From Molecules to Materials

Elena A. Rozhkova • Katsuhiko Ariga
Editors

From Molecules to Materials

Pathways to Artificial Photosynthesis

Springer

Editors
Elena A. Rozhkova
Center for Nanoscale Materials
Argonne National Laboratory
Argonne, IL, USA

Katsuhiko Ariga
Supermolecules Unit, Research
Center for Materials Nanoarchitectonics
National Institute for Materials Science
Tsukuba, Ibaraki, Japan

ISBN 978-3-319-34640-3 ISBN 978-3-319-13800-8 (eBook)
DOI 10.1007/978-3-319-13800-8

Springer Cham Heidelberg New York Dordrecht London

Softcover reprint of the hardcover 1st edition 2015

Printed on acid-free paper

Springer International Publishing AG Switzerland is part of Springer Science+Business Media (www.springer.com)

Preface

This multifaceted book focuses on the various aspects of *artificial photosynthesis* that replicates the natural processes of capturing, conversion, and storage of solar energy. Assisted by *photocatalysts,* artificial photosynthetic schemes allow the direct transformation of sunlight, an infinite energy source, into energy-rich chemical compounds, mostly hydrogen and organic molecules. Motivated by exponentially growing societal demands for clean energy within the last decade, artificial photosynthesis became a multidisciplinary consortium that combines efforts of biologists, chemists, materials scientists, and engineers. Artificial photosynthesis field spans from the understanding of fundamental biological principles of sunlight capturing and splitting of water toward engineered functioning devices. The introduction of nanoscience, nanotechnology, and nanoscale materials gives an additional thrust to the field. The final goal of such multidimensional efforts is to engineer advanced materials with tunable photophysical properties capable of converting solar energy to chemical fuels having high performance in an environmentally friendly and cost-efficient fashion. The building blocks for such advanced energy conversion systems are based on naturally abundant, inexpensive, and stable inorganic materials (semiconductors, carbon materials), organic molecules (dyes), and bioorganic complexes (light harvesting structures, hydrogenases, proton pumps).

The first chapter, by K. Domen and coauthors, introduces the readers to the field, providing a comprehensive overview of the main concepts of the thermodynamics of photocatalysis and photoelectrochemistry, design of visible-light-driven semiconducting materials, as well as the current understanding of the mechanisms of these reactions. Hashimoto and Uragami describe natural tactics applicable in the development of artificial fuel producing systems. The second chapter, by Llansola-Portoles, Palacios, Gust, Thomas Moore, and Ana Moore, demonstrates how knowledge of natural photosynthesis has been employed in building artificial systems, starting from early simple molecular reaction center constructs to the development of more complex systems for the conversion of solar energy to fuel. The following four chapters are dealing with hybrid-based systems that include biological structures as building blocks. In particular, Armstrong, Bachmeier, and Siritanaratkul focus on interfacing hydrogenases with energy materials with the ultimate purpose to "upgrade them to work like enzymes." Braun, Bora, and Gajda-Schrantz summarize works on bio-hybrid electrodes of various types, with particular interest on bio-functionalized hematite photoanodes and synthetic methods for the integration of biomolecules with an electrode. Corronado and coauthors and Aresta and coauthors review the current challenges of photocatalytic CO_2 reduction in two chapters. Next, Ito and D'Souza present the recent advances in photo-sensitized electron-transfer processes on nanocarbon (fullerenes, carbon nanotubes, carbon nanohorns, graphene) surfaces for artificial photosynthesis applications. A chapter presented by Huang and coauthors provides an overview of the plasmonic photocatalysis, one of the most recent developments in the field. The principal feature of this approach is utilization of the localized surface plasmon resonance effects originated from noble metal nanoparticles interfaced with semiconductor particles. The last two chapters (Guo et al.; Bora and Braun) review current status and perspectives of application of synchrotron-based techniques for energy materials characterization.

We are very thankful to all the authors for the generous contribution of their time and for sharing their expertise through excellent comprehensive chapters. We believe that this book will become a desk companion for the multidisciplinary community of students and professionals, for general readers interested in modern technologies, and for everyone who cares for the global future of civilization. We would like to express our gratitude to Ania Levinson and Springer for giving us the opportunity to bring this exciting field to a wider audience and for great cooperation in preparing this book.

Argonne, IL, USA Elena A. Rozhkova
Tsukuba, Ibaraki, Japan Katsuhiko Ariga

Contents

Semiconductors for Photocatalytic and Photoelectrochemical Solar Water Splitting

Guijun Ma, Takashi Hisatomi, and Kazunari Domen

Abstract The fossil energy resources upon which we are heavily dependent will inevitably be depleted in the not-so-distant future. Solar energy may be the best and perhaps only choice for meeting long-term human energy needs. Photocatalytic and photoelectrochemical (PEC) water splitting are ideal approaches to the conversion of solar energy into clean chemical energy H_2, an excellent energy carrier. In this chapter, basic concepts including thermodynamics and experimental methods of photocatalysis and photoelectrochemistry were introduced. As an important aspect of this research, the design of visible-light-driven semiconducting materials was discussed. Then, we summarized recent advancements in photocatalytic and photoelectrochemical water splitting, as well as the current understanding of the mechanisms of these reactions.

Keywords Photocatalysis • Photoelectrochemistry • Semiconductor • Water splitting • Hydrogen

1 Introduction

Over the past 1,000 years, human beings were never forced to seriously consider the issue of energy depletion because of their relatively small population and low energy consumption. Today however, the situation has changed drastically. Exhaustible fossil fuels represent about 80 % of the total world energy supply. It has been estimated that the presently known reserves of oil will last approximately 40 years, while those of natural gas and coal will last approximately 60 and 150 years, respectively [1]. On the other hand, with the rapid increase in global population, energy needs are likely to double within the next 50 years [2]. Moreover, this heavy reliance on fossil fuels is associated with a tremendous production of C, N, S, and P pollution. It is extremely important to develop an abundant and clean energy

G. Ma • T. Hisatomi • K. Domen (✉)
Department of Chemical System Engineering, The University of Tokyo, 7-3-1 Hongo, Bunkyo-ku, Tokyo 113-8656, Japan
e-mail: domen@chemsys.t.u-tokyo.ac.jp

E.A. Rozhkova, K. Ariga (eds.), *From Molecules to Materials*,
DOI 10.1007/978-3-319-13800-8_1

resource as soon as possible. Solar, geothermal, and tidal energies account for >99.9 %, 0.02 %, and 0.002 % of the total energy available on Earth. One of the best choices is to utilize solar energy. Currently, the world's energy consumption rate corresponds to only 0.01 % of the solar energy striking the Earth. Many attempts have been made to explore technologies that can transform solar energy into heat, electricity, or chemical energy. Photocatalytic and photoelectrochemical (PEC) water splitting are ideal approaches to the conversion of solar energy into clean chemical energy H_2, an excellent energy carrier. Thermodynamically, the oxidation of H_2 can provide triple the energy of gasoline or natural gas (per gram). Oxidation of H_2 to H_2O and regeneration of H_2 from H_2O using solar energy are clean and sustainable processes. Furthermore, H_2 is an important chemical reagent in the fine chemical synthesis industry. Since the PEC water-splitting reaction was accomplished on a TiO_2 photoanode in 1972 by Fujishima and Honda, extensive research has been carried out into solar light-driven hydrogen production from water by both photocatalysis and photoelectrochemistry [3].

2 Thermodynamics of Photocatalytic and Photoelectrochemical Water Splitting

Thermodynamically, the process of water splitting to produce H_2 and O_2 is an endothermic reaction, with a positive change in Gibbs free energy of 237 kJ mol^{-1}. Since (1) is a two-electron reaction, water can be electrolyzed under standard conditions using an externally applied voltage of 1.23 eV (according to the equation $-E=\Delta G^{\circ}/nF$, with $n=2$ and $F=96{,}485$ C mol^{-1}). As shown in (2) and (3), the electrolysis of water proceeds via two reactions: proton reduction at a cathode and water oxidation at an anode:

$$H_2O \rightarrow H_2 + 1/2O_2 \quad \Delta G^{\circ} = +237\,\mathrm{kJ\,mol^{-1}} \tag{1}$$

$$2H^+ + 2e^- \rightarrow H_2 \quad E^{\circ} = 0\,\mathrm{V} \tag{2}$$

$$O_2 + 4H^+ + 4e^- \rightarrow 2H_2O \quad E^{\circ} = 1.23\,\mathrm{V} \tag{3}$$

The redox potential of water determines the required band gap and band position for a semiconductor to be used for photocatalytic or PEC water splitting. Since the band properties of semiconductors are usually described as an absolute potential, with the vacuum energy as zero, the correlation between the absolute (abs) potential and the potential of a standard hydrogen electrode (SHE) is shown in (4):

$$E_{(\mathrm{abs})} = -\{E_{(\mathrm{SHE})} + (4.44 \pm 0.02)\}\,\mathrm{V(vs.\ vacuum)} \tag{4}$$

2.1 Photocatalytic Water Splitting on a Semiconductor Particle

According to the traditional definition of catalysis, photocatalytic water splitting is not a catalytic process because it is an uphill reaction. In its early stage, it can also be called photoinduced water splitting on a semiconductor. With the long-term development of this research, the term photocatalytic water splitting has become widely accepted. To differentiate from natural photosynthesis, it has recently also been called artificial photosynthesis.

The key step of the heterogeneous photocatalytic water-splitting reaction is based on the photoelectric effect in semiconductors. Under light irradiation at a wavelength with an energy equal to or higher than the band gap, electrons in the valence band (VB) of a semiconductor can be excited to the conduction band (CB). The top of the VB and the bottom of the CB are often regarded as the highest occupied molecular orbital (HOMO) and the lowest unoccupied molecular orbital (LUMO), respectively, when a limited number of atoms are considered in quantum chemical calculations, borrowing terms from the molecular orbital method. An excited electron in the CB has a reductive potential, while a positively charged hole that has an oxidative potential is produced in the VB. Thus, it can be said that solar energy is first converted to electrical energy after light absorption by a semiconductor. Heterogeneous photocatalytic or PEC water splitting can be taken as the electrolysis of water on the surface of a semiconductor powder or photoelectrode.

Figure 1 shows the water-splitting reaction on a semiconductor particle [4–6]. Under light excitation, electronic transitions occur inside photocatalyst particles, generating reductive electrons in the CB and oxidative holes in the VB. Then, these photoexcited carriers are transferred to the surface of the semiconductor and trapped by hydrogen evolution cocatalyst (HEC) or oxygen evolution cocatalyst (OEC) for

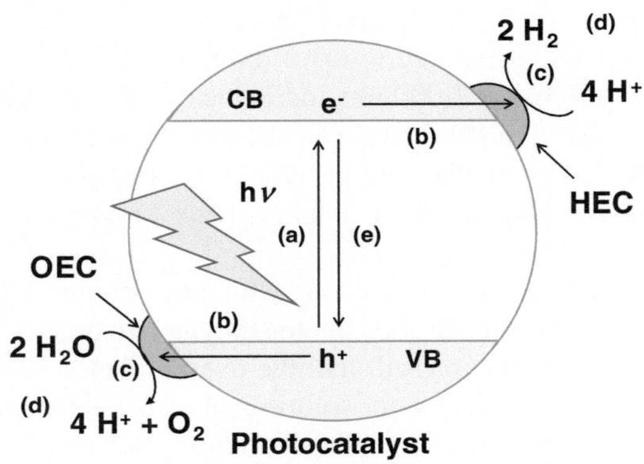

Fig. 1 Photocatalytic water-splitting reactions on a semiconductor particle. (a) Light absorption; (b) charge transfer; (c) redox reactions; (d) adsorption, desorption, and mass diffusion of chemical species; and (e) charge recombination. HEC, OEC, CB, and VB stand for a hydrogen evolution cocatalyst and an oxygen evolution cocatalyst, the conduction band, and the valence band, respectively

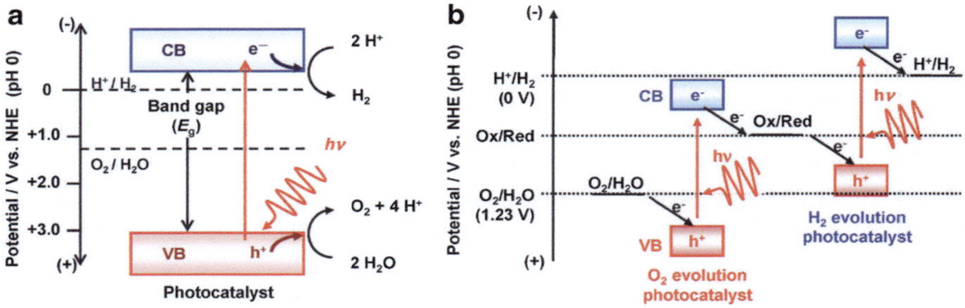

Fig. 2 Energy diagrams for photocatalytic water splitting based on (**a**) one-step excitation and (**b**) two-step excitation (Z-scheme). Reprinted with permission from [6], Copyright © 2014, Royal Society of Chemistry

water reduction or oxidation. Once these electron–hole pairs are generated by light excitation, recombination reactions can occur until the charge carriers are consumed by reactions with water. Thus, there is a competition between charge recombination and photocatalytic water splitting. Unfortunately, charge recombination occurs frequently and rapidly in most semiconductors, which is the primary reason for their low photocatalytic water-splitting activity. The driving force for charge separation is band bending from the semiconductor–water interface to the semiconductor bulk, as will be further discussed in regard to the photoelectrochemistry of semiconductor electrodes.

The reductive and oxidative capabilities of photogenerated electrons and holes are determined by the energy potential of the CB bottom and the VB top, respectively. As shown in Fig. 2a, to achieve overall water splitting, the bottom of the CB must be at a more negative potential than the reduction potential of H^+ to H_2 (0 V vs. NHE at pH 0), while the top of the VB must be more positive than the oxidation potential of H_2O to O_2 (1.23 V vs. NHE) [4–7]. Therefore, the minimum photon energy thermodynamically required to drive the reaction is 1.23 eV, which corresponds to a wavelength of ca. 1,000 nm and is located in the near-infrared region. However, there are energy losses caused by the migration of excited electrons and holes as well as the overpotentials for H_2 and O_2 evolution. The actual minimum required band gap for photocatalytic overall water splitting is estimated to be approximately 2 eV [8].

To maintain charge balance in a semiconductor, photoexcited electrons and holes must be consumed simultaneously in a photocatalytic reaction. Any interruption in photoreduction or photooxidation will halt the overall photoreaction. It is difficult to achieve overall photocatalytic water splitting on a single semiconductor, even if its band gap and band position meet the thermodynamic requirements for water splitting. To separately test the ability of a photocatalyst for water reduction or oxidation, half-reactions for photocatalytic water splitting can be carried out by adding an actively reducible or oxidizable species to the water as a sacrificial reagent to consume photogenerated electrons or holes. By recycling the redoxable sacrificial reagent used as a mediator, overall water splitting can be achieved by combining two half-reactions on two different photocatalysts (Fig. 2b). This system was inspired by natural photosynthesis in green plants and is called the "Z-scheme." The development of Z-scheme water-splitting systems has relied on both finding new semiconductor photocatalysts that efficiently function in the presence of a shuttle redox mediator and creating active sites to promote surface chemical reactions while suppressing backward reactions involving redox mediators [7].

2.2 *PEC Water Splitting on a Thin-Film Semiconductor Photoelectrode*

Not all semiconductors, especially visible-light-responsive materials, meet the band gap and band position requirements for overall water splitting. Figure 3 depicts the band positions of several semiconductors relative to the redox potential for water splitting [2, 9]. It can be seen that the valence bands of CdSe and Si are more negative than the water oxidation potential (1.23 V vs. NHE), while the conduction bands of MoS_2, WO_3, and Fe_2O_3 are more positive than the proton reduction potential (0 V vs. NHE). Therefore, it is impossible for these semiconductors to achieve overall water splitting by photocatalysis. Another issue for most semiconductor photocatalysts is the limited separation and transferability of photogenerated charge carriers, which is the main reason for the poor photocatalytic water-splitting activity. Thus, it is necessary to apply an external bias to the semiconductor to assist the

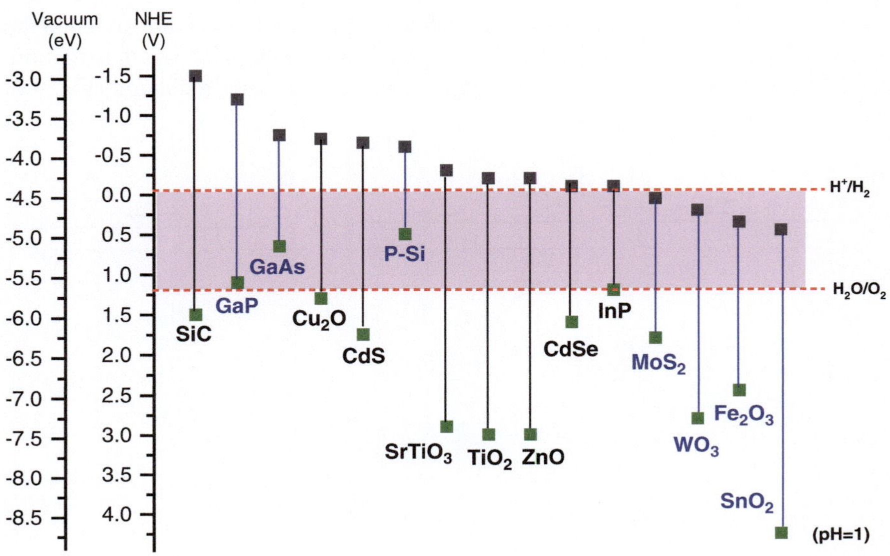

Fig. 3 Relationship between semiconductor band structures and redox potentials for water splitting

separation and migration of photoexcited electrons and holes. To achieve this goal, a semiconductor is deposited on a conductive substrate at the proper thickness as a photoelectrode for PEC water splitting.

When a photoelectrode is immersed in an electrolyte solution, junctions are generally formed between the semiconductor and the solution because of an initial difference in Fermi levels (chemical potential of electrons) between the two phases [10]. The majority charge carriers (electrons in n-type semiconductors and positive holes in p-type semiconductors) are transferred to the solution upon contact, equalizing the Fermi levels and establishing a thermodynamic equilibrium. For an n-type semiconductor, the diffusion of electrons to water produces a positive space charge layer (also called a depletion layer, since the region is depleted of majority charge carriers) in the semiconductor adjacent to the interface with the electrolyte. As a result, the conduction and valence band edges are bent upward, establishing a potential barrier (built-in potential) against further electron transfer into the electrolyte. The inverse but analogous situation occurs with p-type semiconductors that have an initial Fermi level below that of the electrolyte, with downward band bending to generate a built-in potential against further positive hole transfer into the electrolyte. The built-in potential on the surface of a photoelectrode produces important effects such as current rectification (diode behavior) and photovoltaic response (generation of a photovoltage when the junction region is illuminated).

All phenomena associated with photoelectrochemical systems are based on the formation of a semiconductor–electrolyte junction. When a semiconducting photoelectrode is irradiated by light with an energy equal to or greater than the band gap of the semiconductor, excited electron–hole pairs that reach the depletion layer before recombination will be separated by the built-in potential. For n-type semiconductor photoelectrodes (photoanodes), holes (minority carriers) are driven to the semiconductor–water interface by the upward band bending for water oxidation, while electrons are conducted to a counter electrode for water reduction (Fig. 4a).

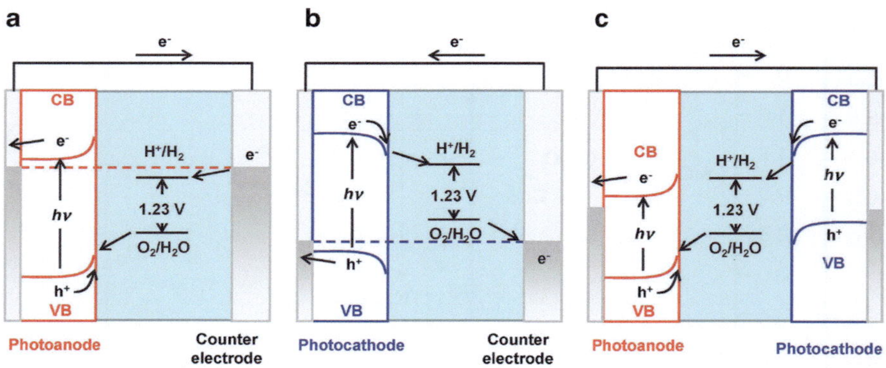

Fig. 4 Energy diagrams for PEC water splitting using (**a**) a photoanode, (**b**) a photocathode, and (**c**) a photoanode and a photocathode in a tandem configuration. Reprinted with permission from [6], Copyright © 2014, Royal Society of Chemistry

Similarly, in p-type semiconductor photoelectrodes (photocathodes), electrons (minority carriers) are driven to the interface by downward band bending to participate in a reduction reaction (Fig. 4b). In summary, band bending results from the diffusion of majority carriers to facilitate the transfer of photoexcited minority carriers from the bulk to the surface of a semiconductor for water splitting.

For semiconductor photoelectrodes whose band gaps do not straddle the reduction and oxidation potentials of water (such as Si, WO_3, and Fe_2O_3, as shown in Fig. 3), an additional external bias voltage can be applied to both the working and counter electrodes to meet the potential requirement for water redox. As a result, in addition to the solar energy absorbed, additional electrical energy is consumed by the PEC reaction system. To develop as many candidate semiconductors as possible, p–n cells can be constructed for overall PEC water splitting as shown in Fig. 4c. Note that the CB potentials of p- and n-type semiconductors are different, which can enhance the available electron–hole pair potential for the net electrolyte reaction when both electrodes are illuminated. An important possible advantage of the p–n cell is that it may allow for the use of smaller band gap semiconductors for the water-splitting reaction. Since the number of solar photons absorbed by a semiconductor increases rapidly with decreasing band gap, semiconductors with small band gap energies could achieve higher energy conversion efficiencies. Recently, after considering energy losses on overpotential, several groups calculated the maximum theoretical solar-to-hydrogen efficiencies of PEC water splitting based on different configurations of photoelectrodes [11–13]. The results indicated that the efficiency of a dual-stacked absorber configuration (ca. 22.8 % with bottom and top photoabsorber band gaps of 1.23 eV and 1.84 eV, respectively) is higher than the efficiency of a dual side-by-side absorber configuration (ca. 15.5 % with photoabsorbers of the same band gap of 1.59 eV) or a single absorber (ca. 11.2 % for a band gap of 2.26 eV) [13].

3 Materials Design of Visible-Light-Driven Photocatalysts

The solar spectrum irradiating the Earth consists of ultraviolet (UV) rays ($\lambda < 400$ nm), visible light (400 nm $< \lambda <$ 800 nm), and infrared rays ($\lambda >$ 800 nm), accounting for 4.6, 54.3, and 41.1 % of the total solar energy, respectively. The relationship between energy and wavelength can be calculated as E (eV) = 1,240/λ (nm). For photocatalytic or PEC water splitting, the band gap of a semiconductor should be as small as possible in order to maximally utilize the solar energy (Fig. 5) [14]. However, a smaller band gap in a semiconductor means smaller offsets between the potentials of the band edges and the redox reactions of the electrolyte solution (see Figs. 2 and 3). Therefore, there is a trade-off between the band gap and redox capability in the design of a suitable semiconductor photocatalyst or photoelectrode for solar light-driven water splitting. Early studies on photocatalysis showed that oxide semiconductors such as TiO_2 (3.2 eV), $K_2La_2Ti_3O_{10}$ (3.4 eV), $K_4Nb_6O_{17}$ (3.4 eV), $Ba_5Nb_4O_{15}$ (3.8 eV), and $NaTaO_3$:La (4.1 eV) performed well in water splitting

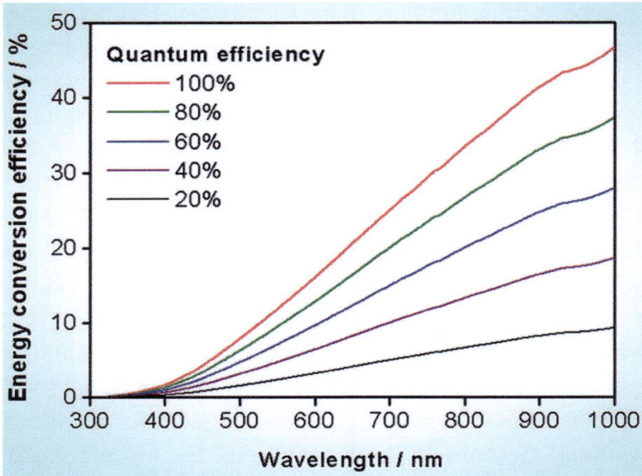

Fig. 5 Calculated solar energy conversion efficiency as a function of wavelength for overall water splitting using photocatalysts with various quantum efficiencies. Solar irradiance used for the calculation was taken from AM 1.5G data. Reprinted with permission from [14], Copyright © 2010, American Chemical Society

under UV light irradiation [9, 15–19]. An optimized NiO(0.2 %)/$NaTaO_3$:La(2 %) photocatalyst had a high activity for water splitting, with an apparent quantum yield of 56 % at 270 nm [19]. However, the wide band gaps of these materials restricted their application to solar energy conversion. Many ideas in band gap engineering, such as doping, solid solution, elemental substitution, and surface sensitization (by dye, quantum dots, or plasma), have been developed to extend the light absorption of semiconductors. Surface sensitization will be introduced in other chapters of this book. Here, we will focus on the other three techniques.

3.1 Doping

Since the report in 1972 of PEC water splitting on TiO_2, a considerable amount of work has been done to test the activity for photocatalytic water splitting on different kinds of semiconductors. Many binary and ternary semiconductor compounds have been evaluated for photocatalytic water splitting [9, 14]. Figure 6 summarizes and assorts the elements most commonly used in semiconductor photocatalysts. Most metal oxide, sulfide, and nitride photocatalysts consist of metal cations with d^0 and d^{10} configurations. The conduction bands for these d^0 and d^{10} photocatalysts are usually composed of d and sp orbitals, respectively. The valence bands of oxide, sulfide, and nitride semiconductors are composed of O 2p, S 3p, and N 2p orbitals, respectively [5, 9, 14]. The idea behind band gap reduction by doping with additional elements is the introduction of a filled energy level above the valence band

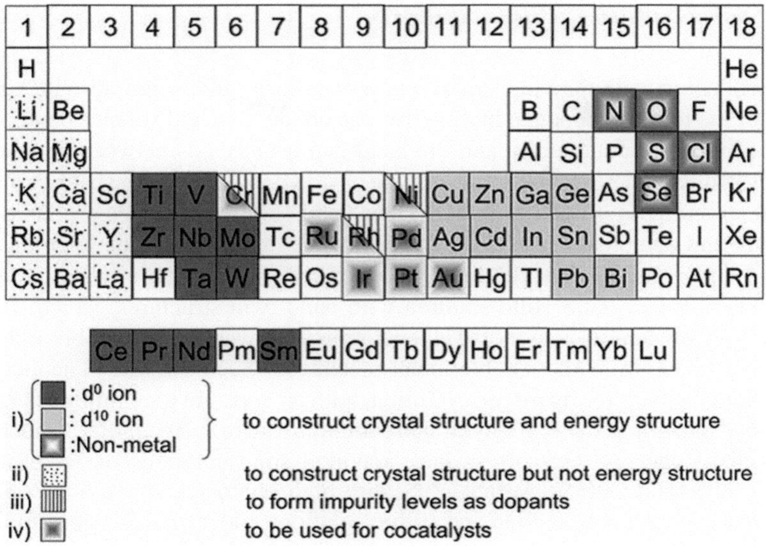

Fig. 6 Elements used to construct heterogeneous photocatalysts. Reprinted with permission from [9], Copyright © 2009, Royal Society of Chemistry

and/or an empty energy level below the conduction band. Among the anions listed in Fig. 6, the O 2p orbital has the most positive potential (ca. 3.0 eV) [20]. That is why most oxide photocatalysts have wide band gaps, capable of absorbing only UV light. Thus, doping experiments predominantly employ oxide photocatalysts. For example, TiO_2 doped with anions (such as C, Cl, S, or N) or cations (such as Cr, Bi, Cu, or Rh) could produce a donor energy level above the valence band composed of the O 2p orbital. Doping can reduce the band gap on one side, but it can also induce lattice defects in the host material on the other side. These lattice defects usually act as recombination centers for electrons and holes [21, 22]. It was demonstrated that the photocatalytic activity of $SrTiO_3$ can be enhanced by doping with a lower valence cation (Na^+ or Ga^{3+}) or decreased by doping with a higher valence cation (Ta^{5+}), because doping with the former cations can introduce oxygen vacancies and decrease the density of Ti^{3+} ions (lattice defects in $SrTiO_3$) [21]. The idea of co-doping was developed in order to increase light absorption while maintaining the charge balance of the host semiconductor. Co-doping of Cr^{3+}/Sb^{5+}, Rh^{3+}/Sb^{5+}, Ni^{2+}/Nb^{5+}, and H^+/N^{3-} into TiO_2 is effective at both reducing the band gap and increasing the photocatalytic activity [9, 23].

The energy levels generated by dopants within the intrinsic band gap are mostly discontinuous, which results in weak and limited light absorption in the extended light absorption region of doped semiconductors. A visible-light-responsive semiconductor with continuous energy levels (energy band) is preferred for solar light utilization, and this can be achieved by solid solution and elemental substitution, as discussed in the following sections.

3.2 Solid Solution Photocatalysts

Semiconductors with the same crystal phases can form solid solutions with variable ratios of the two components. In this case, the conduction and valence bands of the solid solution are continuous, and the band gap usually lies between those of the starting semiconductors. The solid solution of CdS and ZnS, which share a common crystal structure, is a good example. Their solid solution has a controllable band structure and demonstrates excellent performance in photocatalytic hydrogen production under visible-light irradiation [24, 25]. Recently, Guo et al. reported the synthesis of a $Cd_{1-x}Zn_xS$ solid solution with nano-twin structures, which exhibits superior photocatalytic activity for H_2 evolution from water under visible-light irradiation ($\lambda \geq 430$ nm) without noble-metal cocatalysts. Such $Cd_{0.5}Zn_{0.5}S$ nanocrystals have a high activity for hydrogen evolution, with an apparent quantum yield (AQY) of 43 % at 425 nm, achieving a hydrogen evolution rate of 1.79 mmol h^{-1} under solar light [25]. Kudo and co-workers have reported solid solutions of $AgInS_2$–ZnS, $CuInS_2$–ZnS, and $CuInS_2$–$AgInS_2$–ZnS with high photocatalytic activities for H_2 evolution from aqueous sulfide and sulfite solutions under visible-light irradiation [9, 26, 27]. The diffuse reflectance spectra of $CuInS_2$–$AgInS_2$–ZnS solid solutions shift monotonically with changes in the composition of the solid solution, as shown in Fig. 7. Density functional theory (DFT) calculations indicate that the levels of the conduction band consisting of Zn 4s4p and In 5s5p, and of the valence band consisting of Cu 3d, Ag 4d, and S 3p, shift with change in composition. Ru/$Cu_{0.25}Ag_{0.25}In_{0.5}ZnS_2$ demonstrated an excellent activity for H_2 evolution in a solar simulator (AM-1.5), and this activity was much higher than that for the starting precursors, which were ZnS, $CuInS_2$, and $AgInS_2$.

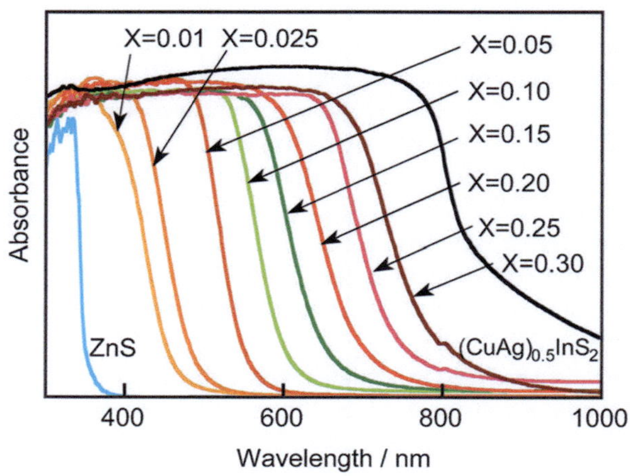

Fig. 7 Diffuse reflection spectra of $(CuAg)_xIn_{2x}Zn_{2(1-2x)}S_2$ solid solution. Reprinted with permission from [9], Copyright © 2009, Royal Society of Chemistry

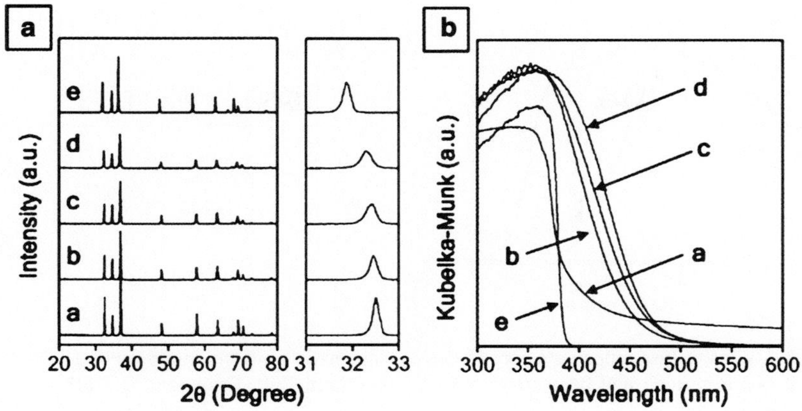

Fig. 8 (**a**) Powder XRD pattern and (**b**) UV-visible diffuse reflectance spectra for (*a*) GaN (ref), (*b*) GaN–ZnO (Zn 3.4 at%), (*c*) GaN–ZnO (Zn 6.4 at%), (*d*) GaN–ZnO (Zn 13.3 at%), and (*e*) ZnO. Reprinted with permission from [28], Copyright © 2005, American Chemical Society

The authors' group has reported oxynitride solid solutions of GaN–ZnO and $ZnGeN_2$–ZnO consisting of wurtzite structures with similar lattice parameters [5, 14, 28, 29]. Taking into account the large band gap energies of GaN (3.4 eV) and ZnO (3.2 eV), it had been considered that the band gap of the solid solution should exceed 3 eV, although the band gaps of $(GaN)_{1-x}$–$(ZnO)_x$ are 2.4 ~ 2.8 eV (Fig. 8). DFT calculations indicate that the bottom of the conduction band for GaN–ZnO is mainly composed of Ga 4s and 4p orbitals, while the top of the valence band consists of N 2p orbitals followed by Zn 3d orbitals. It was hypothesized that p-d repulsion in the GaN–ZnO solid solution (i.e., N 2p–Zn 3d repulsion) might cause the top of the valence band formed by N 2p atomic orbitals to be raised to a higher potential energy, resulting in a narrower band gap for GaN. Until now, GaN–ZnO solid solutions loaded with proper cocatalysts have demonstrated the highest activity for photocatalytic overall water splitting under visible-light irradiation.

3.3 Elemental Substitution of O by S and N

Since the valence bands of most oxide semiconductors are composed of deep O 2p orbitals, partly or completely replacing O atoms with an element that has a lower potential should effectively reduce the band gap of the semiconductor. The potentials of the N 2p and S 3p orbitals are more negative than the O 2p orbital, which is why most (oxy)nitrides and (oxy)sulfides have a smaller band gap than the corresponding oxides. (Oxy)sulfide or (oxy)nitride production is usually carried out by calcining the related oxide precursor at high temperature under a flow of H_2S or NH_3 gas, respectively. These processes can also be understood as the high-temperature

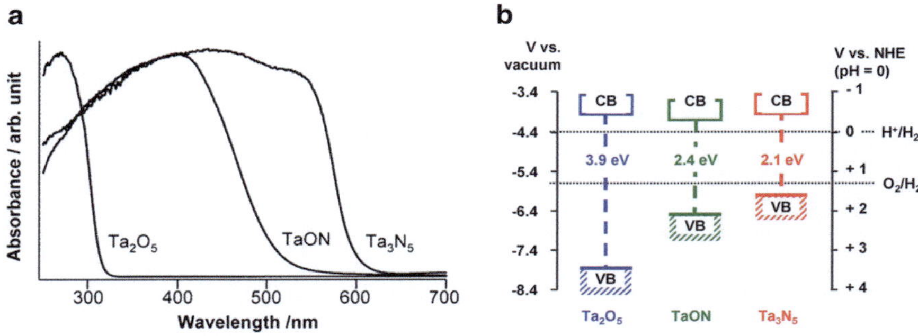

Fig. 9 UV-vis diffuse reflectance spectra (**a**) and schematic band structures (**b**) of Ta_2O_5, TaON, and Ta_3N_5. Reprinted and adapted with permission from [30], Copyright © 2003, American Chemical Society

reduction of oxide semiconductors by reductive H_2S or NH_3 gases [28–35]. Thus, the extent of sulfurization or nitridation should be carefully controlled to prevent the production of over-reduced products. Unlike doped samples, these prepared (oxy) sulfides and (oxy)nitrides have continuous energy levels in the VB for visible-light absorption. Furthermore, the crystal phases of (oxy)sulfides and (oxy)nitrides are sometimes different from those of their oxide precursors, which means that a new type of semiconductor is prepared by the elemental substitution.

Figure 9 shows the light absorption and band structures of Ta_2O_5, TaON, and Ta_3N_5 [30]. The preparation of TaON and Ta_3N_5 is accomplished by holding the Ta_2O_5 powder in NH_3 flow at ca. 850 °C for appropriate times. As the nitridation reaction proceeds, the O^{2-} anions in the precursor Ta_2O_5 are gradually replaced by N^{3-} anions from the NH_3. Further nitridation of Ta_3N_5 can produce more reduced by-products such as TaN. The top of the valence band of Ta_2O_5 consists of O 2p orbitals and is located at ca. +3.4 V vs. NHE. In contrast, the N 2p orbitals of TaON and Ta_3N_5 are incorporated into the valence bands, shifting the top of the valence bands to ca. +2.0 V vs. NHE for TaON and +1.5 V vs. NHE for Ta_3N_5. Consequently, the light absorption of Ta_2O_5 is extended from 320 to 500 nm for TaON and 600 nm for Ta_3N_5. Using this reaction process, many visible-light-responsive oxynitride photocatalysts can be prepared by the nitridation of oxide precursors such as $LaTiO_2N$ (from $La_2Ti_2O_7$), $BaTaO_2N$ (from mixture of $BaCO_3$ and Ta_2O_5), $BaNbO_2N$ (from $Ba_5Nb_4O_{15}$), and the previously mentioned GaN–ZnO solid solutions (from a mixture of Ga_2O_3 and ZnO) [28, 29, 31–33]. By changing the NH_3 to H_2S gas, oxysulfides such as $Sm_2Ti_2S_2O_5$ can be prepared from an $Sm_2Ti_2O_7$ precursor under similar procedures [34, 35]. Another way to prepare oxysulfides is to calcine vacuum-sealed precursors of oxide and sulfide compounds such as $Sm_2Ti_2S_2O_5$ (from precursors of Sm_2S_3, Sm_2O_3, and TiO_2) and $La_5Ti_2CuS_5O_7$ (from precursors of La_2O_3, La_2S_3, TiO_2, and Cu_2S) [36, 37]. In addition to the increased light absorption, another important priority for these oxysulfides and oxynitrides is for their valence bands to remain positive enough to oxidize water, despite their narrowed band gaps.

4 Measurement of Water-Splitting Activity

4.1 Apparatus

A variety of photocatalytic and PEC systems have been considered in the pursuit of efficient, scalable, and inexpensive overall water splitting [13, 38, 39]. This subsection will focus on setups for water-splitting reactions and activity evaluation with technical remarks and recommendations on how to obtain reliable results. Photocatalytic water-splitting reactions can be carried out in a closed-circulation system equipped with a vacuum system or a flow system at atmospheric pressure [6]. An example of a closed-circulation system is shown in Fig. 10. In such a system, a photocatalyst sample is dispersed in a reaction solution, and the system is evacuated to eliminate dissolved air from the solution. Enough inert gas is introduced into the reaction system to keep the gas composition homogeneous by circulation and to avoid boiling of the reaction solution under illumination at a reduced pressure. Alternatively, an open-flow system can be used for photocatalytic water splitting, with a continuous inert gas flow through the headspace of a reactor. The advantage of closed circulation is that product gas accumulates in the reaction system. Therefore, the photocatalytic activity can be estimated even if the total amount of products is small or if the gas evolution is discontinuous because of the use of sacrificial reagents or deactivation of the photocatalyst. On the other hand, an open-flow system allows for better control of the temperature and reactor atmosphere. Hydrogen and oxygen accumulation in the reaction system, which would enhance the recombination of hydrogen and oxygen into water, does not occur. In both cases, gaseous products were analyzed with a gas chromatograph integrated with the reaction system to avoid air contamination.

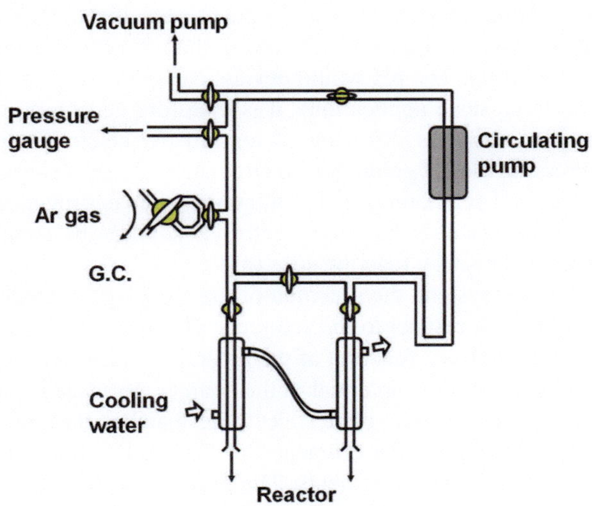

Fig. 10 Schematic of a closed-circulation system for photocatalytic reaction

In practice, the pressure inside the reaction system affects the photocatalytic activity [40]. At a high pressure, the photocatalytic activity for overall water splitting generally decreases because of slower bubble formation and desorption of product gases, which induces side reactions such as oxygen reduction and water formation from the produced hydrogen and oxygen. These side reactions are critical problems in the practical operation of solar water splitting at ambient pressure. A hydrogen evolution cocatalyst that is highly selective for the desired reaction, such as $Rh_{2-y}Cr_yO_3$ [40] or Rh–Cr_2O_3 core/shell composite [41], is needed.

PEC water splitting is performed using a single- or double-compartment cell equipped with a flat optical window [6, 42]. The PEC cell has several ports for electrodes, purging, and gas sampling. Two-electrode and three-electrode systems are used to measure the efficiency of a whole cell and the performance of a working electrode as a half-cell, respectively. In a two-electrode system, the photocurrent and dark current are recorded as a function of voltage against a counter electrode. The potential of the counter electrode is ideally the same as the potential of the redox reaction occurring on the counter electrode. However, in reality, the potential of the counter electrode is not at the equilibrium potential of the redox reaction because of the reaction overpotential, which in turn depends on the current density. Thus, the current–voltage curves are affected by the properties of the counter electrode. Consequently, the potential of the working electrode cannot be measured exactly in the simple two-electrode configuration. The purpose of the two-electrode measurement is to assess the efficiency of a PEC cell, taking overpotential loss in both working and counter electrodes into account. It is important to suppress the polarization of the electrolyte solution in order to maintain a high PEC activity. The pH of the electrolyte solution near the (photo) anode and (photo) cathode decreases and increases, respectively, as a result of the oxygen and hydrogen evolution reactions. This difference in local pH generates a chemical bias that hinders water splitting. In this situation, an externally applied voltage greater than that theoretically required for water electrolysis is needed, and this lowers the efficiency of the PEC water-splitting cell significantly. The pH would deviate easily in a neutral aqueous solution. In practical large-scale applications, it is therefore necessary to design PEC cells that allow for sufficient diffusion of ions in the electrolyte solution. One method of suppressing the pH gradient is to circulate buffered electrolyte solutions. Recently, a stable, self-regulating, and continuous solar hydrogen generator was demonstrated in a near-neutral buffered electrolyte solution by using a controlled recirculating stream across the reaction sites [43].

In a three-electrode system, the potential of the working electrode is controlled using a potentiostat with respect to the reference electrode. The photocurrent and dark current are recorded as a function of the potential against the electrolyte solution. It should be noted that the potential of the counter electrode is adjusted so that the current passing through the counter electrode matches that generated on the working electrode. Therefore, the efficiency of PEC cells could not be assessed from these three-electrode measurements. The purpose of the three-electrode system is to define the potential of the working electrode against the electrolyte solution, independent of the properties of the counter electrode. A silver–silver chloride electrode, a saturated calomel electrode, and a mercury–mercury oxide electrode are commonly used in PEC water splitting. These reference electrodes can indicate the known potentials with respect to electrolyte solutions containing different redox species, temperatures, and concentrations. The potential of the working electrode measured with respect to a reference electrode can be converted to its potential vs. the reversible hydrogen electrode (RHE) using the following equation:

$$E_{\mathrm{RHE}} = E_{\mathrm{ref}} + E^{\mathrm{o}}_{\mathrm{ref}} + \frac{RT \ln 10}{F} pH, \tag{5}$$

where E_{RHE}, E_{ref}, and E°_{ref} denote, respectively, the potential of the working electrode against the electrolyte solution with respect to the RHE, the potential of the working electrode measured with respect to a reference electrode, and the potential of the reference electrode with respect to the normal hydrogen electrode (NHE), if the potential of the reference electrode used is independent of pH. *R*, *T*, and *F* represent the gas constant, temperature, and Faraday constant, respectively. The potentials at 0 and +1.23 V vs. RHE indicate the potentials at which hydrogen and oxygen are in equilibrium with the actual electrolyte solution.

Various sources can be used for illumination, including Hg lamps, Xe lamps, halogen lamps, and light-emitting diodes. Hg lamps and light-emitting diodes generate emission line spectra, while Xe and halogen lamps mostly generate continuous spectra. Xe lamps equipped with appropriate filters can be used as a solar simulator. One must consider the power spectrum (i.e., intensity at different wavelengths) of a light source at different distances, positions, and times, because the photon and energy fluxes are important in the evaluation of quantum efficiencies and energy conversion efficiencies, respectively. Moreover, heating of the reaction solution under illumination should be avoided because the water-splitting rate depends on the reaction temperature [4]. Furthermore, fluctuations in vapor pressure could cause inhomogeneity of the gaseous components in the reactor.

4.2 Efficiency

The benchmark efficiency for solar hydrogen production via water splitting and the diagnostic efficiencies for the investigation of material performance were recently reviewed [42]. The solar-to-hydrogen efficiency (STH) can be calculated from the product of the hydrogen production rate (r_{H2}) and the increase in Gibbs free energy (ΔG) of 237 kJ mol^{-1} [9, 42]:

$$STH = \frac{\text{Output energy as } \mathrm{H_2}}{\text{Energy of incident solar light}} = \frac{r_{\mathrm{H_2}} \times \Delta G}{P_{\mathrm{sun}} \times S}, \tag{6}$$

where P_{sun} is the energy flux of sunlight and *S* is the area of the reactor. Considering ASTM-G173 AM1.5 global tilt, solar irradiation has an energy flux of 1.0×10^3 W m^{-2} and a well-defined power spectrum.

When an external voltage is applied to a PEC system, the electrical energy must be subtracted from the energy gain. For two-electrode measurements, the applied bias photon-to-current efficiency (ABPE) is frequently used [42]. ABPE is defined as

$$ABPE = \frac{|j| \times (V_{th} - V_{bias})}{P_{sun}}, \tag{7}$$

where j, V_{th}, and V_{bias} are the photocurrent density, the theoretical water electrolysis voltage (1.23 V), and the applied voltage, respectively. One concern regarding ABPE as an alternative expression of STH is that the Faradaic efficiency (η_{F}) is not considered explicitly [42]. Thus, the photocurrent is not necessarily attributed to water splitting, unless the production of hydrogen and oxygen is confirmed at a Faradaic efficiency of unity. The Faradaic efficiency for water splitting is the ratio of the current contributing to water splitting into hydrogen and oxygen to the observed current. In the ideal scenario, four electrons correspond to two hydrogen molecules and one oxygen molecule. If the Faradaic efficiency is below unity, a part of the current should be attributed to redox processes other than water splitting.

Because of the ambiguous definition and naming of ABPE either as energy conversion efficiency or quantum efficiency, applied-bias-compensated solar-to-hydrogen efficiency (AB-STH) was suggested as a general expression for STH via PEC water splitting under sunlight [6]:

$$AB - STH = \frac{|j| \times \eta_{\mathrm{F}} \times (V_{th} - V_{bias})}{P_{sun}}. \tag{8}$$

STH corresponds to a special case of AB-STH such that η_{F} is unity and V_{bias} is zero. Note that current–voltage curves often have a significant hysteresis in cyclic voltammetry because of charging–discharging of trap states, cocatalysts, and photoelectrode surface effects. To accurately estimate AB-STH, it is necessary to employ a two-electrode system to measure the photocurrent at a static applied voltage. It is obviously necessary to quantify the amounts of evolved H_2 and O_2 in order to determine η_{F}. PEC cells should be purged with an inert gas when side reactions involving oxygen and hydrogen are a concern. Moreover, even if the value of AB-STH is high, the performance is considered to be low when the potential gain is much smaller than 1.23 V. Such systems split water mostly by electrical energy.

For a photoanode (photocathode) measured in the three-electrode configuration, the product of the potential gain and the photocurrent can be regarded as the hypothetical half-cell solar-to-hydrogen efficiency (HC-STH) [6]. The gain in potential by PEC water splitting can be approximated as the difference between the potential of a photoanode (photocathode) and the oxygen (hydrogen) equilibrium potential:

$$HC - STH = \begin{cases} \dfrac{|j| \times (E_{\mathrm{O_2/H_2O}} - E_{\mathrm{RHE}})}{P_{\mathrm{sun}}} & \text{(for photoanodes)} \\ \dfrac{|j| \times (E_{\mathrm{RHE}} - E_{\mathrm{H^+/H_2}})}{P_{\mathrm{sun}}} & \text{(for photocathodes)} \end{cases}, \tag{9}$$

where E_{O_2/H_2O} and E_{H^+/H_2} represent the equilibrium potentials of oxygen evolution (+1.23 V vs. RHE) and hydrogen evolution (0 V vs. RHE), respectively. Provided that η_F for PEC water splitting is unity and that a counter electrode drives the counter reaction without any overpotential, HC-STH is a convenient estimate of the maximum AB-STH of a half-cell, independent of the counter electrode properties. In reality, however, the potential of the counter electrode will not be in equilibrium with the potential of the redox reaction on the counter electrode, as mentioned in the previous section. Therefore, HC-STH should not be regarded as AB-STH.

STH can be used as a universal and practical standard for the performance of photocatalysts and photoelectrodes, although it conveys too much information to understand the performance of photocatalyst and photoelectrode materials in water splitting. To understand the performance of materials and perform reliable measurements, it is necessary to understand diagnostic efficiencies. The activities of photocatalysts and photoelectrodes are often reported as gas evolution rates and photocurrents in photocatalytic and PEC water splitting, respectively, measured using various light sources for convenience. However, it is practically impossible to compare the activities measured in different reaction systems because of variations in the power spectra of light sources. Hence, to discuss performance on a kinetic basis, it is necessary to compare the extents to which photons contribute to the photocatalytic and PEC reactions. The apparent quantum yield (AQY) and the incident photon-to-current efficiency (IPCE) can be used as standard measures of activity. The AQY is defined as

$$AQY = \frac{nR}{I}, \tag{10}$$

where n, R, and I denote the number of electrons involved in the photocatalytic reaction, the molecular production rate, and the rate of incident photons, respectively. In overall water splitting using a single photocatalyst, the constant n for the hydrogen and oxygen evolution rates is equal to two and four, respectively. IPCE is defined as the ratio of the number of electrons passing through an external circuit as a photocurrent to the number of photons irradiating a photoelectrode. Thus, IPCE corresponds to the external quantum efficiency (EQY). If the total absorptance of a photoelectrode is available, the absorbed photon-to-current efficiency (APCE) can be calculated by dividing IPCE by the absorptance [42]. It should be again noted that the photocurrent used for the calculation of IPCE and APCE is not necessarily attributable to the water-splitting reaction unless the production of hydrogen and oxygen is confirmed at a Faradaic efficiency of unity.

AQY, IPCE, and APCE are dependent on the wavelengths of photons, typically decreasing as the irradiation wavelength approaches the absorption edge because of the lower absorption coefficients and longer migration distances for photoexcited carriers. Therefore, the above efficiencies are usually determined under monochromatic irradiation. Note that in most cases, AQY, IPCE, and APCE depend on the light intensity. A number of previous studies have reported that AQY decreased with increasing light intensity because, in the high-intensity region, charge recombination was a second-order reaction with respect to light intensity [4]. If the AQY and IPCE spectra are independent of the photon flux, it should be possible to reproduce the observed water-splitting rates and photocurrents from the power spectrum of a light source and the AQY and IPCE spectra, respectively.

4.3 *Test Reaction with Sacrificial Reagents*

Photocatalytic water splitting remains challenging because it is an uphill reaction associated with the rapid recombination of photoexcited electrons and holes. Indeed, the kinds of photocatalysts that are capable of steady overall water splitting under visible light are very limited [5–7, 9]. An electron donor such as methanol and an electron acceptor such as the silver cation are often added to the reaction solution to measure the photocatalytic activity for the hydrogen and oxygen reactions, respectively [5, 44]. These additives are more reactive than water, consumed irreversibly, and are called sacrificial reagents. For example, the standard electrode potential for the reduction of carbonic acid to methanol ($E^{o}_{H_2CO_3(aq)/CH_3OH}$, +0.04 V) is more negative than that for the reduction of oxygen to water ($E^{o}_{O_2/H_2O}$, +1.23 V). That is, methanol is thermodynamically more easily oxidized than water. Similarly, the standard electrode potential for the reduction of a proton to hydrogen ($E^{o}_{H^+/H_2}$, 0 V by definition) is more negative than that for the reduction of a silver cation to metallic silver ($E^{o}_{Ag^+/Ag}$, +0.80 V). Therefore, the silver cation is more easily reduced than the proton, from a thermodynamic perspective. Undesirable recombination reactions can be suppressed in the presence of sacrificial reagents because photoexcited electrons and holes can be consumed by the sacrificial reagents irreversibly and rapidly. The overall reactions in the presence of methanol or silver ions can be expressed as follows:

$$CH_3OH + H_2O \rightarrow CO_2 + 3H_2 \tag{11}$$

$$4Ag^+ + 2H_2O \rightarrow O_2 + 4Ag^0 + 4H^+ \tag{12}$$

Clearly, the energy gains for the above reactions are much smaller than that for water splitting. Thus, these reactions are not intended for solar energy conversion. Sacrificial photocatalytic water splitting can proceed as long as the thermodynamic requirement for hydrogen or oxygen evolution is satisfied. The bottom of the conduction band should be more negative than the hydrogen evolution potential when hydrogen is evolved from an aqueous methanol solution under light irradiation. Likewise, the top of the valence band should be more positive than the oxygen evolution potential when oxygen is evolved from an aqueous silver ion solution. Note that a photocatalyst is not necessarily active for overall water splitting, even if it is active for both the sacrificial hydrogen and oxygen evolution reactions, because charge separation is strongly supported by sacrificial reagents. Moreover, the determinants of the photocatalytic activity for sacrificial hydrogen and oxygen evolution are not necessarily identical [44]. However, this does not present a problem for the estimation of the charge separation efficiency of photoelectrodes, where either the reduction or the oxidation reaction proceeds on the surface. Given that sacrificial reagents consume all of the photoexcited carriers reaching the surface, a photocurrent observed in the presence of sacrificial reagents reflects the flux of minority carriers that have escaped recombination with majority carriers. If the absorptance of a photoelectrode is available, one can estimate the efficiencies of charge separation inside the photoelectrode and charge injection on the surface, as described in a later subsection.

5 Recent Advancements in Photocatalytic Water Splitting

5.1 Overall Water Splitting with a Single Photocatalyst

5.1.1 Oxide Photocatalysts Active Under UV Light Illumination

Some transition metal oxides with a d^0 electronic configuration and typical metal oxides with a d^{10} electronic configuration are active for photocatalytic water splitting under UV light illumination. TiO_2 [45] and $SrTiO_3$ [46, 47] modified with cocatalysts were reported in 1980 as the first reliable materials to be active for photocatalytic overall water splitting. The photocatalytic activity of unmodified $SrTiO_3$ for overall water splitting is negligible due to a lack of hydrogen evolution sites [48]. In fact, oxide photocatalysts generally exhibit decent activity for the sacrificial oxygen evolution reaction but not for the sacrificial hydrogen evolution reaction in the absence of cocatalysts. This suggests that the surface of oxide photocatalysts is essentially inactive for hydrogen evolution. Pt is a well-known hydrogen evolution catalyst. However, modification of photocatalysts with Pt as a hydrogen evolution cocatalyst is not effective in photocatalytic overall water splitting, because Pt catalyzes the formation of water from hydrogen and oxygen, the product gases of the photocatalytic reaction. In the earliest studies, water vapor was split into hydrogen and oxygen, and the formation of water (back reaction) on cocatalysts was avoided by wetting the photocatalysts with a small amount of water, presumably to suppress exposure of the cocatalyst surface [45], or by using cocatalysts that were less active for water formation such as NiO [46]. Subsequent studies focused primarily on photocatalytic water splitting in the aqueous phase, because the rate of liquid water splitting was generally higher than that of water vapor splitting [49]. It was revealed that Ni/NiO core/shell particles (Fig. 11a) were more effective as hydrogen evolution sites for water splitting using $SrTiO_3$ than NiO [48, 50], presumably because the Ni metal between NiO and $SrTiO_3$ facilitated electron transfer between the photocatalyst and the cocatalyst.

Subsequently, certain layered oxides were found to split water efficiently. $K_4Nb_6O_{17}$ is a layered oxide with a unique ion-exchange capability. Two alternating types of interlayer spaces are formed by $Nb_6O_{17}^{4-}$ niobate layers, and K^+ cations are

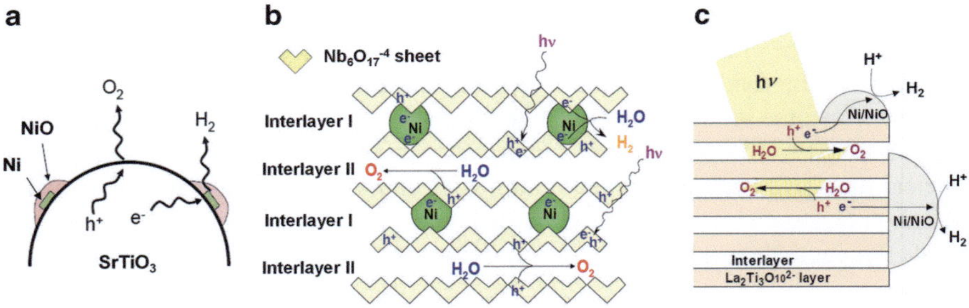

Fig. 11 Several oxide photocatalysts active for overall water splitting under UV light. (**a**) NiO/Ni/$SrTiO_3$, (**b**) Ni/$K_4Nb_6O_{17}$, and (**c**) NiO/Ni/$Rb_2La_2Ti_3O_{10}$. Reprinted with permission from [16], Copyright © 1997, American Chemical Society; [50], Copyright © 1986, American Chemical Society; and [51], Copyright © 1989, Elsevier

distributed in the interlayer spaces. K^+ cations are readily replaced by Ni^{2+} cations in the first type of interlayer space (interlayer space I), but not in the second (interlayer space II). Therefore, it is possible to selectively load $K_4Nb_6O_{17}$ with Ni metal particles in interlayer space I. The proposed reaction mechanism for overall water splitting using Ni/$K_4Nb_6O_{17}$ is shown in Fig. 11b [51]. Photoexcited electrons migrate to the Ni metal particles in interlayer space I and reduce H^+ to hydrogen, while photoexcited holes react with water in interlayer space II to form oxygen. Since the sites for hydrogen and oxygen evolution are physically separated by niobate layers, it is believed that the formation of water from the gaseous products is suppressed. Moreover, hydrated water molecules in the interlayer spaces are involved in the water-splitting reaction, which is advantageous in that the migration distance for photoexcited carriers from the photocatalyst bulk to the surface reaction sites is reduced. Since the discovery of the beneficial effects of the layered structure of $K_4Nb_6O_{17}$ on the photocatalytic activity for overall water splitting, many other semiconductor materials with unique crystal structures have been investigated [52, 53]. For example, $Rb_2La_2Ti_3O_{10}$ is a Ruddlesden–Popper layered perovskite oxide consisting of $La_2Ti_3O_{10}^{2-}$ perovskite layers containing Rb^+ cations in the interlayer space. The interlayer is hydrated to form $Rb_2La_2Ti_3O_{10}{\cdot}H_2O$ under ambient conditions. $Rb_2La_2Ti_3O_{10}$ shows good photocatalytic activity for overall water splitting under UV light irradiation when the external surface is modified with a Ni/NiO core/shell cocatalyst [16]. It was suggested that the reaction sites for photoexcited electrons and holes were separated at the external surface and the interlayer spaces, as shown in Fig. 11c, and that hydrated water was involved in the water-splitting reaction, similar to the case for the Ni/$K_4Nb_6O_{17}$ photocatalyst.

Today, $NaTaO_3$ doped with La [19] and Ga_2O_3 doped with Zn [54] exhibit the highest known water-splitting rates after modification with NiO and $Rh_{2-y}Cr_yO_3$, respectively, the latter being a solid solution of Rh_2O_3 and Cr_2O_3 as described later, as hydrogen evolution cocatalysts. As with $SrTiO_3$, cocatalyst loading is essential for achieving efficient photocatalytic water splitting with these photocatalysts. Loading with NiO and $Rh_{2-y}Cr_yO_3$ significantly improves the hydrogen evolution activity and thus enables overall water splitting. In addition, a carrier dynamics study based on transient absorption spectroscopy found that photoexcited electrons in the conduction band were quenched by the loading of a hydrogen evolution cocatalyst [55]. This indicates that electrons are trapped by the cocatalyst and separated from photoexcited holes. Therefore, cocatalysts can facilitate both charge separation and surface kinetics. It is likely preferable that an ohmic contact is formed at the interface of a photocatalyst and a hydrogen evolution cocatalyst so that electrons can be injected preferentially into the cocatalyst. Otherwise, the cocatalyst would also collect photoexcited holes and function as a recombination center.

5.1.2 (Oxy)nitride Photocatalysts for Visible-Light-Driven Water Splitting

The types of non-oxide photocatalysts that are active for overall water splitting are limited due to the relatively low reactivity of the photoexcited carriers because of their lower band gap energies and difficulties in preparing materials of adequate quality. However, certain oxynitride photocatalysts show activity for overall water splitting after modification with hydrogen evolution cocatalysts. For example, $(Ga_{1-x}Zn_x)(N_{1-x}O_x)$ and $(Zn_{1+x}Ge)(N_2O_x)$ loaded with hydrogen evolution cocatalysts can split water [5]. In particular, $(Ga_{1-x}Zn_x)(N_{1-x}O_x)$ modified with $Rh_{2-y}Cr_yO_3$ has the highest AQY for overall water splitting using a single photocatalyst under visible light (5.1 % at 410 nm) [14]. To analyze the roles of the respective elements in the $Rh_{2-y}Cr_yO_3$ cocatalyst, the photocatalytic activity of $(Ga_{1-x}Zn_x)(N_{1-x}O_x)$ modified with various amounts of rhodium and chromium oxides was investigated, as shown in Table 1 [56]. $(Ga_{1-x}Zn_x)(N_{1-x}O_x)$ loaded with 1 wt% rhodium oxide alone generated hydrogen at a relatively slow rate. When chromium oxide was coloaded with 1 wt% rhodium oxide, the water-splitting rate drastically increased, peaking at a coloading of 1.5 wt% chromium oxide, and stoichiometric water splitting occurred. The cocatalyst particles were 10–30 nm in size and consisted of trivalent rhodium

Table 1 Effect of the composition of the $Rh_{2-y}Cr_yO_3$ cocatalyst on the photocatalytic activity of $(Ga_{1-x}Zn_x)(N_{1-x}O_x)$ [56]

Amount of loaded metal oxides/wt%[a]		Rate of gas evolution/mmol h^{-1} [b]	
Rh_2O_3	Cr_2O_3	H_2	O_2
1.0	0	<0.1	0
1.0	0.5	1.2	0.6
1.0	1.0	2.2	1.1
1.0	1.5	3.8	1.9
1.0	2.0	2.9	1.4
1.0	2.5	2.7	1.4
0	1.5	0	0

[a]Metallic content
[b]Reaction conditions: catalyst (0.3 g); distilled water (370–400 mL); light source: a 450 W high-pressure mercury lamp ($\lambda > 300$ nm)

Table 2 Photocatalytic activities of $(Ga_{1-x}Zn_x)(N_{1-x}O_x)$ in the presence of sacrificial reagents

		Activity/mmol h^{-1}[a]	
Cocatalyst	Reaction solution	H_2[b]	O_2[c]
None	10 vol% CH_3OH aq	0	–
$Rh_{2-y}Cr_yO_3$	10 vol% CH_3OH aq	0.36	–
RuO_2	10 vol% CH_3OH aq	0.04	–
Rh	10 vol% CH_3OH aq	0.28	–
None	10 mM $AgNO_3$ aq	–	0.55
$Rh_{2-y}Cr_yO_3$	10 mM $AgNO_3$ aq	–	0.31
RuO_2	10 mM $AgNO_3$ aq	–	0.60

[a]Reaction conditions: catalyst (0.3 g); reaction solution (370 mL); light source: a 450 W high-pressure mercury lamp equipped with an aqueous $NaNO_2$ solution filter ($\lambda > 400$ nm)
[b]Steady rate of gas evolution
[c]Initial rate of gas evolution. Reprinted from [40] with permission.

and chromium oxides, although the composition varied somewhat among the particles [57]. Therefore, the $Rh_{2-y}Cr_yO_3$ cocatalyst is usually regarded as a mixed oxide of corundum-type Rh_2O_3 and Cr_2O_3.

The function of the $Rh_{2-y}Cr_yO_3$ cocatalyst was confirmed using a sacrificial electron donor (methanol) and an acceptor (silver cation). The photocatalytic activity for the hydrogen evolution and oxygen evolution reactions in the presence of the sacrificial reagents is listed in Table 2 [40]. $(Ga_{1-x}Zn_x)(N_{1-x}O_x)$ without cocatalyst loading showed no activity for the photocatalytic hydrogen evolution reaction. However, hydrogen was evolved after loading $Rh_{2-y}Cr_yO_3$ at a rate comparable to that for the case in which metallic Rh was loaded. The rates for hydrogen evolution on these photocatalysts were roughly ten times higher than on RuO_2-loaded samples, demonstrating the high performance of $Rh_{2-y}Cr_yO_3$ for H_2 evolution. On the other hand, loading $Rh_{2-y}Cr_yO_3$ lowered the oxygen evolution activity, indicating that $Rh_{2-y}Cr_yO_3$ did not enhance oxygen evolution, unlike RuO_2. The $Rh_{2-y}Cr_yO_3$ loading probably prevented light absorption by the photocatalyst and/or blocked the excitation light. This experiment indicated that $Rh_{2-y}Cr_yO_3$ functions as an efficient hydrogen evolution site. Another important feature of the $Rh_{2-y}Cr_yO_3$ cocatalyst is its high selectivity for the H_2 evolution reaction [40]. The water-splitting rate is significantly reduced in the presence of oxygen when RuO_2 is used as a cocatalyst, because the photoreduction of oxygen competes with the hydrogen evolution reaction on RuO_2. In contrast, the water-splitting rate for $(Ga_{1-x}Zn_x)(N_{1-x}O_x)$ modified with $Rh_{2-y}Cr_yO_3$ is largely independent of the partial pressure of oxygen in the reaction system. Therefore, the high selectivity of $Rh_{2-y}Cr_yO_3$ for hydrogen evolution contributes to the high activity of $Rh_{2-y}Cr_yO_3/(Ga_{1-x}Zn_x)(N_{1-x}O_x)$. It is also noteworthy that the $Rh_{2-y}Cr_yO_3$ cocatalyst is generally applicable to various other photocatalysts [56].

As mentioned earlier, noble-metal nanoparticles cannot be applied as a cocatalyst for overall water splitting because side reactions such as the formation of water

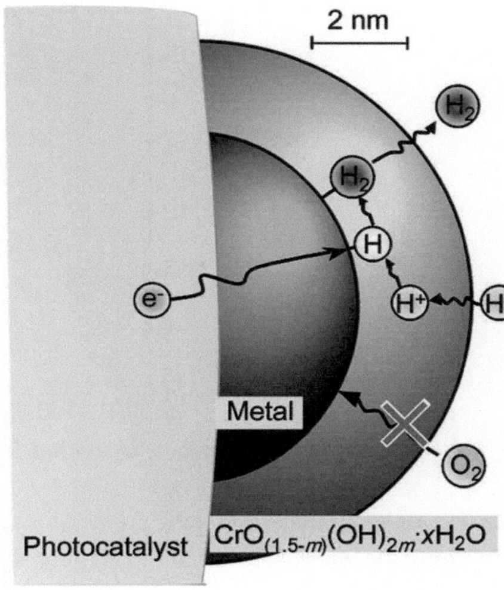

Fig. 12 Schematic model of the H_2 evolution reaction on a core/shell noble-metal/Cr_2O_3 cocatalyst particle for photocatalytic overall water splitting. Reprinted from [58] with permission. Copyright © 2009 American Chemical Society

from hydrogen and oxygen and oxygen reduction occur readily on noble metals. However, undesirable side reactions can be suppressed by using a core/shell-structured cocatalyst, in which noble metal particles are covered by a thin layer of metal oxide. Such core/shell-type nanocomposites can be deposited on $(Ga_{1-x}Zn_x)(N_{1-x}O_x)$ by sequential photodeposition [41]. To form an Rh–Cr_2O_3 core/shell cocatalyst, Rh nanoparticles are first photodeposited in a solution containing Rh cations such as Na_3RhCl_6. Subsequently, a Cr_2O_3 shell is photodeposited in a solution of hexavalent chromium species such as K_2CrO_4. Stepwise photodeposition results in the formation of an Rh–Cr_2O_3 core/shell nanocomposite with a chromium oxide shell ca. 2-nm thick. $(Ga_{1-x}Zn_x)(N_{1-x}O_x)$ modified with the Rh–Cr_2O_3 core/shell cocatalyst shows photocatalytic activity for overall water splitting under visible-light irradiation. Note that the addition of Rh/$(Ga_{1-x}Zn_x)(N_{1-x}O_x)$ without a Cr_2O_3 shell to the reaction solution halts photocatalytic water splitting because of immediate water formation on the Rh metal particles. It is therefore clear that a Cr_2O_3 shell can effectively prevent the backward reaction. Electrochemical analysis revealed that the Cr_2O_3 shell suppressed water formation on the Rh nanoparticles because of the selective permeability of the ultrathin hydrated chromia layer to protons and gaseous hydrogen. Since oxygen molecules cannot access the Rh surface, the reductive side reactions involving oxygen molecules are negligible [58], as shown in Fig. 12.

5.1.3 Coloading of Hydrogen and Oxygen Evolution Cocatalysts

The loading of oxygen evolution cocatalysts has a weaker effect on the photocatalytic activity than the loading of hydrogen evolution cocatalysts. For example, the loading of Mn_3O_4 alone as an oxygen evolution cocatalyst on $(Ga_{1-x}Zn_x)(N_{1-x}O_x)$ does not allow for overall water splitting because of the lack of hydrogen evolution

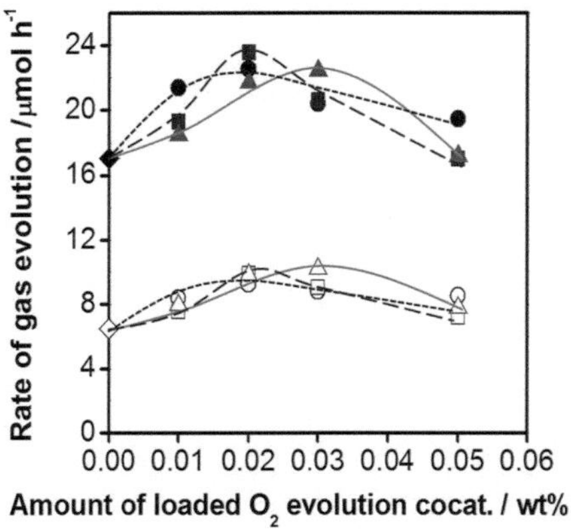

Fig. 13 Photocatalytic activity of GaN–ZnO coloaded with different O_2 evolution cocatalysts and Rh–Cr_2O_3 for water splitting under visible light (λ>420 nm). *Circles*, *triangles*, and *squares* indicate the loading of Mn_3O_4, IrO_2, and RuO_2, respectively. *Closed* and *open symbols* denote H_2 and O_2, respectively. Reprinted from [60] with permission. ©2014 Wiley-VCH Verlag GmbH & Co. KGaA, Weinheim

sites [59]. The loading of Mn_3O_4, RuO_2, and IrO_2 as oxygen evolution cocatalysts is effective when they are coloaded on the photocatalyst along with a hydrogen evolution cocatalyst such as an Rh–Cr_2O_3 core/shell composite [59, 60]. However, the optimal loading amounts of oxygen evolution cocatalysts were below 0.05 wt% (see Fig. 13), whereas the typical loading amount of $Rh_{2-y}Cr_yO_3$ as a hydrogen evolution cocatalyst was 1 wt% (for Rh alone) or larger [60]. Moreover, coloading with oxygen evolution cocatalysts improved the water-splitting rates by a factor of 1.4 at most, regardless of which cocatalyst was used. These results suggest that the photocatalytic activity was limited by the hydrogen evolution process. However, coloading with oxygen evolution cocatalysts significantly improved the durability of the non-oxide photocatalysts, because the photooxidation of water and the photocatalyst itself compete on the surface. It was reported that coloading RuO_2 on $(Ga_{1-x}Zn_x)(N_{1-x}O_x)$ along with Rh–Cr_2O_3 suppressed the loss of nitrogen on the photocatalyst during photocatalytic water splitting [61]. As a result, the deactivation of the photocatalyst was also suppressed.

Recently, ZrO_2-modified TaON (ZrO_2/TaON) was also reported to be active for overall water splitting when coloaded with cocatalysts for both hydrogen and oxygen evolution. This was the first report of overall water splitting by a transition metal oxynitride [62]. TaON is known to generate hydrogen and oxygen under visible-light illumination in the presence of methanol and silver cations, respectively. TaON exhibited a decent AQY for the sacrificial oxygen evolution reaction, but not for the sacrificial hydrogen evolution reaction, even with cocatalyst modifications. In addition, TaON generated only a small amount of hydrogen and no oxygen when it was applied to overall water splitting. These results suggest that photoexcited electrons did not migrate to cocatalysts effectively because of a high

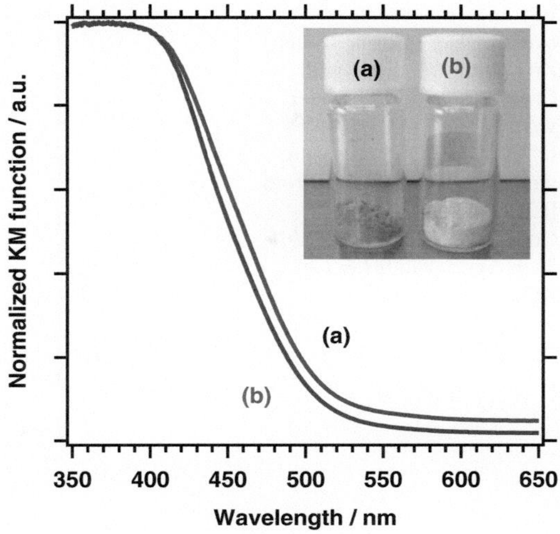

Fig. 14 Diffuse reflectance spectra of (*a*) TaON and (*b*) ZrO_2/TaON. Adapted from [63] with permission. Copyright © 2008 The Chemical Society of Japan

defect density in TaON and/or because photoexcited holes were consumed by the self-oxidation of TaON rather than water oxidation. Therefore, it was necessary to improve the TaON synthesis conditions and the cocatalyst loading methods.

Modifying Ta_2O_5 with ZrO_2 prior to nitridation effectively suppressed the reduction of pentavalent Ta ions during the nitridation [63]. Figure 14 shows absorption spectra of TaON and ZrO_2-modified TaON [63]. Unmodified TaON exhibits light absorption attributable to reduced Ta species at a wavelength longer than 500 nm, the absorption edge wavelength of TaON. Such background light absorption is significantly suppressed for ZrO_2-modified TaON. This is likely because ionic Zr^{4+} is not significantly reduced to generate anion vacancies during the nitridation. Photoluminescence spectroscopy and PEC measurements revealed that the n-type semiconducting character of TaON was moderated as a result of the ZrO_2 modification. The photocatalytic activity of TaON for sacrificial hydrogen evolution improved as a result of the modification with ZrO_2. Furthermore, the activity of a Z-scheme water-splitting system based on TaON as a hydrogen evolution photocatalyst, WO_3 as an oxygen evolution photocatalyst, and IO_3^-/I^- as a redox shuttle also improved when ZrO_2-modified TaON was employed instead of conventional TaON [64].

Subsequently, the coloading of a core/shell-type hydrogen evolution cocatalyst and an oxygen evolution cocatalyst was found to enable overall water splitting using ZrO_2/TaON [62]. ZrO_2/TaON was modified with an RuO_x/Cr_2O_3 core/shell-type hydrogen evolution cocatalyst and subsequently with IrO_2 as an oxygen evolution cocatalyst. ZrO_2/TaON modified with RuO_x/Cr_2O_3 showed some activity for overall water splitting under UV illumination, although the gas evolution rates decreased over time because of deactivation of the photocatalyst. When IrO_2 was coloaded as an oxygen evolution cocatalyst on ZrO_2/TaON with RuO_x/Cr_2O_3, overall water splitting proceeded steadily. By optimizing the preparation conditions for

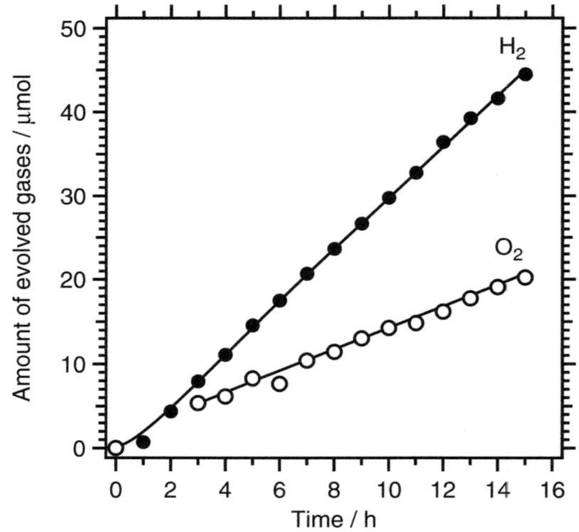

Fig. 15 Time course of overall water splitting using ZrO_2/TaON modified with Cr_2O_3/RuO_x and IrO_2 under visible light irradiation ($\lambda > 400$ nm, a high-pressure mercury lamp). Adapted from [62] with permission. Copyright © 2013 WILEY-VCH Verlag GmbH & Co. KGaA, Weinheim

the photocatalyst/cocatalyst composite, overall water splitting was achieved, even under visible-light irradiation, as shown in Fig. 15. Note that coloading with RuO_x and IrO_2 did not lead to oxygen evolution. These results highlight the importance of activation and stabilization of the photocatalyst by the coloading of hydrogen and oxygen evolution cocatalysts and the suppression of side reactions by the ultra-thin chromia layer.

5.2 Z-Scheme Water Splitting

Z-scheme water splitting is advantageous over one-step photoexcitation water splitting in that a wider range of visible wavelengths can be used, as semiconductors with either water reduction or oxidation potentials can be employed. In fact, WO_3 [64] and $BiVO_4$ [65, 66] have been applied as oxygen evolution photocatalysts in Z-scheme water splitting. Furthermore, $BaTaO_2N$, which has an absorption edge wavelength of 660 nm, is applicable as a hydrogen evolution photocatalyst [67]. On the other hand, the number of photons required to generate a given amount of hydrogen is double of that required for one-step splitting. Furthermore, the Z-scheme requires a balance between the photocatalytic activity of the hydrogen and oxygen evolution photocatalysts under shared pH conditions and redox mediator concentrations.

The Z-scheme system was inspired by photosynthesis in green plants and similarities in the excitation and transfer processes of photoexcited electrons. The idea of Z-scheme water splitting was originally suggested in the late 1970s by Bard et al. [68]. Since the conception of the Z-scheme, semiconductor photocatalysts applicable to Z-scheme water splitting have been extensively investigated. As recently reviewed by Maeda [7], earlier work focused on the half-reactions of Z-scheme water splitting, that is, either hydrogen evolution or oxygen evolution in the presence of reversible redox couples. If redox couples are regenerated spontaneously via photochemical, photocatalytic, or photoelectrochemical processes, the cycle of Z-scheme water splitting is complete. Fujihara et al. constructed a Z-scheme water-splitting system using rutile-type TiO_2 and two redox mediators (Br_2/Br^- and Fe^{3+}/Fe^{2+}) [69]. However, stoichiometric evolution of hydrogen and oxygen was not achieved, supposedly because of side reactions involving the redox mediators.

Stoichiometric Z-scheme water splitting was first demonstrated by Sayama et al. in 1997 using RuO_2-loaded WO_3 powder suspended in an aqueous $FeSO_4$ solution under UV irradiation ($\lambda > 200$ nm) [70]. In this system, Fe^{2+} is excited by UV light, followed by photochemical reduction of water to form H_2 and Fe^{3+}, while WO_3 photocatalyzes water oxidation using Fe^{3+} as an electron acceptor. A later Z-scheme system consisted of Pt-modified anatase TiO_2 as the H_2 evolution photocatalyst and bare rutile TiO_2 as the O_2 evolution photocatalyst under UV irradiation ($\lambda > 300$ nm) in the presence of an iodate/iodide (IO_3^-/I^-) shuttle redox mediator [71]. This Z-scheme for water splitting employed semiconductor photocatalysts for both H_2 and O_2 evolution. Since the band-edge potentials and band gap energies of semiconductors can be modified by changing their composition, various photocatalysts may be applicable to Z-scheme water splitting. Z-scheme water splitting under visible light ($\lambda > 420$ nm) was achieved using suspended particles of Pt-loaded $SrTiO_3$ doped with Cr and Ta as a hydrogen evolution photocatalyst and Pt-loaded WO_3 as an oxygen evolution photocatalyst with an IO_3^-/I^- redox pair [72]. These findings established the principles of Z-scheme water splitting under visible light using various H_2 and O_2 evolution photocatalysts in the presence of IO_3^-/I^- or Fe^{3+}/Fe^{2+} redox mediators.

The highest AQY of 6.3 % at 420 nm was achieved with a Z-scheme system based on Pt-loaded ZrO_2/TaON as a hydrogen evolution photocatalyst, Pt-loaded WO_3 as an oxygen evolution photocatalyst, and IO_3^-/I^- as a redox couple (Fig. 16) [64]. To achieve efficient Z-scheme water splitting, it is critical to suppress backward

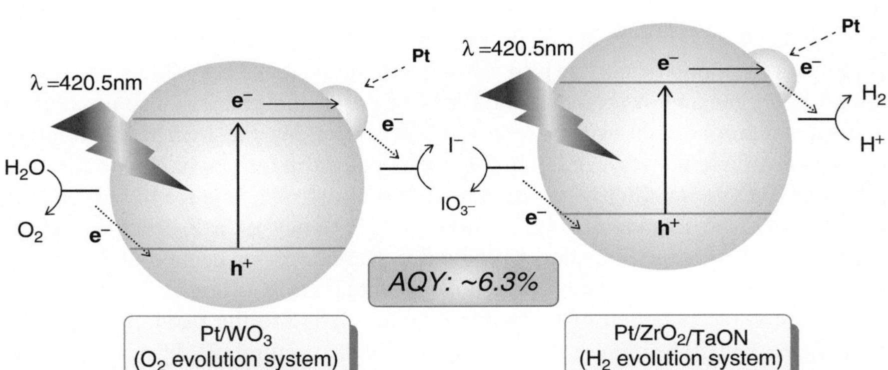

Fig. 16 Schematic reaction mechanism of Z-scheme water splitting using Pt/ZrO_2/TaON, Pt/WO_3, and IO_3^-/I^- as the hydrogen evolution photocatalyst, oxygen evolution photocatalyst, and reversible redox mediator, respectively. Reprinted with permission from [64]. ©2010 American Chemical Society

Table 3 Effect of the weight ratio of Pt/ZrO_2/TaON–Pt/WO_3–NaI in the reaction solution on the water-splitting activity under monochromatic light irradiation ($\lambda = 420.5$ nm)

Pt/ZrO_2/TaON–Pt/ WO_3–NaI/mg	Concentration of NaI/mM	Loading amt of Pt on ZrO_2/TaON/wt%	Gas evolution rate/μmol h^{-1}		AQY/%
			H_2	O_2	
50–100–15	1.0	1.0	3.3	1.6	2.7
25–50–7.5	0.5	1.0	5.5	2.7	4.3
10–20–3.0	0.2	1.0	4.6	2.1	3.5
10–20–15	1.0	1.0	4.3	1.4	2.3
25–50–7.5	0.5	0.5	7.5	3.8	6.3
25–50–7.5	0.5	0.1	<1	n.d.	

reactions involving redox mediators that are thermodynamically more favorable than water splitting. Forward reactions should proceed efficiently to avoid side reactions. I^- ions as a donor and IO_3^- ions as an acceptor react efficiently on Pt-loaded ZrO_2/TaON and Pt-loaded WO_3, respectively. As a result, undesirable backward reactions such as the oxidation of I^- ions on Pt-loaded WO_3 and the reduction of IO_3^- on Pt-loaded ZrO_2/TaON are minimized. As mentioned previously, modifying TaON with ZrO_2 was essential to render TaON active for hydrogen evolution. In addition, reaction conditions such as Pt loading and photocatalyst and redox mediator concentrations should be optimized so that photoexcited electrons are consumed in hydrogen evolution instead of the reduction of IO_3^- ions, as shown in Table 3 [64].

A combination of Ru-loaded $SrTiO_3$ doped with Rh ($SrTiO_3$:Rh) for hydrogen evolution and $BiVO_4$ for oxygen evolution enables efficient Z-scheme water splitting in the presence of various reversible redox couples [73]. In particular, the use of $[Co(bpy)_3]^{3+/2+}$ as a redox couple exhibited an excellent AQY (2.1 % at 420 nm) for overall water splitting, although this redox couple was not effective with other combinations of photocatalysts [66]. This Z-scheme system works stably and shows decent hydrogen and oxygen evolution under natural sunlight. It is also possible to generate hydrogen and oxygen separately by dispersing the hydrogen evolution photocatalyst and the oxygen evolution photocatalyst in a two-chamber cell separated by a membrane filter penetrating the redox couple. The choice of reaction pH plays a key role in maximizing the reaction efficiency of a Z-scheme system, because the band-edge potentials of the oxide photocatalysts depend on the pH of the reaction solution, while the redox potential of $[Co(bpy)_3]^{3+/2+}$ does not. Accordingly, the potential of the redox couple relative to the band edges is also dependent on pH. The highest activity was observed at neutral pH. In an acidic solution, the reduction of $Co(bpy)^{3+}$ on $BiVO_4$ for oxygen evolution is suppressed because the potential of $[Co(bpy)_3]^{3+/2+}$ shifts to a more negative value than the band edges of $SrTiO_3$:Rh and $BiVO_4$. It was also reported that oxidation of $Co(bpy)^{2+}$ strongly competes with oxygen evolution on $BiVO_4$. As a result, the activity for Z-scheme water splitting decreased drastically with decreasing pH. On the other hand, the oxidation of $Co(bpy)^{2+}$ for hydrogen evolution on Ru-loaded $SrTiO_3$:Rh was suppressed in a basic solution, leading to a lower activity for water splitting.

Interestingly, a mixture of Ru-loaded $SrTiO_3$:Rh and $BiVO_4$ generated hydrogen and oxygen in the stoichiometric ratio, even in the absence of redox mediators [65]. The activity for overall water splitting was sensitive to the pH of the reaction solution, peaking sharply at pH 3.5. Optical microscopy of suspensions of Ru-loaded $SrTiO_3$:Rh and $BiVO_4$ revealed that particles of both photocatalysts aggregated best at pH 3.5, whereas there was little aggregation at pH 7. This was due to the sign of the electrostatic charge of the photocatalyst particles. The isoelectric point of Ru-loaded $SrTiO_3$:Rh was approximately 4, indicating that the powder is negatively charged at pH>4 and positively charged at pH<4. On the other hand, $BiVO_4$ was negatively charged even at pH 2. Therefore, Ru-loaded $SrTiO_3$:Rh and $BiVO_4$ had an attractive electrostatic force in aqueous solution at pH<4, forming aggregates composed of both photocatalysts. Note that water splitting did not proceed when the two photocatalysts were physically separated. This indicates that Z-scheme water splitting took place via interparticle electron transfer, and physical contact between the two photocatalysts was necessary to complete the process (Fig. 17) [65].

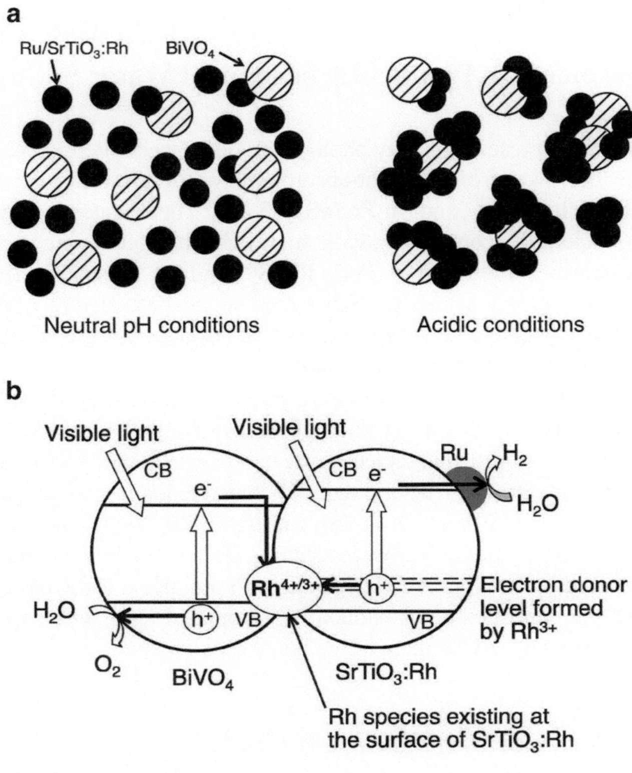

Fig. 17 Mechanism of water splitting using a Z-scheme photocatalysis system driven by electron transfer between H_2- and O_2-photocatalysts. (**a**) Suspension of Ru/$SrTiO_3$:Rh and $BiVO_4$ in neutral and acidic conditions. (**b**) Scheme of photocatalytic water splitting. Reprinted with permission from [65] ©2009 American Chemical Society

Z-scheme water splitting via interparticle electron transfer can have advantages over conventional Z-scheme water splitting with reversible redox couples because thermodynamically favorable backward reactions involving reversible redox couples do not occur. Therefore, (oxy)nitride photocatalysts that have a low selectivity for the forward reaction can be applied to Z-scheme water splitting without elaborate surface modification. For example, a suspension of Ru-loaded $SrTiO_3$:Rh and Ir/CoO_x-coloaded Ta_3N_5 shows activity for Z-scheme water splitting and potentially utilizes photons with wavelengths up to 600 nm for oxygen evolution on Ta_3N_5 [74]. Unlike $BiVO_4$, the use of Ta_3N_5 did not lead to steady water splitting, although Ta_3N_5 is active for oxygen evolution as long as photoexcited electrons are scavenged. One major challenge in this Z-scheme water-splitting system was the facilitation of interparticle electron transfer between $SrTiO_3$:Rh and Ta_3N_5. It was found that Ir loaded on Ta_3N_5 effectively captured the photoexcited electrons of Ta_3N_5, facilitated electron transfer to $SrTiO_3$:Rh via physical contact, and thus enabled Z-scheme water splitting. In addition, coloading with CoO_x as an oxygen evolution cocatalyst was found to improve the durability of the Z-scheme system, similarly to overall water splitting on $(Ga_{1-x}Zn_x)(N_{1-x}O_x)$ [61] and TaON [62].

6 Advancements in Photoelectrochemical Water Splitting

In PEC water-splitting research, many photoelectrodes, especially photocathode materials, have been borrowed from the light-absorbing layer of thin-film solar cells such as Si, GaAs, GaP, Cu(In,Ga)Se_2, and Cu_2ZnSnS_4 [75–80]. These materials have high light absorption coefficients (except for Si), wide solar light absorption ranges, and efficient charge carrier densities. Furthermore, As^{3-}, P^{3-}, S^{2-}, and Se^{2-} anions in these materials would be vulnerable to (photo)oxidative corrosion but are protected by a negative photocurrent. The difference between cells in solar panels and photoelectrodes is that the latter does not require a flat surface and a panel substrate to conduct and circulate electricity from the semiconductor. The PEC water reduction or oxidation reaction can take place on photoelectrodes with any morphology, such as nanotubes, nanorods, nanotrees [77, 81]. The substrates of the photoelectrode can be a panel, belt, rod, or any shape to which the semiconductor can be affixed. We have reported the preparation of a Cu_2ZnSnS_4 photocathode using a Mo mesh substrate [80]. In this research, the metal precursors of a Zn/Sn/Cu multilayer were electrodeposited on the individual wires of a Mo mesh, followed by heating in a H_2S gas flow to produce a Cu_2ZnSnS_4 photoelectrode. Because the deposited semiconductors surrounded their wire substrates, they could not easily peel away from the wire without breaking apart.

6.1 Photoelectrode Preparation Methods

To prepare a photoelectrode, semiconductor precursors can be directly deposited on a conductive substrate, followed by posttreatment. Another route is to synthesize the semiconductor powder first and then deposit it on a substrate. Figure 18 lists several

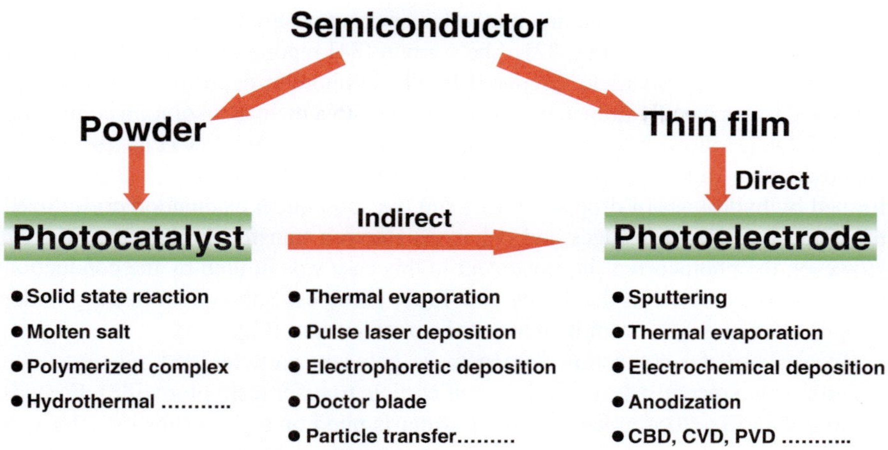

Fig. 18 Photocatalyst and photoelectrode preparation methods

different methods of fabricating photoelectrodes. Thin-film photoelectrodes can be prepared via different routes. For example, it was reported that $CuInS_2$ photocathodes can be prepared by sputtering, evaporation, and electrodeposition, and Ta_3N_5 photoanodes can be prepared by electrophoretic deposition and anodization [82–86]. The technology used to prepare a photoelectrode depends upon the chemical properties of the photoelectrode components. For example, a $CuInS_2$/Mo thin-film photoelectrode can be prepared by electrodepositing Cu and In metals on a Mo substrate in aqueous solution, but it is difficult to prepare $CuGaS_2$ by the same method because the reduction potential of the Ga^{3+} cation is more negative than that of H_2O [82].

It is preferable to prepare a photoelectrode by a direct route, as shown in Fig. 18, because the direct growth of a semiconductor on a conductive substrate usually produces a good contact at the semiconductor–substrate interfaces, which is crucial for the transfer of photogenerated charge carriers. Recently, the authors' group reported the preparation of Ta_3N_5 nanorods grown vertically on a Ta substrate via anodization of an Al/Ta metal plate followed by nitridation in NH_3 flow [85, 86]. After doping with Ba, a maximum HC-STH of 1.5 % was achieved on the prepared Ta_3N_5/Ta photoanode [86].

It is difficult to prepare certain semiconductors by any direct method, especially bi-transitional metal oxynitrides or oxysulfides such as $LaTiO_2N$, $BaTaO_2N$, and $Sm_2Ti_2S_2O_5$. Therefore, these semiconductor-based photoelectrodes must be fabricated through an indirect route, such as by synthesizing highly crystallized powder precursors and then depositing them on a conductive substrate by a doctor blade (paste) or electrophoresis. Unfortunately, stacked semiconductor particles on a substrate contain many particle–particle and particle–substrate boundaries, which greatly inhibit charge transfer from particle to substrate. Since the poor charge transfer capability is due to the loose physical stacking of particles, there are two potential methods of overcoming this disadvantage: the first is to combine the particles by chemical bonds, and the second is to increase contacting compactness between particles and substrate. In the first method, a necking posttreatment is applied to the fabricated photoelectrode [84, 87]. Abe's group [84] reported that the photocurrent in a TaON/FTO photoanode (prepared by electrophoretic deposition) was greatly improved by repeatedly immersing the electrode into a methanol solution containing 10 mM $TaCl_5$ followed by ordinal calcination in air and NH_3 flow (Fig. 19). It was supposed that Ta_2O_5 bridges between the TaON particles on the FTO were first formed by hydrolysis of dropped $TaCl_5$, and the subsequent calcination crystallized the amorphous Ta_2O_5 bridges and enhanced electron transfer through the bridges. However, the photocurrent improvement in this case was limited by the conduction of the interface between the TaON and Ta_2O_5 particles. Subsequent heating under NH_3 resulted in a significant increase of the photocurrent (Fig. 19d).

To increase the contacting compactness between particles and substrate, the authors' group recently developed a novel particle transfer technology [88]. First, as illustrated in Fig. 20, semiconducting powder is piled on a glass substrate. Then, a thin metal layer (approximately 100–300 nm) is deposited by sputtering as a contact layer. The choice of metal is considered to be crucial in establishing favorable electrical contact between the semiconductor particles and the metal layer. Next, a continuous metal film, several micrometers thick, is deposited by sputtering to provide adequate conductivity and mechanical strength. The resulting metal film is transferred to another substrate using a resin, and the primary substrate is peeled off.

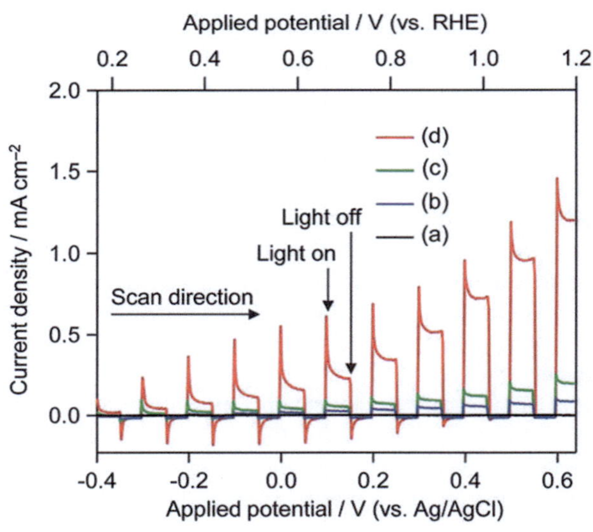

Fig. 19 Current–potential curves over TaON electrodes in aqueous 0.1 M Na_2SO_4 solution (pH 6) under chopped visible-light irradiation (λ>400 nm). (*a*) as prepared, (*b*) dropped with $TaCl_5$ and dried at room temperature, (*c*) dropped with $TaCl_5$ and calcined at 673 K in air, (*d*) heated in NH_3 at 773 K after treatment (*c*). Reprinted with permission from [84], Copyright © 2011, Royal Society of Chemistry

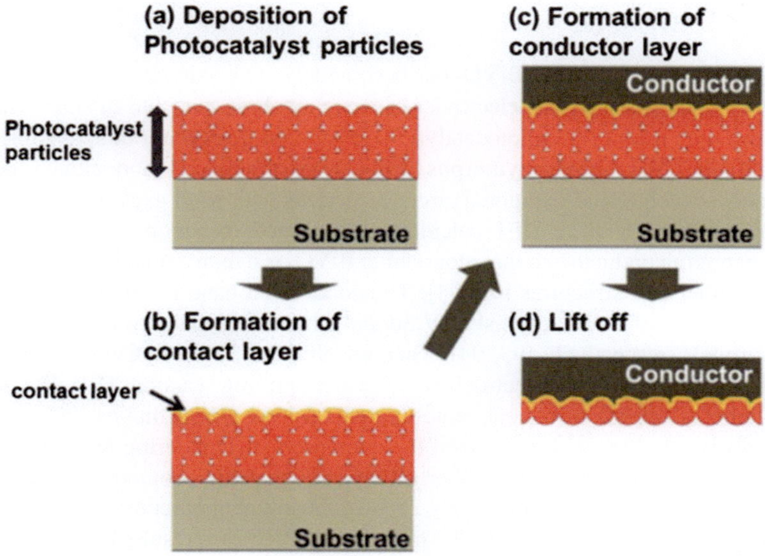

Fig. 20 Schematic diagram of the particle transfer (PT) method. Reprinted with permission from [88], Copyright © 2013, Royal Society of Chemistry

The excess powder is removed by ultrasonication in water. This method ensures adequate mechanical and electrical contact between the semiconducting particles and the metal layer. Furthermore, since any semiconductor particles that are only loosely attached to the electrodes are removed, a one-particle-thick layer covers the metal conductor film. It should be noted that this process is applicable to a variety of powdered semiconductors that can be used in the fabrication of photoelectrodes. To increase the light absorption capability of the photoelectrode, a highly crystallized, large-particle-size precursor may be preferable because only one layer of particles is deposited on the metal substrate by this method. The optimal size of the semiconductor material is dependent on its light absorption coefficient and charge separation efficiency.

6.2 *Earth-Abundant Photoelectrode Materials*

As new photoelectrochemical water-splitting systems are developed, issues relevant to practical application become increasingly important and must be considered when planning and carrying out research. Semiconductors with appropriate band properties and low cost are preferred. In recent years, semiconductor materials that contain abundant and cheap components such as $BiVO_4$, Fe_2O_3, WO_3, and Cu_2O have attracted the most attention.

6.2.1 $BiVO_4$

As an n-type semiconductor, $BiVO_4$ has been widely studied in recent years for use in both photocatalysts and photoelectrodes, since monoclinic scheelite $BiVO_4$ was found to show a high activity for photocatalytic O_2 evolution from water under visible-light irradiation [89–91]. Three polymorphs of $BiVO_4$ exist in nature: monoclinic scheelite, tetragonal scheelite, and tetragonal zircon structures with band gaps of 2.4, 2.34, and 2.9 eV, respectively [92]. DFT calculations indicated that monoclinic $BiVO_4$ is a direct-gap semiconductor, so that monoclinic $BiVO_4$ is a more suitable photo absorber than the other two structures [93, 94]. The conduction band minimum and valence band maximum of $BiVO_4$ consist of V 3d and a hybridization between the O 2p and Bi 6s orbitals, respectively [93, 94]. Theoretically, the STH of $BiVO_4$ can reach as high as 9.1 % in the photoelectrochemical water-splitting reaction. However, a pure $BiVO_4$ photoelectrode would encounter two difficulties. One is that its IPCE at lower potentials is very low, which is one of the main problems hindering its application in solar hydrogen production. The other difficulty is that the photocurrent stability of $BiVO_4$ is quite poor. In one study, only 25 % of the initial photocurrent remained after half an hour of illumination [95]. The photocurrent decrease resulted from V^{5+} dissolution into solution and oxidation products on the surface of the photoelectrode such as H_2O_2 or O_2, which act as recombination centers [92].

Nonisovalent substitutional doping often increases the carrier concentration of $BiVO_4$. Zou's group reported that the IPCE of $BiVO_4$ can be increased from 9 % to 41 % at 420 nm and 1.0 V RHE by doping 3 % Mo into the V site of $BiVO_4$ [96]. When Mo^{6+} substitutes for V^{5+}, it acts as an electron donor in the $BiVO_4$ lattice. The carrier concentration of 3 % Mo-doped $BiVO_4$ photoelectrodes increases 80-fold, compared with pure $BiVO_4$ photoelectrodes. They also investigated the influence of other dopant ions (Mo^{6+}, W^{6+}, Ta^{5+}, Zr^{4+}, Si^{4+}, Ti^{4+}, La^{3+}, Fe^{3+}, Sr^{2+}, Zn^{2+}, and Ag^{+}) on the PEC properties and found that only Mo^{6+} or W^{6+} doping significantly enhanced the PEC performance.

Electrocatalysts such as Co_3O_4, RhO_2, and cobalt–phosphate (Co–Pi) can be loaded onto the surface of $BiVO_4$ photoanodes to decrease the overpotential and improve the PEC stability [96–99]. Gamelin's group reported that the modification of W-doped $BiVO_4$ photoanode surfaces with Co–Pi resulted in a very large cathodic shift of ~440 mV in the onset potential of sustained PEC water oxidation at pH = 8 [99]. In an attempt to mend the water oxidation kinetics of these $BiVO_4$ photoelectrodes, a layer of FeOOH was deposited on the $BiVO_4$ surface as an oxygen evolution catalyst. The IPCE of bare $BiVO_4$ photoelectrodes reaches a maximum of 9 % at 350 nm and 1.2 VRHE, while that of $BiVO_4$/FeOOH achieves 44 % at the same wavelength [100].

6.2.2 Hematite (α-Fe_2O_3)

Hematite (a-Fe_2O_3), which has a band gap of 1.9–2.2 eV, has emerged as a promising photoelectrode material due to its significant light absorption, chemical stability in aqueous environments, and ample abundance (6.3 % by weight in the Earth's crust)

[101, 102]. However, its performance as a water-oxidizing photoanode has been limited by its poor optoelectronic properties, which lead to both a low light-harvesting efficiency and a large requisite overpotential for photoelectrochemical water oxidation [101]. The hole diffusion length in hematite ranges from 2 to 4 nm, shorter than that for TiO_2 (on the order of 1 μm) and WO_3 (on the order of 100 nm), which results in an ultrafast recombination of photogenerated carriers (time constants on the order of 10 ps) as well as a low minority charge mobility (0.2 $cm^2\ V^{-1}\ s^{-1}$) [92, 101–104]. The conduction band minimum of hematite consists of Fe 3d e_g, while the valence band maximum is composed of Fe 3d t_{2g} (the main part) and O 2p [102]. The absorption coefficient (α) of pure hematite increases abruptly at the band gap energy. Still higher in the electromagnetic spectrum, α continues to increase, and additional absorption features have been observed, centered on 2.4, 3.2, and 5.8 eV (with α on the order of $10^5\ cm^{-1}$) in samples of polycrystalline hematite [104, 105].

Because of their poor charge carrier transferability, hematite photoanodes are usually prepared from nanometer-scale particles so that photogenerated holes can migrate to the surface before recombination. Another method of increasing the photocurrent is to modify the surface of hematite with a proper cocatalyst to reduce the overpotential for water oxidation. Figure 21 shows that a maximal photocurrent of 12 mA cm^{-2} and an onset potential of 0.4 V vs. RHE can theoretically be achieved on a hematite photoanode under solar light irradiation [101]. Grätzel and co-workers reported the preparation of ultrathin hematite films (with a thickness around 9 nm) on Nb_2O_5 (2 nm)-covered FTO, achieving an APCE as high as 40 % at 400 nm and +1.43 V vs. RHE. The photocurrent of $Fe_2O_3/Nb_2O_5/FTO$ became saturated when the thickness of Fe_2O_3 reached 16 nm [106]. Based on the flat band potential usually reported for hematite, an external bias of only 0.3–0.4 V vs. RHE would be necessary to initiate the water-splitting reaction. However, the onset of water oxidation

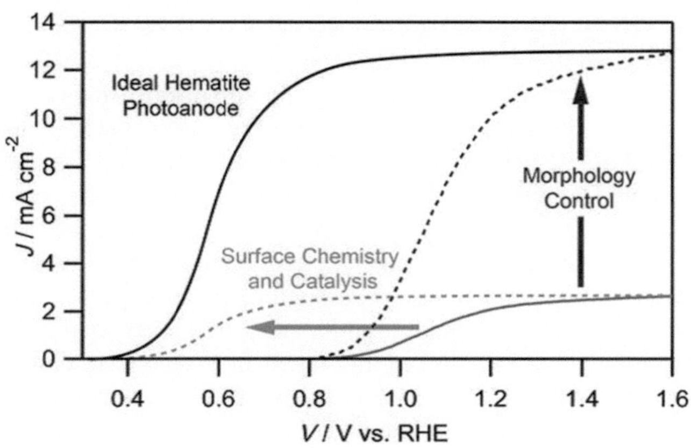

Fig. 21 A two-part strategy for improving hematite performance is shown with respect to the photocurrent density *J* vs. voltage for an idealized hematite photoanode (*solid black trace*) compared to the typical performance (*solid grey trace*) under AM1.5G 100 mW cm^{-2} simulated sunlight and the expected effects of improving the surface chemistry and morphology. Reprinted with permission from [101], Copyright © 2011, John Wiley and Sons

photocurrent is usually not observed until 0.8–1.0 V vs. RHE, even at a high pH of 13.6 (1 M NaOH) and for single-crystal electrodes [107]. Surface modifications of hematite with CoO, IrO_2, and Co–Pi have resulted in obvious reductions of onset potential [81, 108]. In 2011, a photo-assisted electrodeposition approach was used to deposit Co–Pi onto dendritic mesostructures of α-Fe_2O_3. A comparison between this approach, the electrodeposition of Co–Pi, and Co^{2+} wet impregnation showed that photo-assisted electrodeposition of Co–Pi yields superior α-Fe_2O_3 photoanodes for photoelectrochemical water oxidation. Stable photocurrent densities of 1.0 mA cm^{-2} at 1.0 V and 2.8 mA cm^{-2} at 1.23 V vs. RHE, measured under standard illumination in basic conditions, were achieved [108].

6.2.3 WO_3

WO_3, an n-type semiconductor with an experimental direct gap of 2.6 eV, is well known for its photosensitivity and stability against photocorrosion in acidic aqueous solutions [109]. DFT calculations have shown that, as expected for a metal-oxide semiconductor, the valence band of WO_3 mainly consists of O 2p bands, whereas its conduction band has a predominantly W 5d character [110]. The conduction band edges of WO_3 are more positive than the water reduction potential. With a deep valence band, WO_3 has a high activity for the photocatalytic oxidation of water or organics [111, 112].

Tungsten trioxide thin films can be synthesized by many different physicochemical methods, and there are a numerous examples of interesting nanoscale materials prepared using techniques such as vapor deposition, sputtering, electrodeposition, and a variety of sol–gel methods [113].

Theoretically, the STH efficiency of a WO_3 photoanode can reach ca. 4.5 %, based on the band gap energy of 2.7 eV. Given their relatively positive Fermi level (ca. 0.5 V vs. NHE at pH 0), WO_3 photoelectrodes are generally characterized by a low STH, as they require a rather positive potential (>1.2 V vs. NHE) to approach the photoanodic plateau, which is limited by the rate of diffusion and/or migration of minority carriers at the interface [114]. As a consequence, in the final application, i.e., a complete PEC cell water splitting without an externally applied bias, WO_3 photoanodes should be serially connected to p-type photoelectrodes or to other voltage sources capable of providing an appropriate voltage for the successful operation of the photoanode. Lee's group fabricated heterojunction electrodes using layer-by-layer deposition of WO_3 and $BiVO_4$ on a conducting glass and investigated their photoelectrochemical water oxidation under simulated solar light [115]. An electrode with the optimal composition of four layers of WO_3 covered by a single layer of $BiVO_4$ demonstrated photoactivity enhancements of 74 % relative to bare WO_3 and 730 % relative to bare $BiVO_4$. Although the conduction band of WO_3 is too low in energy to reduce water, it can receive an external bias from an economical PV device such as a dye-sensitized solar cell (DSC), which can harvest solar photons not absorbed by the semitransparent photoanode [2, 116]. This photoanode/DSC concept is shown in Fig. 22. Utilizing this concept, a WO_3 photoanode was used to demonstrate zero-bias overall water splitting under solar light irradiation [116].

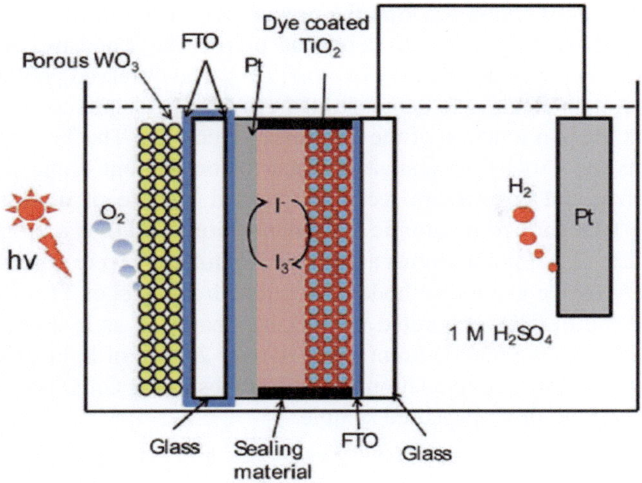

Fig. 22 Schematic diagram of a photovoltaic/photoelectrochemical tandem cell with a DSSC and a WO_3 photoelectrode. Reprinted with permission from [116], Copyright © 2011, Royal Society of Chemistry

6.2.4 Cu_2O

Cuprous oxide has received relatively little attention as a material for the conversion of solar energy into electrical or chemical energy. Cu_2O is a p-type semiconductor with a direct band gap of 2 eV, which should make it highly suitable for solar applications. The conduction band position of Cu_2O is approximately 0.7 V more negative than the reduction potential of hydrogen, so direct generation of hydrogen at the photocathode surface is possible [117]. Furthermore the position of the Fermi level suggests the possibility of constructing a standalone device for water splitting by coupling Cu_2O with an n-type semiconductor electrode, such as n-Fe_2O_3 [118]. One potential problem is the limited stability of Cu_2O in aqueous solutions. Cu_2O is stable only in a limited pH range. The redox potentials for the reduction and oxidation of monovalent copper oxide lie well within the band gap, making decomposition reactions with photogenerated electrons or holes thermodynamically possible [117]. For polycrystalline Cu_2O, it can be assumed that the Cu^+-terminated (111) surface is more stable than the O^{2-}-terminated or O^{2-}- and Cu^+-terminated (211) and (311) surfaces, respectively, in PEC water reduction because the former can stabilize O_2 reduction but not H^+-assisted Cu_2O reduction [118, 119]. Recently, p-type copper oxide nanowires were synthesized on a copper foil as a photocathode via a template route through the thermal transformation of one-dimensional $Cu(OH)_2$. PEC measurements indicated that a highly active photocathode of copper oxide nanowires can achieve a photocurrent of 1.3 mA cm^{-2} at a potential of 0.4 V vs. Ag/AgCl (AM 1.5) [120]. To improve its stability, the surface of a Cu_2O photocathode can be covered with other conductive or semiconductive transparent materials to protect the Cu species from photocorrosion. In 2003,

McFarland and co-workers reported the preparation of a thin-film heterojunction photocathode consisting of electrodeposited p-type Cu_2O and n-type TiO_2 (100 nm) deposited by physical vapor deposition [121]. The observed photoresponse of the electrode in an electrolyte demonstrated efficient light-induced charge carrier generation at the p–n junction of the Cu_2O/TiO_2 electrode. The TiO_2 film successfully protects the Cu_2O film against photocorrosion without diminishing performance. Grätzel and co-workers recently reported the use of ultrathin films of Al:ZnO and TiO_2 formed by atomic layer deposition (ALD) as a protective layer for Cu_2O [122, 123]. Photocurrents as high as 7.6 mA cm^{-2} were observed (0 V vs. RHE, AM 1.5) for Cu_2O photocathodes covered with a bilayer of 20 nm of Al-doped ZnO and 10 nm of TiO_2 decorated with Pt nanoparticles as hydrogen evolution catalysts. Although the photocurrent decayed over 20 min of light irradiation at 0 V vs. RHE, the chemical degradation of the light-absorbing Cu_2O was ruled out by XPS analysis of the PEC-measured sample.

7 Mechanistic Aspects of the Water-Splitting Reaction

7.1 Introduction

Photocatalytic water splitting has been regarded as a potential technology for the large-scale production of renewable hydrogen. Currently, some oxide photocatalysts can very efficiently split water to hydrogen and oxygen in the stoichiometric ratio under UV irradiation. For example, the AQY for overall water splitting is above 50 % at 270 nm when $NaTaO_3$:La modified with NiO is used for the reaction [19]. In order to achieve an STH efficiency higher than 5 % via one-step photoexcitation, it is necessary to split water at an AQY of 30 % for the wavelength of up to 600 nm or longer [14]. Obviously, it is necessary to harvest visible sunlight for water splitting, although the AQY for overall water splitting still remains several percent at 420 nm [14]. Therefore, it is necessary to improve the photocatalytic activity under visible-light irradiation.

In earlier studies, photocatalytic activity has mostly been discussed in terms of band gap, crystallinity, particle morphology of photocatalytic materials, and cocatalyst dispersivity [52, 124]. However, because photocatalysis is a non-equilibrium phenomenon caused by photoexcited carriers, it is necessary to consider the photocatalytic activity on the basis of information acquired during photocatalytic reactions [4, 125]. As shown in Fig. 1, a photoexcited electron and a hole are generated in a photocatalyst particle when a photon is absorbed. Photoexcited electrons and holes migrate to surface active sites and are used for the reduction and oxidation of water, respectively. Mass transfer of reactants and products occur simultaneously at the corresponding rates in the reaction solution, and the recombination of electrons and holes proceeds in parallel. It is generally difficult to specify which factor dominates the photocatalytic activity, because photocatalytic reactions occur via a sequence of competing multistep processes. However, it may be possible to estimate how much reaction processes occurring inside and on the surface of the photocatalyst and in the bulk solution limit the overall reaction rate on the basis of kinetic investigation. In fact, understanding photocatalytic water splitting from a kinetic perspective is essential for the formation of rational strategies for the development of efficient photocatalysts. This chapter presents recent attempts to understand the kinetic aspects of photocatalytic water splitting. Additionally, the use of sacrificial reagents in analyzing PEC water splitting is also discussed.

7.2 *Kinetic Aspects of Photocatalytic Water Splitting*

7.2.1 Light Intensity and Cocatalyst Loading

The rate of a photocatalytic reaction increases with increasing excitation intensity, although it is not necessarily proportional to the excitation intensity. Some reaction models suggest that the reaction order for light intensity decreases from unity to one-half as the light intensity increases [125]. This is because the recombination of photoexcited carriers is second order with respect to carrier concentrations (proportional to both electron and hole concentrations). On the other hand, under low light intensities at which the concentration of photoexcited carriers is negligible with respect to the intrinsic majority carrier concentration, it is a reasonable approximation to assume that only the minority carrier concentration depends on the excitation intensity and that the majority carrier concentration is constant. As a result, the recombination reaction is approximated as a quasi-first-order reaction with respect to the minority carrier concentration generated by photoexcitation, and the photocatalytic reaction rate becomes proportional to the light intensity. Accordingly, the reaction order for light intensity can be an indirect measure of how many photoexcited carriers exist in photocatalyst particles under illumination.

Figure 23 shows the light intensity dependence of the water-splitting rates obtained using $(Ga_{1-x}Zn_x)(N_{1-x}O_x)$ modified with various amounts of $Rh_{2-y}Cr_yO_3$ as a hydrogen evolution cocatalyst [125]. The $(Ga_{1-x}Zn_x)(N_{1-x}O_x)$ was prepared by nitriding a mixture of ZnO and Ga_2O_3 under an NH_3 flow, and $Rh_{2-y}Cr_yO_3$ was loaded by an impregnation method using Na_3RhCl_6 and $Cr(NO_3)_3$ as precursors. The AQY of the photocatalytic overall water splitting using the obtained $Rh_{2-y}Cr_yO_3/(Ga_{1-x}Zn_x)(N_{1-x}O_x)$ was 0.5 %. The reaction rate was proportional to light intensity when the $(Ga_{1-x}Zn_x)(N_{1-x}O_x)$ was modified with a sufficient amount of $Rh_{2-y}Cr_yO_3$. However, the reaction order for light intensity decreased as the cocatalyst loading amount was reduced, and the water-splitting rate decreased. The steady-state concentration of photoexcited carriers under photoexcitation is determined by a balance in the rates of carrier generation by photoexcitation and carrier consumption by surface reactions and recombination.The above result suggests that charge recombination was enhanced by reducing the cocatalyst loading because the hydrogen evolution process became a bottleneck. When excessive cocatalyst was loaded, on the other hand, the reaction rate slowed but

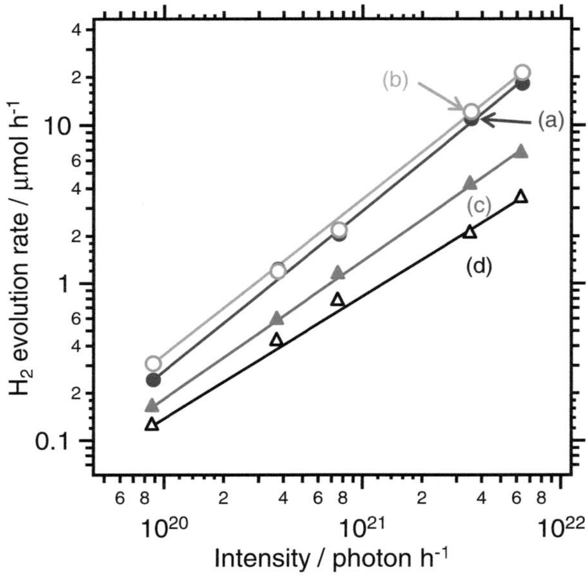

Fig. 23 Effect of light intensity and loading amount of $Rh_{2-y}Cr_yO_3$ cocatalyst on the photocatalytic activity of $Rh_{2-y}Cr_yO_3/(Ga_{1-x}Zn_x)(N_{1-x}O_x)$ for water splitting. The amount of $Rh_{2-y}Cr_yO_3$ loaded was (*a*) Rh 3.0 wt%, Cr 4.5 wt%; (*b*) Rh 1.0 wt%, Cr 1.5 wt%; (*c*) Rh 0.2 wt%, Cr 0.3 wt%; (*d*) Rh 0.1 wt%, Cr 0.15 wt%. The reactions were performed under Xe lamp illumination (300 nm $< \lambda <$ 500 nm). Adapted with permission from [125]. Copyright 2009 American Chemical Society

remained proportional to the light intensity. In this situation, a low concentration of photoexcited electrons would be expected as the hydrogen evolution processes were facilitated, resulting in the observed proportionality of the water-splitting rate to the light intensity. However, excessive cocatalyst loading could instead cause the aggregation of cocatalyst particles, photocatalyst shading, and/or blockage of oxygen evolution sites on the photocatalyst surfaces. As a result, the water-splitting rate was decreased by the increased loading of the hydrogen evolution cocatalyst. Therefore, it is important to optimize the cocatalyst loading amounts at the light intensity intended for actual applications in order to optimize the performance of photocatalytic systems.

The activities of most photocatalysts follow the aforementioned light intensity dependency; that is, the reaction order changes from unity to one-half as the light intensity increases. This means that the AQY monotonically decreases with increasing light intensity. However, in some cases, the AQY of photocatalytic water splitting increases with increasing light intensity under weak excitation conditions. In order to achieve a high quantum efficiency, it is probably necessary to saturate certain trap states with photoexcited carriers by generating photoexcited carriers more frequently than the charge recombination mediated by the trap states so that the photoexcited carriers have a high mobility.

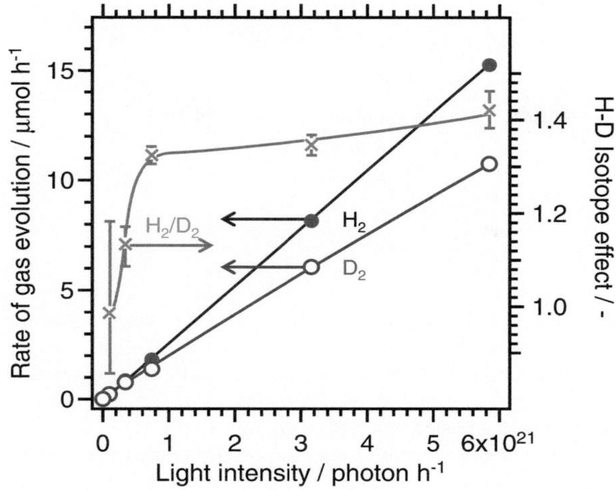

Fig. 24 Influence of light intensity on the H-D isotope effect in the water-splitting reaction using $Rh_{2-y}Cr_yO_3/(Ga_{1-x}Zn_x)(N_{1-x}O_x)$. The reactions were performed under Xe lamp illumination (300 nm $<\lambda<$ 500 nm). Reprinted with permission from [125]. Copyright 2009 American Chemical Society

7.2.2 Hydrogen–Deuterium Isotope Effect

The hydrogen–deuterium (H-D) isotope effect results from reaction processes involving hydrogen (deuterium) atoms at the interface of a photocatalyst and a reaction solution and is expected to reflect the reaction mechanism of the rate-determining step. However, when processes occurring inside a photocatalyst particle have a significant influence on the overall reaction rate, the H-D isotope effect should be small because hydrogen atoms are not involved in the rate-determining step.

The H-D isotope effect was investigated using $(Ga_{1-x}Zn_x)(N_{1-x}O_x)$ and Zn-added Ga_2O_3 (Ga_2O_3:Zn) modified with $Rh_{2-y}Cr_yO_3$. Figure 24 shows rates of light water (H_2O) and heavy water (D_2O) splitting and the H-D isotope effect using $Rh_{2-y}Cr_yO_3/(Ga_{1-x}Zn_x)(N_{1-x}O_x)$ under various light intensities, with the H-D isotope effect defined as the ratio of H_2O and D_2O splitting rates [125]. The H-D isotope effect for the photocatalytic water-splitting reaction was 1.4 at most, which was smaller than that for typical chemical reactions. In addition, the H-D isotope effect did not change significantly when various sacrificial electron donors or acceptors were added, as shown in Table 4 [125]. The H-D isotope effect would change drastically after adding sacrificial reagents that react more readily than water if it originated from surface reactions. Therefore, these results suggest that it was mostly independent of surface reactions. The rates of light water and heavy water splitting were the same when the light intensity was extremely weak. This result suggests that the water-splitting rate on $Rh_{2-y}Cr_yO_3/(Ga_{1-x}Zn_x)(N_{1-x}O_x)$ was primarily controlled by processes within the photocatalyst such as photoexcitation and/or migration of photoexcited carriers to the surface, rather than surface redox reactions involving hydrogen atoms.

Table 4 H-D isotope effect and apparent activation energy for photocatalytic water splitting using $Rh_{2-y}Cr_yO_3/(Ga_{1-x}Zn_x)(N_{1-x}O_x)$ in the presence of sacrificial reagents[125][a]

Sacrificial reagent	Reaction	Isotope effect[b]	E_a[c]/kJ mol^{-1}
None[d]	Water splitting	1.4	8
CH_3OH[e]	H_2 evolution	1.5	10
CH_3CH_2OH[f]	H_2 evolution	–	9
$C_2O_4^{2-}$, $HC_2O_4^-$[g]	H_2 evolution	1.2	–
HCO_2^-[h]	H_2 evolution	1.8	–
Ag^+[i]	O_2 evolution	1.3	9

[a]Reaction conditions: $Rh_{2-y}Cr_yO_3/(Ga_{1-x}Zn_x)(N_{1-x}O_x)$, 0.10 g; H_2O (D_2O) containing the sacrificial reagents, 140 mL; light source, 300 W Xe lamp (300 nm$<\lambda<$500 nm)
[b]The ratio of the gas evolution rates from H_2O solutions to those from D_2O solutions
[c]Apparent activation energy
[d]The pH of the solutions was adjusted to 4.5 by H_2SO_4
[e]10 and 80 vol% CH_3OH aqueous solutions for measurements of the isotope effect and the apparent activation energy, respectively
[f]80 vol% CH_3CH_2OH aqueous solutions
[g]45 mM $(COONa)_2$ and 5 mM $(COOH)_2$ aqueous solutions
[h]50 mM HCOONa aqueous solutions
[i]50 mM $AgNO_3$ aqueous solutions

Table 5 Effect of reaction conditions on the H-D isotope effect for photocatalytic water splitting using $Rh_{2-y}Cr_yO_3/Ga_2O_3$:Zn [126]

Ga_2O_3:Zn /g	$Rh_{2-y}Cr_yO_3$/wt%	Light source[a]	Water-splitting rate/ mmol h^{-1}	Isotope effect[b]
0.4	Rh 0.5–Cr 0.75	Xe	0.11	1.0
0.4	Rh 0.5–Cr 0.75	Hg	11.2	1.2
0.4	Rh 0.05–Cr 0.075	Hg	7.3	1.4
0.04	Rh 0.05–Cr 0.075	Hg	1.0	1.9

[a]Xe: 300 W Xe lamp (200 nm$<\lambda<$500 nm), Hg: 450 W Xe lamp ($\lambda>$200 nm)
[b]Ratio of H_2O and D_2O splitting rates. Reprinted with permission from [126]. Copyright © 2010, Elsevier

$Rh_{2-y}Cr_yO_3/Ga_2O_3$:Zn, which can split water with an AQY of over 10 %, was used to further study the isotope effect in photocatalytic water splitting, because the photocatalytic activity of $Rh_{2-y}Cr_yO_3/(Ga_{1-x}Zn_x)(N_{1-x}O_x)$ could be effectively controlled by charge recombination due to the relatively low AQY. The effects of the reaction conditions on the water-splitting rate and isotope effect for $Rh_{2-y}Cr_yO_3/Ga_2O_3$:Zn are presented in Table 5 [126]. The H-D isotope effect became stronger as the excitation frequency per photocatalyst particle was raised by increasing the light intensity and decreasing the amount of photocatalyst. The H-D isotope effect also became stronger when the amount of loaded cocatalyst was decreased to suppress the hydrogen evolution reaction. However, it should be noted that the H-D isotope effect was at most two, even when the photocatalytic water splitting was carried out using $Rh_{2-y}Cr_yO_3/Ga_2O_3$:Zn at decent rates. These results suggest that in most cases, the rate of photocatalytic water splitting is mainly determined by bulk processes inside the photocatalyst particles.

7.2.3 Activation Energy

The effect of reaction temperature on the photocatalytic activity for water splitting was investigated using $(Ga_{1-x}Zn_x)(N_{1-x}O_x)$ and Ga_2O_3:Zn modified with various cocatalysts. It is natural to expect that the apparent activation energy of photocatalytic water splitting reflects the activation energy of the slowest reaction step and that reaction processes involving the cleavage and formation of chemical bonds have higher activation energies than physical processes such as charge migration.

The apparent activation energies for photocatalytic water splitting using $Rh_{2-y}Cr_yO_3/(Ga_{1-x}Zn_x)(N_{1-x}O_x)$ and $Rh_{2-y}Cr_yO_3/Ga_2O_3$:Zn were 8 kJ mol^{-1} from 298 to 323 K [125, 126]. This is significantly lower than that for the electrochemical hydrogen evolution reaction on Rh electrodes (31 kJ mol^{-1}) [127]. Moreover, the apparent activation energies for heavy water splitting, water splitting in the presence of sacrificial reagents, and gas phase water vapor splitting were very similar when $Rh_{2-y}Cr_yO_3/(Ga_{1-x}Zn_x)(N_{1-x}O_x)$ was used [125, 128]. This indicates that the apparent activation energy is associated with processes occurring inside the photocatalyst, rather than the formation and dissociation of chemical bonds involving hydrogen or the diffusion, adsorption, and desorption of chemical species. This is most likely because only electrons that have successfully escaped recombination with photoexcited holes can contribute to the photocatalytic hydrogen evolution reaction on the surface, whereas the electrons needed to drive the hydrogen evolution can be supplied immediately depending on the potential of the electrode in the electrochemical hydrogen evolution reaction.

The apparent activation energy depends on the cocatalyst type. For example, loading Ni instead of $Rh_{2-y}Cr_yO_3$ on Ga_2O_3:Zn increased the apparent activation energy from 8 to 15 kJ mol^{-1} while lowering the water-splitting rate at room temperature to 40 % [126]. This result suggests that reaction processes involving both the photocatalyst and the cocatalysts, such as electron transfer from the photocatalyst to the cocatalyst, contributed to the apparent activation energy. In fact, it was revealed by surface-enhanced infrared spectroscopy under potential control that a potential barrier for electron migration existed at the interface between an n-type GaN single crystal and deposited Pt particles [129]. The slightly higher activation energy of Ni could also be associated with surface electrochemical reactions, taking the relatively low electrochemical activity of Ni into account. The apparent activation energy for the hydrogen evolution reaction on Ni was reported to be 56 kJ mol^{-1} [130]. However, the apparent activation energy is often much lower for photocatalytic reactions than for electrochemical hydrogen evolution reactions using corresponding electrodes, because bulk processes have a dominant effect on the apparent activation energy.

Note that the apparent activation energy decreased from 8 to 0 kJ mol^{-1} and the water-splitting rate decreased to 40 % when $(Ga_{1-x}Zn_x)(N_{1-x}O_x)$ was modified with RuO_2 instead of $Rh_{2-y}Cr_yO_3$ [4]. This change in the water-splitting rate cannot be explained by the apparent activation energy alone. It is known that side reactions such as oxygen reduction compete with hydrogen evolution on RuO_2, while $Rh_{2-y}Cr_yO_3$ is selectively active for hydrogen evolution [40]. As a result, the water-splitting rate on $RuO_2/(Ga_{1-x}Zn_x)(N_{1-x}O_x)$ decreases drastically in the presence of oxygen. Such competing reactions could significantly complicate the kinetics of photocatalysis.

7.3 Charge Separation Efficiency and Change Injection Efficiency in Photoelectrochemical Water Splitting

In PEC water splitting, only minority carriers react on the surface, unlike the case of water splitting on particulate photocatalysts. Furthermore, it is possible to control the potential of photoelectrodes. The number of photons absorbed by a semiconductor photoelectrode can also be measured if the photoelectrodes are grown on a transparent conductive electrode such as fluorine-doped tin oxide (FTO). Therefore, kinetic investigation of PEC water splitting is easier than that of photocatalytic water splitting. If sacrificial reagents are used, minority carriers that reach the surface can be consumed very efficiently by redox reactions. Therefore, the photocurrent obtained in the presence of sacrificial reagents reflects the maximum rate at which photoexcited carriers reach the surface. This also provides an estimate of how much the PEC activity can be enhanced by facilitating the surface redox reaction by loading catalysts. This method was applied by Dotan et al. to hematite photoanodes in 2011 [131] and subsequently to various other photoelectrodes.

J_{abs}, the photocurrent when the absorbed photon-to-current efficiency (APCE) is 100 %, is expressed by the following equation:

$$J_{abs} = e\int_{\infty}^{0} \%A(\lambda) \times I(\lambda) d\lambda, \tag{13}$$

where e, $\%A(\lambda)$, and $I(\lambda)$ stand for the elementary charge, absorptance, and photon flux from a light source at a given wavelength, respectively. J_{abs} corresponds to the number of absorbed photons converted to the photocurrent. $\%A(\lambda)$ can be calculated from the total reflectance and total transmittance measured using a spectrometer equipped with an integration sphere. The photocurrent resulting from PEC water splitting, J_{H2O}, can be expressed as

$$J_{H_2O} = J_{abs} \times \phi_{sep} \times \phi_{inj} \tag{14}$$

using the probability that photoexcited minority carriers reach the surface or, in other words, the charge separation efficiency ϕ_{sep} and the probability that minority carriers on the surface are injected into water ϕ_{inj} (the charge injection efficiency). For photoanodes, when H_2O_2, $[Fe^{II}(CN)_6]^{-4}$ or SO_3^{2-}, which can be oxidized more readily than water, is added to an electrolyte solution, all of the minority carriers that reach the photoelectrode surface may be injected into the electrolyte solution. In other words, ϕ_{inj} can be regarded as unity. With this assumption, the photocurrent in the presence of a sacrificial reagent, J_{sac}, is

$$J_{sac} = J_{abs} \times \phi_{sep}. \tag{15}$$

J_{H2O}, J_{sac}, and J_{abs} can be obtained from PEC measurements and spectroscopy. Therefore, ϕ_{sep} and ϕ_{inj} can be calculated.

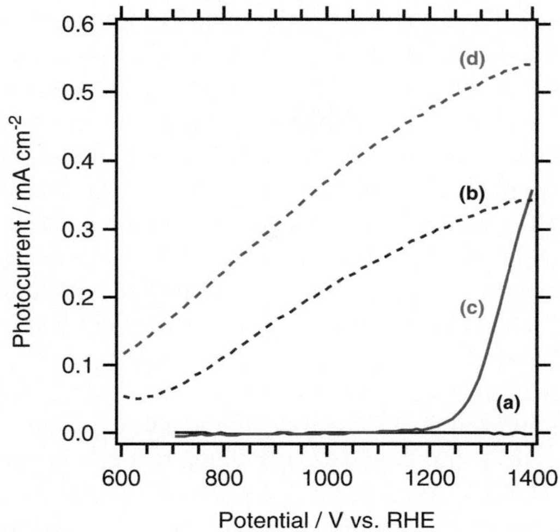

Fig. 25 Photocurrent–potential curves for 9-nm-thick hematite photoanodes (*a*, *b*) without and (*c*, *d*) with a 2-nm-thick Nb_2O_5 underlayer under a simulated sunlight in 1 M NaOH aqueous solution (*a*, *c*) without and (*b*, *d*) with 0.5 M H_2O_2. Adapted with permission from [106]. Copyright © 2012 WILEY-VCH Verlag GmbH & Co. KGaA, Weinheim

Current–potential curves for 9-nm-thick hematite photoanodes in a 1 M NaOH aqueous solution are shown in Fig. 25 [106]. An ultrathin hematite photoanode was prepared by spray pyrolysis according to a previously reported method [132]. The ultrathin hematite photoanode generated no photocurrent. However, when H_2O_2 was added as a sacrificial reagent, a photoanodic current was observed at +0.6 V vs. RHE and increased as the potential became more positive, indicating that ϕ_{sep} also became higher. ϕ_{sep} was estimated to be approximately 30 % at +1.4 V vs. RHE, since J_{abs} of the hematite electrode was 1.1 mA cm^{-2} under AM1.5G irradiation. The absence of photoanodic current in a 1 M NaOH aqueous solution suggests that holes reaching the surface recombined with electrons before contributing to water oxidation. It is likely that photoexcited holes at the surface interact with FTO electron clouds and recombine with electrons because the hematite layer is too thin. Extremely thin hematite layers near FTO are generally inactive for PEC water splitting and are often called dead layers [132].

Incorporating a thin oxide layer, which can pass electrons from hematite to FTO but not in the opposite direction, between the ultrathin hematite layer and the FTO may suppress interactions between photoexcited holes trapped on the surface of the hematite and electrons from the FTO. This approach has been successfully used in photovoltaic applications [133]. Therefore, a 2-nm-thick Nb_2O_5 layer was deposited by atomic layer deposition on FTO prior to deposition of ultrathin hematite [106]. This process resulted in an ultrathin hematite photoelectrode that had an ultrathin Nb_2O_5 layer between the FTO and the hematite. Figure 25 shows current–potential curves of hematite photoanodes deposited on FTO coated with a 2-nm-thick Nb_2O_5 layer. Unlike the unmodified FTO case, a photoanodic current was clearly observed at +1.2 V vs. RHE and above. Furthermore, over the measured potential range, the photocurrent in the presence of H_2O_2 increased significantly. Consequently, ϕ_{sep} became 1.5 to 2 times higher. Moreover, ϕ_{inj} was also improved, probably because photoexcited holes and intermediates of the oxygen evolution reaction trapped on the surface could avoid recombination with electrons. In an earlier work, it was suggested that intermediates that trapped photoexcited holes might exist during PEC water oxidation using hematite thin films [134]. Such oxidation intermediates would be reduced back to inactive states if the FTO were to provide electrons to the thin hematite layer. Such deactivation processes involve the recombination of electrons and holes and must be suppressed. By incorporating Nb_2O_5 as an intermediate layer, electron injection from FTO into the hematite layer can be suppressed at any potential more positive than the conduction band edge. As for photoexcited carriers in the hematite, it is likely that electrons will be injected from the hematite to the FTO through the ultrathin Nb_2O_5 layer, as the conduction band offset is as small as 0.2 eV. Holes will not be similarly injected because of the large valence band offset (approximately 1.0 eV). For these reasons, the incorporation of the Nb_2O_5 layer suppressed charge recombination and improved the PEC activity of the ultrathin hematite photoanode.

It is important to evaluate the rectification behavior at the interface between the FTO and the hematite photoanode in order to discuss charge separation and PEC activity. Reversible redox couples such as $[Fe^{III}(CN)_6]^{3-}/[Fe^{II}(CN)_6]^{4-}$ can be used to semiquantitatively estimate how the ultrathin Nb_2O_5 layer suppresses the contribution of electrons from a conductive substrate. Note that the hematite photoanodes were illuminated from the FTO side because $[Fe^{III}(CN)_6]^{3-}/[Fe^{II}(CN)_6]^{4-}$ absorbs visible light. The current–potential curves shown in Fig. 26 show that the cathodic dark current was suppressed by incorporating the Nb_2O_5 layer [106]. This is because electron injection from FTO to the thin hematite film and the electrolyte solution

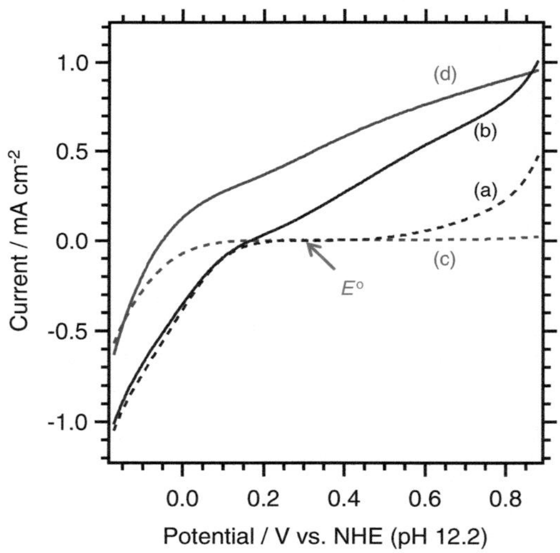

Fig. 26 Current–potential curves for 9-nm-thick hematite photoanodes (*a*, *b*) without and (*c*, *d*) with a 2-nm-thick Nb_2O_5 underlayer acquired (*a*, *c*) under darkness and (*b*, *d*) under a simulated sunlight in an aqueous solution containing 200 mM $K_4Fe(CN)_6$ and 20 mM $K_3Fe(CN)_6$ adjusted to pH 12.2 with an aqueous KOH solution. E° corresponds to the potential of the $[Fe^{III}(CN)_6]^{3-}/[Fe^{II}(CN)_6]^{4-}$ redox couple employed. Adapted with permission from [106]. Copyright © 2012 WILEY-VCH Verlag GmbH & Co. KGaA, Weinheim

was suppressed. In this context, the Nb_2O_5 layer acted as an electron blocking layer. The anodic dark current was also suppressed, indicating that the surface of the FTO was totally coated with a 2-nm-thick Nb_2O_5 layer.

7.4 Summary

Kinetic aspects of water splitting on semiconductors were discussed on the basis of recent studies using a $(Ga_{1-x}Zn_x)(N_{1-x}O_x)$ photocatalyst with visible-light activity, a Ga_2O_3:Zn photocatalyst with high UV light activity, and an ultrathin hematite photoelectrode. When particulate photocatalysts were used for the water-splitting reaction, the kinetic isotope effect was weak and the apparent activation energy was small because the number of photoexcited carriers available for the surface reaction was limited. This is because the majority of photoexcited carriers recombined before contributing to the surface reactions. Therefore, it is important to promote processes occurring within the photocatalyst. As for n-type semiconductors, it was suggested that lowering defect densities and moderating n-type semiconducting properties could effectively improve the photocatalytic activity [125]. It may also be effective to promote the oxygen evolution reaction in order to lower the selectivity of holes' reaction toward charge recombination [125]. In fact, there is experimental evidence supporting the effectiveness of both approaches for $(Ga_{1-x}Zn_x)(N_{1-x}O_x)$ [59, 60, 135], $SrTiO_3$ [21], and TaON [62]. Furthermore, the importance of charge separation and injection efficiency in the analysis and design of photoelectrodes was presented, using hematite photoanodes as an example. The concepts of charge injection and separation efficiency have been applied to other photoelectrodes such as $BiVO_4$ and have contributed to the development of highly active photoelectrodes [100, 136]. Therefore, understanding the kinetic aspects of the water-splitting reaction is highly beneficial for the development of highly active photocatalysts and photoelectrodes.

8 Summary

The fossil energy resources upon which we are heavily dependent will inevitably be depleted in the not-so-distant future, and the timescales for coal, oil, and natural gas regeneration are far too long. Solar energy may be the best and perhaps only choice for meeting long-term human energy needs. As an important technology for converting and storing solar energy, photocatalytic and photoelectrochemical H_2 production from water has made significant progress in the past 40 years. The range of solar light absorption has been successfully extended from the UV to the visible region, enhancing the utilization of solar energy. The highest reported STH efficiency has increased from less than 0.1 % (on a TiO_2 photoanode in 1972) to as high as 12.4 % (on a GaAs–GaInP_2 photocathode in 1998) [3, 75]. In recent years, research on photocatalysis and photoelectrochemistry has begun to address important factors other than efficiency, and the Earth abundance, cost, and reaction stability of the semiconductor materials employed are now being taken into account.

In this chapter, basic concepts including thermodynamics and experimental photocatalysis and photoelectrochemistry were introduced. As an important aspect of this research, the design of visible-light-driven semiconducting materials was discussed. Then we summarized recent advancements in photocatalytic and photoelectrochemical water splitting, as well as the current understanding of the mechanisms of these reactions. In the numerous recent efforts to develop new inorganic photocatalysts and photoelectrodes, most semiconductors composed of binary, ternary, and even quaternary elements have been evaluated. In recent years, research on photocatalysis and photoelectrochemistry has mainly focused on the optimization of known photocatalysts through bulk and surface modifications such as doping with alien elements, preparation via new methods, constructing heterojunctions, and cocatalyst loading.

Both photocatalysis and photoelectrochemistry are based on the photoelectric effect in semiconductors, although the former experiments are carried out with semiconductor powders and the latter are carried out with thin films. The difference in reaction technologies leads to different ideas for the design and optimization of photocatalysts and photoelectrodes, as follows:

1. *Particle size*. A photocatalytic reaction is carried out by dispersing semiconductor particles into an appropriate amount of aqueous solution sealed in a reactor with a transparent window. It is not necessary for one photocatalyst particle to absorb all of the light irradiating upon it, because any penetrating or scattered light can be absorbed by other nearby particles in the solution. A highly crystallized nanoscale semiconductor is preferred as a photocatalyst, in order to achieve better dispersivity for light absorption. PEC reactions are carried on thin-film semiconductors, so there is only one chance for the semiconductor to capture irradiated light. Photons that are not absorbed by the thin film will be reflected and/or scattered. The optimal thickness of the photoelectrode depends upon the light absorption coefficient of the semiconductor, and usually a thickness of hundreds or thousands of nanometers is required to absorb most of the irradiated light. An alternative method of increasing the light absorption is to fabricate a multilayer (tandem) semiconductor film while creating architectures that can trap photons within the electrodes, although the preparation of such electrodes is often very complicated. In the field of photoelectrochemistry, nanoscale inorganic semiconductors can also be used to prepare quantum-dot-sensitized photoelectrodes for enhanced light absorption.
2. *Surface modification*. The oxidation and reduction reactions both occur on the same particle in a photocatalytic reaction, whereas only one of the two reactions proceeds on an electrode in a photoelectrochemical reaction. It is preferable to load both oxidation and reduction cocatalysts on different sites of the photocatalyst because photogenerated electrons and holes must be trapped and consumed at the same time to maintain charge balance on the particle. In contrast, only one kind of cocatalyst is needed for a photoelectrode, as the other can be loaded on the counter electrode.

3. *Reaction driving force*. The reaction driving force for the photocatalyst is provided only by the band bending close to the semiconductor–solution and semiconductor–cocatalyst junctions. All modifications to the semiconductor photocatalyst must be made before it is suspended in an aqueous solution. Compared with photocatalysis, photoelectrochemistry is more controllable because an additional bias can be applied to the photoelectrode to adjust the redox potential of the semiconductor. With the application of an applied external voltage, the PEC reaction offers a much wider selection of semiconductor materials than photocatalysis for overall water splitting.
4. *Cost*. The cost of hydrogen production from water by photocatalysis and photoelectrochemistry had been calculated by the U.S. Department of Energy (DOE) H2A model for a 10 t per day production scale with a plant gate output at 300 psi [13]. All capital expenditures and operating costs for the reactors and auxiliaries (such as compressors, control systems) were considered. The final cost of photocatalytic and PEC hydrogen production is $1.60 (with an STH efficiency of 10 % and a lifetime of 5 years) and $10.40 (with an STH efficiency of 10 % and lifetime of 10 years) per kg of H_2, respectively. By using a solar concentrator and tracking components, the cost of operating a photoelectrochemical reaction system can be reduced to $4.1 per kg of H_2. Photocatalytic reaction systems have significantly lower capital costs, but there is greater uncertainty associated with their operation, such as safety concerns over the cogeneration of H_2 and O_2. This work clearly demonstrates that if technical progress is made to meet materials performance targets and after appropriate plant-scale engineering, direct solar hydrogen can be produced by PEC water splitting at a cost that meets the DOE target of $2.00–$4.00 per kg of H_2.

Acknowledgement This work was supported by Grant-in-Aids for Specially Promoted Research (no. 23000009) and for Young Scientists (B) (no. 25810112) of the Japan Society for the Promotion of Science (JSPS) and the Artificial Photosynthesis Project of the Ministry of Economy, Trade and Industry (METI) of Japan.

References

1. Goldemberg J (2007) Ethanol for a sustainable energy future. Science **315**:808–10.
2. Grätzel M (2001) Photoelectrochemical cells. Nature **414**:338–44.
3. Fujishima A, Honda K (1972) Electrochemical photolysis of water at a semiconductor electrode. Nature **238**:37–8.
4. Hisatomi T, Minegishi T, Domen K (2012) Kinetic assessment and numerical modeling of photocatalytic water splitting toward efficient solar hydrogen production. Bull Chem Soc Jpn **85**:647–55.
5. Maeda K, Domen K (2007) New non-oxide photocatalysts designed for overall water splitting under visible light. J Phys Chem C **111**:7851–61.
6. Hisatomi T, Kubota J, Domen K (2014) Recent advances in semiconductors for photocatalytic and photoelectrochemical water splitting. Chem Soc Rev 43:7520–35
7. Maeda K (2013) Z-scheme water splitting using two different semiconductor photocatalysts. ACS Catal **3**:1486 – 503.

8. Murphy A B, Barnes P R F, Randeniya L K, Plumb I C, Grey I E, Horne M D, Glasscock J A (2006) Efficiency of solar water splitting using semiconductor electrodes. Int J Hydrogen Energy **31**:1999–2017.
9. Kudo A, Miseki Y (2009) Heterogeneous photocatalyst materials for water splitting. Chem Soc Rev **38**:253–78.
10. Nozik A J (1978) Photoelectrochemistry: applications to solar energy conversion. Ann Rev Phys Chem **29**:189–222.
11. Prévot M S, Kevin Sivula K (2013) Photoelectrochemical tandem cells for solar water splitting. J Phys Chem C **117**:17879 – 93.
12. Hu S, Xiang C, Haussener S, Bergercd A D, Lewis N S (2013) An analysis of the optimal band gaps of light absorbers in integrated tandem photoelectrochemical water-splitting systems. Energy Environ Sci **6**:2984–93.
13. Pinaud B A, Benck J D, Seitz L C, Forman A J, Chen Z, Deutsch T G, James B D, Baum K N, Baum G N, Ardo S, Wang H, Miller E, Jaramillo T F (2013) Technical and economic feasibility of centralized facilities for solar hydrogen production via photocatalysis and photoelectrochemistry. Energy Environ Sci **6**:1983–2002.
14. Maeda K, Domen K (2010) Photocatalytic water splitting: recent progress and future challenges. J Phys Chem Lett **1**:2655–61.
15. Sayama K, Arakawa H (1997) Effect of carbonate salt addition on the photocatalytic decomposition of liquid water over Pt–TiO_2 catalyst. J Chem Soc Faraday Trans **93**:1647–54.
16. Takata T, Furumi Y, Shinohara K, Tanaka A, Hara M, Kondo J N, Domen K (1997) Photocatalytic decomposition of water on spontaneously hydrated layered perovskites. Chem Mater **9**:1063–4.
17. Domen K, Kudo A, Shinozaki A, Tanaka A, Maruya K, Onishi T (1986) Photodecomposition of water and hydrogen evolution from aqueous methanol solution over novel niobate photocatalysts. J Chem Soc Chem Commun **4**:356–7.
18. Miseki Y, Kato H, Kudo A (2006) Water splitting into H_2 and O_2 over $Ba_5Nb_4O_{15}$ photocatalysts with layered perovskite structure prepared by polymerizable complex method. Chem Lett **35**:1052–53.
19. Kato H, Asakura K, Kudo A (2003) Highly efficient water splitting into H_2 and O_2 over lanthanum-doped $NaTaO_3$ photocatalysts with high crystallinity and surface nanostructure. J Am Chem Soc **125**:3082–9.
20. Scaife D E (1980) Oxide semiconductor in photoelectrochemical conversion of solar energy. Sol Energy **25**:41–54.
21. Takata T, Domen K (2009) Defect engineering of photocatalysts by doping of aliovalent metal cations for efficient water splitting. J Phys Chem C **113**:19386–8.
22. Pan X, Yang M, Fu X, Zhang N, Xu Y (2013) Defective TiO_2 with oxygen vacancies: synthesis properties and photocatalytic applications. Nanoscale **5**:3601–14.
23. Hoang S, Berglund S P, Hahn N T, Bard A J, Mullins C B (2012) Enhancing visible light photo-oxidation of water with TiO_2 nanowire arrays via cotreatment with H_2 and NH_3: synergistic effects between Ti^{3+} and N. J Am Chem Soc **134**:3659–62.
24. Wang L, Wang W, Shang M, Yin W, Sun S, Zhang L (2010) Enhanced photocatalytic hydrogen evolution under visible light over $Cd_{1-x}Zn_xS$ solid solution with cubic zinc blend phase. Int J Hydrogen Energy **35**:19–25.
25. Liu M, Wang L, Lu G, Yao X, Guo L (2011) Twins in $Cd_{1-x}Zn_xS$ solid solution: Highly efficient photocatalyst for hydrogen generation from water. Energy Environ Sci **4**:1372–8.
26. Tsuji I, Kato H, Kobayashi H, Kudo A (2004) Photocatalytic H_2 evolution reaction from aqueous solutions over band structure-controlled $(AgIn)_xZn_{2(1-x)}S_2$ solid solution photocatalysts with visible-light response and their surface nanostructures. J Am Chem Soc **126**:13406–13.
27. Tsuji I, Kato H, Kudo A (2005) Visible-light-induced H_2 evolution from an aqueous solution containing sulfide and sulfite over a ZnS–$CuInS_2$–$AgInS_2$ solid-solution photocatalyst. Angew Chem Int Ed **44**:3565–8.

28. Maeda K, Takata T, Hara M, Saito N, Inoue Y, Kobayashi H, Domen K (2005) GaN:ZnO solid solution as a photocatalyst for visible-light-driven overall water splitting. J Am Chem Soc **127**:8286–7.
29. Lee Y, Terashima H, Shimodaira Y, Teramura K, Hara M, Kobayashi H, Domen K, Yashima M (2007) Zinc germanium oxynitride as a photocatalyst for overall water splitting under visible light. J Phys Chem C **111**:1042–8.
30. Chun W, Ishikawa A, Fujisawa H, Takata T, Kondo J N, Hara M, Kawai M, Matsumoto Y, Domen K (2003) Conduction and valence band positions of Ta_2O_5, TaON, and Ta_3N_5 by UPS and electrochemical methods. J Phys Chem B **107**:1798–803.
31. Zhang F, Yamakata A, Maeda K, Moriya Y, Takata T, Kubota J, Teshima K, Oishi S, Domen K (2012) Cobalt-modified porous single-crystalline $LaTiO_2N$ for highly efficient water oxidation under visible light. J Am Chem Soc **134**:8348–51.
32. Maeda K, Lu D, Domen K (2013) Oxidation of water under visible-light irradiation over modified $BaTaO_2N$ photocatalysts promoted by tungsten species. Angew Chem Int Ed **52**:1–5.
33. Hisatomi T, Katayama C, Moriya Y, Minegishi M, Katayama M, Nishiyama H, Yamadad T, Domen K (2013) Photocatalytic oxygen evolution using $BaNbO_2N$ modified with cobalt oxide under photoexcitation up to 740 nm. Energy Environ Sci **6**:3595–9.
34. Ishikawa A, Yamada Y, Takata T, Kondo J N, Hara M, Kobayashi H, Domen K (2003) Novel synthesis and photocatalytic activity of oxysulfide $Sm_2Ti_2S_2O_5$. Chem Mater **15**:4442–6.
35. Zhao W, Maeda K, Zhang F, Hisatomia T, Domen K (2014) Effect of post-treatments on the photocatalytic activity of $Sm_2Ti_2S_2O_5$ for the hydrogen evolution reaction. Phys Chem Chem Phys **16**:12051–6.
36. Ishikawa A, Takata T, Kondo J N, Hara M, Kobayashi H, Domen K (2002) Oxysulfide $Sm_2Ti_2S_2O_5$ as a stable photocatalyst for water oxidation and reduction under visible light irradiation ($\lambda \leq 650$ nm). J Am Chem Soc **124**:13547–53.
37. Suzuki T, Hisatomi T, Teramura K, Shimodaira Y, Kobayashi H, Domen K (2012) A titanium-based oxysulfide photocatalyst: $La_5Ti_2MS_5O_7$ (M = Ag, Cu) for water reduction and oxidation. Phys Chem Chem Phys **14**:15475–81.
38. Xing Z, Zong X, Pan J, Wang L (2013) On the engineering part of solar hydrogen production from water splitting: Photoreactor design. Chem Eng Sci **104**:125–46.
39. Huang Z, Xiang C, Lewerenz H, Lewis N S, (2014) Two stories from the ISACS 12 conference: solar-fuel devices and catalyst identification. Energy Environ Sci **7**:1207–11.
40. Maeda K, Teramura K, Masuda H, Takata T, Saito N, Inoue Y, Domen K (2006) Efficient overall water splitting under visible-light irradiation on $(Ga_{1-x}Zn_x)(N_{1-x}O_x)$ dispersed with Rh − Cr mixed-oxide nanoparticles: effect of reaction conditions on photocatalytic activity. J Phys Chem **110**:13107–12.
41. Maeda K, Teramura K, Lu D, Saito N, Inoue Y, Domen K (2006) Noble-metal/Cr_2O_3 core/shell nanoparticles as a cocatalyst for photocatalytic overall water splitting. Angew Chem Int Ed **45**:7806–9.
42. Chen Z, Jaramillo T F, Deutsch T G, Kleiman-Shwarsctein A, Forman A J, Gaillard N, Garland R, Takanabe K, Heske C, Sunkara M, McFarland E W, Domen K, Miller E L, Turner J A, Dinh H N (2010) Accelerating materials development for photoelectrochemical hydrogen production: Standards for methods, definitions, and reporting protocols. J Mater Res **25**:3–16.
43. Modestino M A, Walczak K A, Berger A, Evans C M, Haussener S, Koval C, Newman J S, Ager J W, Segalman R A (2014) Robust production of purified H_2 in a stable, self-regulating, and continuously operating solar fuel generator. Energy Environ Sci **7**:297–301.
44. Maeda K, Hashiguchi H, Masuda H, Abe R, Domen K (2008) Photocatalytic Activity of $(Ga_{1-}xZnx)(N_{1-x}O_x)$ for Visible-Light-Driven H_2 and O_2 Evolution in the Presence of Sacrificial Reagents. J Phys Chem C **112**:3447–52.
45. Sato S, White J M (1980) Photodecomposition of water over Pt/TiO_2 catalysts. Chem Phys Lett **72**:83–6.
46. Domen K, Naito S, Soma M, Onishi T, Tamaru K (1980) Photocatalytic decomposition of water vapour on an NiO–$SrTiO_3$ catalyst. J Chem Soc Chem Commun 543–4.

47. Lehn J M, Sauvage J P, Ziessel R (1980) Photochemical water splitting.- Continuous generation of hydrogen and oxygen on irradiation of aqueous suspensions of metal loaded strontium titanate. Nouv J Chim **4**:623.
48. Domen K, Kudo A, Onishi T (1986) Mechanism of photocatalytic decomposition of water into H_2 and O_2 over NiO/$SrTiO_3$. J Catal **102**:92–8.
49. Domen K, Naito S, Onishi T, Tamaru K (1982) Study of the photocatalytic decomposition of water vapor over a NiO–SrTiO catalyst. J Phys Chem **86**:3657–61.
50. Domen K, Kudo A, Onishi T, Kosugi N, Kuroda H (1986) Photocatalytic decomposition of water into hydrogen and oxygen over nickel(II) oxide-strontium titanate ($SrTiO_3$) powder. 1. structure of the catalysts. J Phys Chem **90**:292–5.
51. Kudo A, Sayama K, Tanaka A, Asakura K, Domen K, Maruya K, Onishi T (1989) Nickel-loaded $K_4Nb_6O_{17}$ photocatalyst in the decomposition of H_2O into H_2 and O_2: structure and reaction mechanism. J Catal **120**:337–52.
52. Osterloh F E (2008) Inorganic materials as catalysts for photochemical splitting of water, Chem Mater **20**:35–54.
53. Inoue Y (2009) Photocatalytic water splitting by RuO_2-loaded metal oxides and nitrides with d^{0-} and d^{10-} related electronic configurations. Energy Environ Sci **2**:364–86.
54. Sakata Y, Matsuda Y, Nakagawa T, Yasunaga R, Imamura H, Teramura K (2011) Remarkable improvement of the photocatalytic activity of Ga_2O_3 towards the overall splitting of H_2O. ChemSusChem **4**:181–4.
55. Yamakata A, Ishibashi T, Kato H, Kudo A, Onishi H (2003) Photodynamics of $NaTaO_3$ catalysts for efficient water splitting. J Phys Chem B **107**:14383–7.
56. Maeda K, Teramura K, Saito N, Inoue Y, Domen K (2006) Improvement of photocatalytic activity of $(Ga_{1-x}Zn_x)(N_{1-x}O_x)$ solid solution for overall water splitting by co-loading Cr and another transition metal. J Catal **243**:303–8.
57. Maeda K, Teramura K, Lu D, Takata T, Saito N, Inoue Y, Domen K (2006) Characterization of Rh–Cr mixed-oxide nanoparticles dispersed on $(Ga_{1-x}Zn_x)(N_{1-x}O_x)$ as a cocatalyst for visible-light-driven overall water splitting. J Phys Chem B **110**:13753–8.
58. Yoshida M, Takanabe K, Maeda K, Ishikawa A, Kubota J, Sakata Y, Ikezawa Y, Domen K (2009) Role and function of noble-metal/Cr-layer core/shell structure cocatalysts for photocatalytic overall water splitting studied by model electrodes. J Phys Chem C **113**:10151–7.
59. Maeda K, Xiong A, Yoshinaga T, Ikeda T, Sakamoto N, Hisatomi T, Takashima M, Lu D, Kanehara M, Setoyama T, Teranishi T, Domen K (2010) Photocatalytic overall water splitting promoted by two different cocatalysts for hydrogen and oxygen evolution under visible light. Angew Chem Int Ed **49**:4096–9.
60. Xiong A, Yoshinaga T, Ikeda T, Takashima M, Hisatomi T, Maeda K, Setoyama T, Teranishi T, Domen K (2014) Effect of hydrogen and oxygen evolution cocatalysts on photocatalytic activity of GaN:ZnO. Eur J Inorg Chem **4**:767–72.
61. Ohno T, Bai L, Hisatomi T, Maeda K, Domen K (2012) Photocatalytic water splitting using modified GaN:ZnO solid solution under visible light: long-time operation and regeneration of activity. J Am Chem Soc **134**:8254–9.
62. Maeda K, Lu D, Domen K (2013) Direct water splitting into hydrogen and oxygen under visible light by using modified TaON photocatalysts with d^0 electronic configuration. Chem Eur J **19**:4986–91.
63. Maeda K, Terashima H, Kase K, Higashi M, Tabata M, Domen K (2008) Surface modification of TaON with monoclinic ZrO_2 to produce a composite photocatalyst with enhanced hydrogen evolution activity under visible light, Bull Chem Soc Jpn **81**:927–37.
64. Maeda K, Higashi M, Lu D, Abe R, Domen K (2010) Efficient nonsacrificial water splitting through two-step photoexcitation by visible light using a modified oxynitride as a hydrogen evolution photocatalyst. J Am Chem Soc **132**:5858–68.
65. Sasaki Y, Nemoto H, Saito K, Kudo A (2009) Solar water splitting using powdered photocatalysts driven by z-schematic inter-particle electron transfer without an electron mediator. J Phys Chem C **113**:17536–42.

66. Sasaki Y, Kato H, Kudo A (2013) $[Co(bpy)_3]^{3+/2+}$ and $[Co(phen)_3]^{3+/2+}$ electron mediators for overall water splitting under sunlight irradiation using z-scheme photocatalyst system. J Am Chem Soc **135**:5441–9.
67. Higashi M, Abe R, Takata T, Domen K (2009) Photocatalytic overall water splitting under visible light using $ATaO_2N$ (A=Ca, Sr, Ba) and WO_3 in a IO^{3-}/I^- shuttle redox mediated system. Chem Mater **21**:1543–9.
68. Bard A J (1979) Photoelectrochemistry and heterogeneous photo-catalysis at semiconductors. J Photochem **10**:59–75.
69. Fujihara K, Ohno T, Matsumura M (1998) Splitting of water by electrochemical combination of two photocatalytic reactions on TiO_2 particles. J Chem Soc Faraday Trans **94**:3705–9.
70. Sayama K, Yoshida R, Kusama H, Okabe K, Abe Y, Arakawa H (1997) Photocatalytic decomposition of water into H_2 and O_2 by a two-step photoexcitation reaction using a WO_3 suspension catalyst and an Fe^{3+}/Fe^{2+} redox system. Chem Phys Lett **277**:387–91.
71. Abe R, Sayama K, Domen K, Arakawa H (2001) A new type of water splitting system composed of two different TiO_2 photocatalysts (anatase, rutile) and a IO^{3-}/I^- shuttle redox mediator. Chem Phys Lett **344**:339–44.
72. Sayama K, Mukasa K, Abe R, Abe Y, Arakawa H (2001) Stoichiometric water splitting into H_2 and O_2 using a mixture of two different photocatalysts and an IO_3^-/I^- shuttle redox mediator under visible light irradiation. Chem Commun 2416–7.
73. Sasaki Y, Iwase A, Kato A, Kudo A (2008) The effect of co-catalyst for Z-scheme photocatalysis systems with an Fe^{3+}/Fe^{2+} electron mediator on overall water splitting under visible light irradiation. J Catal **259**:133–7.
74. Ma S S K, Maeda K, Hisatomi T, Tabata M, Kudo A, Domen K (2013) A redox-mediator-free solar-driven z-scheme water-splitting system consisting of modified Ta_3N_5 as an oxygen-evolution photocatalyst. Chem Eur J **19**:7480–6.
75. Khaselev O, Turner J A (1998) A monolithic photovoltaic-photoelectrochemical device for hydrogen production via water splitting. Science **280**:425–7.
76. Reece S Y, Hamel J A, Sung K, Jarvi T D, Esswein A J, Pijpers J J H, Nocera D G (2011) Wireless solar water splitting using silicon-based semiconductors and earth-abundant catalysts. Science **334**:645–8.
77. Liu C, Tang J, Chen H M, Liu B, Yang P (2013) A fully integrated nanosystem of semiconductor nanowires for direct solar water splitting. Nano Lett **13**:2989–92.
78. Kaneshiro J, Gaillard N, Rocheleau R, Miller E (2010) Advances in copper-chalcopyrite thin films for solar energy conversion. Sol Energy Mater Sol Cells **94**:12–6.
79. Moriya M, Minegishi T, Kumagai H, Katayama M, Kubota J, Domen K (2013) Stable hydrogen evolution from CdS-modified $CuGaSe_2$ photoelectrode under visible-light irradiation. J Am Chem Soc **135**:3733–5.
80. Ma G, Minegishi T, Yokoyama D, Kubota J, Domen K (2011) Photoelectrochemical hydrogen production on Cu_2ZnSnS_4/Mo-mesh thin-film electrodes prepared by electroplating. Chem Phys Lett **501**:619–22.
81. Tilley S D, Cornuz M, Sivula K, Grätzel M (2010) Light-induced water splitting with hematite: improved nanostructure and iridium oxide catalysis. Angew Chem Int Ed **49**:6405–8.
82. Lee S M, Ikeda S, Otsuka Y, Septina W, Harada T, Matsumura M (2012) Homogeneous electrochemical deposition of In on a Cu–covered Mo substrate for fabrication of efficient solar cells with a $CuInS_2$ photoabsorber. Electrochim Acta **79**:189–96.
83. Bandyopadhyaya S, Chaudhuri S, Pal A K (2000) Synthesis of $CuInS_2$ films by sulphurization of Cu/In stacked elemental layers. Sol Energy Mater Sol Cells **60**:323–39.
84. Higashi M, Domen K, Abe R (2011) Fabrication of efficient TaON and Ta_3N_5 photoanodes for water splitting under visible light irradiation. Energy Environ Sci **4**:4138–47.
85. Li Y, Takata T, Cha D, Takanabe K, Minegishi T, Kubota J, Domen K (2013) Vertically aligned Ta_3N_5 nanorod arrays for solar-driven photoelectrochemical water splitting. Adv Mater **25**:125–31.

86. Li Y, Zhang L, Torres-Pardo A, González-Calbet J M, Ma Y, Oleynikov P, Terasaki O, Asahina S, Shima M, Cha D, Zhao Lan, Takanabe K, Kubota J, Domen K (2013) Cobalt phosphate-modified barium-doped tantalum nitride nanorod photoanode with 1.5% solar energy conversion efficiency. Nature Commun **4**:2566.
87. Nishimura N, Maeda K, Takata T, Lu D, Kubota J, Domen K (2013) Fabrication of photoelectrodes from $LaTiO_2N$ particles for photoelectrochemical water splitting. Bull Chem Soc Jpn **86**:540–6.
88. Minegishi T, Nishimura N, Kubota J, Domen K (2013) Photoelectrochemical properties of $LaTiO_2N$ electrodes prepared by particle transfer for sunlight-driven water splitting. Chem Sci **4**:1120–4.
89. Kudo A, Omori K, Kato H (1999) A novel aqueous process for preparation of crystal form-controlled and highly crystalline $BiVO_4$ powder from layered vanadates at room temperature and its photocatalytic and photophysical properties. J Am Chem Soc **121**:11459–67.
90. Jia Q, Iwashina K, Kudo A (2012) Facile fabrication of an efficient $BiVO_4$ thin film electrode for water splitting under visible light irradiation. Proc Natl Acad Sci U S A **109**:11564–9.
91. Li R, Zhang F, Wang D, Yang J, Li M, Zhu J, Zhou X, Han H, Li C (2013) Spatial separation of photogenerated electrons and holes among {010} and {110} crystal facets of $BiVO_4$. Nature Commun **4**:1432.
92. Li Z, Luo W, Zhang M, Feng J, Zou Z (2013) Photoelectrochemical cells for solar hydrogen production: current state of promising photoelectrodes, methods to improve their properties, and outlook. Energy Environ Sci **6**:347–70.
93. Yin W, Wei S, Al-Jassim M M, Turner J, Yan Y (2011) Doping properties of monoclinic $BiVO_4$ studied by first-principles density-functional theory. Phys Rev B **83**:155102.
94. Zhao Z, Luo W, Li Z, Zou Z (2010) Density functional theory study of doping effects in monoclinic clinobisvanite $BiVO_4$. Phys Lett A **374**:4919–27.
95. Sayama K, Nomura A, Zou Z, Abe R, Abe Y, Arakawa H (2003) Photoelectrochemical decomposition of water on nanocrystalline $BiVO_4$ film electrodes under visible light. Chem Commun 2908–9.
96. Luo W, Yang Z, Li Z, Zhang J, Liu J, Zhao Z, Wang Z, Yan S, Yu T, Zou Z (2011) Solar hydrogen generation from seawater with a modified $BiVO_4$ photoanode. Energy Environ Sci **4**:4046–51.
97. Long M, Cai W, Kisch H (2008) Visible light induced photoelectrochemical properties of n-$BiVO_4$ and n-$BiVO_4$/p-Co_3O_4. J Phys Chem C **112**:548–54.
98. Wang D, Li R, Zhu J, Shi J, Han J, Zong X, Li C (2012) Photocatalytic water oxidation on $BiVO_4$ with the electrocatalyst as an oxidation cocatalyst: essential relations between electrocatalyst and photocatalyst. J Phys Chem C **116**:5082–9.
99. Zhong D K, Choi S, Gamelin D R (2011) Near-complete suppression of surface recombination in solar photoelectrolysis by "Co-Pi" catalyst-modified W:$BiVO_4$. J Am Chem Soc **133**:18370–7.
100. Seabold J A, Choi K (2012) Efficient and stable photo-oxidation of water by a bismuth vanadate photoanode coupled with an iron oxyhydroxide oxygen evolution catalyst. J Am Chem Soc **134**:2186–92.
101. Sivula K, Formal F L, Grätzel M (2011) Solar water splitting: progress using hematite (a-Fe_2O_3) photoelectrodes. ChemSusChem **4**:432–49.
102. Sivula K (2013) Metal oxide photoelectrodes for solar fuel production, surface traps, and catalysis. J Phys Chem Lett **4**:1624 – 33.
103. Cherepy N J, Liston D B, Lovejoy J A, Deng H, Zhang J Z (1998) Ultrafast studies of photoexcited electron dynamics in γ- and α-Fe_2O_3 semiconductor nanoparticles. J Phys Chem B **102**:770–6.
104. Gardner R F G, Sweett F, Tanner D W (1963) The electrical properties of alpha ferric oxide—II: Ferric oxide of high purity. J Phys Chem Solids **24**:1183–6.
105. Barroso M, Pendlebury S R, Cowan A J, Durrant J R (2013) Charge carrier trapping, recombination and transfer in hematite (α-Fe_2O_3) water splitting photoanodes. Chem Sci **4**:2724–34.

106. Hisatomi T, Dotan H, Stefik M, Sivula K, Rothschild A, Grätzel M, Mathews N (2012) Enhancement in the performance of ultrathin hematite photoanode for water splitting by an oxide underlayer. Adv Mater **24**:2699–702.
107. Brillet J, Cornuz M, Formal F L, Yum J, Grätzel M, Sivula K (2010) Examining architectures of photoanode–photovoltaic tandem cells for solar water splitting. J Mater Res **25**:17–24.
108. Zhong D K, Cornuz M, Sivula K, Grätzel M, Gamelin D R (2011) Photo-assisted electrodeposition of cobalt–phosphate (Co–Pi) catalyst on hematite photoanodes for solar water oxidation. Energy Environ Sci **4**:1759–64.
109. Bak T, Nowotny J, Rekas M, Sorrell C C (2002) Photo-electrochemical hydrogen generation from water using solar energy. Materials-related aspects. Int J Hydrogen Energy **27**:991–1022.
110. Huda M N, Yan Y, Moon C, Wei S, Al-Jassim M M (2008) Density-functional theory study of the effects of atomic impurity on the band edges of monoclinic WO_3. Phys Rev B: Condens Matter Mater Phys **77**:195102.
111. Liu G, Wang X, Wang X, Han H, Li C (2012) Photocatalytic H_2 and O_2 evolution over tungsten oxide dispersed on silica. J Catal **293**:61–6.
112. Abe R, Takami H, Murakami N, Ohtani B (2008) Pristine simple oxides as visible light driven photocatalysts: highly efficient decomposition of organic compounds over platinum-loaded tungsten oxide. J Am Chem Soc **130**:7780–1.
113. Yang B, Zhang Y, Drabarek E, Barnes P R F, Luca V (2007) Enhanced photoelectrochemical activity of sol-gel tungsten trioxide films through textural control. Chem Mater **19**:5664–72.
114. Bignozzi C A, Caramori S, Cristino V, Argazzi R, Medac L, Taccac A (2013) Nanostructured photoelectrodes based on WO_3: applications to photooxidation of aqueous electrolytes. Chem Soc Rev **42**:2228–46.
115. Hong S J, Lee S, Jang J S, Lee J S (2011) Heterojunction $BiVO_4/WO_3$ electrodes for enhanced photoactivity of water oxidation. Energy Environ Sci **4**:1781–7.
116. Kim J K, Shin K, Cho S M, Leeb T, Park J H (2011) Synthesis of transparent mesoporous tungsten trioxide films with enhanced photoelectrochemical response: application to unassisted solar water splitting. Energy Environ Sci **4**:1465–70.
117. Jongh P E, Vanmaekelbergh D, Kelly J J (2000) Photoelectrochemistry of Electrodeposited Cu_2O. J Electrochem Soc **147**:486–9.
118. Engel C J, Polson T A, Spado J R, Bell J M, Fillinger A (2008) Photoelectrochemistry of Porous p-Cu_2O Films. J Electrochem Soc, **155**:37–42.
119. Amano F, Ebina T, Ohtani B (2014) Enhancement of photocathodic stability of p-type copper(I) oxide electrodes by surface etching treatment. Thin Solid Films **550**:340–6.
120. Hsu Y, Yu C, Lin H, Chen Y, Lin Y (2013) Template synthesis of copper oxide nanowires for photoelectrochemical hydrogen generation. J Electroanal Chem **704**:19–23.
121. Siripala W, Ivanovskaya A, Jaramillo T F, Baeck S, McFarland E W (2003) A Cu_2O/TiO_2 heterojunction thin film cathode for photoelectrocatalysis. Sol Energy Mater Sol Cells **77**:229–37.
122. Paracchino A, Laporte V, Sivula K, Grätzel M, Thimsen E (2011) Highly active oxide photocathode for photoelectrochemical water reduction. Nature Mater **10**:456–61.
123. Paracchino A, Mathews N, Hisatomi T, Stefik M, Tilley S D, Grätzel M (2012) Ultrathin films on copper(I) oxide water splitting photocathodes: a study on performance and stability. Energy Environ Sci **5**:8673–81.
124. Maeda K (2011) Photocatalytic water splitting using semiconductor particles: History and recent developments. J Photochem Photobiol C **12**:237–68.
125. Hisatomi T, Maeda K, Takanabe K, Kubota J, Domen K (2009) Aspects of the water splitting mechanism on $(Ga_{1-x}Zn_x)(N_{1-x}O_x)$ Photocatalyst Modified with $Rh_{2-y}Cr_yO_3$ cocatalyst. J Phys Chem C **113**:21458–66.
126. Hisatomi T, Miyazaki K, Takanabe K, Maeda K, Kubota J, Sakata Y, Domen K (2010) Isotopic and kinetic assessment of photocatalytic water splitting on Zn-added Ga_2O_3 photocatalyst loaded with $Rh_{2-y}Cr_yO_3$ cocatalyst. Chem Phys Lett **486**:144–6.
127. Wrona P K, Lasia A, Lessard M, Ménard H (1992) Kinetics of the hydrogen evolution reaction on a rhodium electrode. Electrochim Acta **37**:1283–94.

128. Dionigi F, Vesborg P C K, Pedersen T, Hansen O, Dahl S, Xiong A, Maeda K, Domen K, Chorkendorff I (2011) Gas phase photocatalytic water splitting with $Rh_{2-y}Cr_yO_3$/GaN:ZnO in μ-reactors, Energy Environ Sci **4**:2937–42.
129. Yoshida M, Yamakata A, Takanabe K, Kubota J, Osawa M, Domen K (2009) ATR-SEIRAS investigation of the Fermi level of Pt cocatalyst on a GaN photocatalyst for hydrogen evolution under irradiation. J Am Chem Soc **131**:13218–9.
130. Correia A N, Machado S A S (1998) Hydrogen evolution on electrodeposited Ni and Hg ultramicroelectrodes. Electrochim Acta **43**:367–73.
131. Dotan H, Sivula K, Grätzel M, Rothschild A, Warren S C (2011) Probing the photoelectrochemical properties of hematite (α-Fe_2O_3) electrodes using hydrogen peroxide as a hole scavenger. Energy Environ Sci **4**:958–64.
132. Formal F L, Grätzel M, Sivula K (2010) Controlling photoactivity in ultrathin hematite films for solar water-splitting. Adv Funct Mater **20**:1099–107.
133. Xia J, Masaki N, Jiang K, Yanagida S (2007) Sputtered Nb_2O_5 as a novel blocking layer at conducting glass/TiO_2 interfaces in dye-sensitized ionic liquid solar cells. J Phys Chem C **111**:8092–7.
134. Klahr B, Gimenez S, Fabregat-Santiago F, Bisquert J, Hamann T W (2012) Electrochemical and photoelectrochemical investigation of water oxidation with hematite electrodes, Energy Environ Sci **5**:7626–36.
135. Maeda K, Teramura K, Domen K (2008) Effect of post-calcination on photocatalytic activity of $(Ga_{1-x}Zn_x)(N_{1-x}O_x)$ solid solution for overall water splitting under visible light. J Catal **254**:198–204.
136. Abdi F F, Han L, Smets A H M, Zeman M, Dam B, Krol R V (2013) Efficient solar water splitting by enhanced charge separation in a bismuth vanadate-silicon tandem photoelectrode. Nat Commun **4**:2195.

Artificial Photosynthesis Producing Solar Fuels: Natural Tactics of Photosynthesis

Hideki Hashimoto and Chiasa Uragami

Abstract 70 % of mankind's current energy needs are met by burning fossil fuels. This is already problematic since oil and gas supplies are limited and because of the adverse environmental effects of rising levels of carbon dioxide in the atmosphere. Moreover this situation is set to get worse as current predictions estimate that our energy needs will double by 2050. Mankind is, therefore, facing a major challenge to find new ways of creating clean, renewable fuels. One potentially abundant source of energy is solar radiation. More energy strikes the earth surface every hour than mankind uses each year. The problem is how to harness such an abundant yet diffuse source of energy. Photosynthesis has evolved mechanisms to achieve this. Conceptually photosynthesis can be divided into the following partial reactions: light harvesting (concentration), charge separation (conversion of solar energy into chemical energy), and finally multi-electron catalysis that takes electrons from water and uses them to reduce carbon dioxide to carbohydrates (fuel). Any proposed strategies that set out to mimic this process in order to make solar fuels must begin with a light-harvesting or light-concentration step. We know a great deal about the structure and function of photosynthetic light-harvesting complexes as individual molecules, but rather little is known about how their supramolecular organization within their photosynthetic membranes affects their overall function in vivo. We wish to understand how the supramolecular arrangement of the light-harvesting apparatus relates to overall light-harvesting efficiency and to be able to translate this information to inform the design of robust artificial light-harvesting arrays that can, in the long term, be used in devices for producing solar fuels. Photosynthetic antenna complexes are organized on the nanoscale, and a major question is how to translate this information into the design of meso- to macroscale light-harvesting devices. This chapter will outline how photosynthesis achieves "solar to fuel" conversion concentrating on the general principles involved. Recent progress on understating the molecular details of the key reactions in the photosynthetic process has been remarkable. We are now at the stage where it is realistic to

---

H. Hashimoto (✉) • C. Uragami
The Osaka City University Advanced Research Institute for Natural Science and Technology (OCARINA), Osaka City University, 3-3-138 Sugimoto, Sumiyoshi-ku, Osaka 558-8585, Japan
e-mail: hassy@ocarina.osaka-cu.ac.jp

E.A. Rozhkova, K. Ariga (eds.), *From Molecules to Materials*,
DOI 10.1007/978-3-319-13800-8_2

start to use this “biological blueprint” to begin to construct devices that have the capability to mimic the key steps in the natural process. This is one of the grand scientific challenges of our time.

Keywords Photosynthesis • Artficial photosynthesis • Solar fuel

1 Introduction

Figure 1 visualizes the ratio of the source of world primary energy supply in the year 2008, which was reported by the IEA [1]. The fossil fuels such as oil, coal, and natural gas occupy genuinely 80 % of all the energy supply. We human beings use 80 % (60 % of the total) of the fossil fuels in their original form, e.g., transportation, fire power plant, chemicals, etc. If we estimate the amount of remaining fossil fuels based on a predicted consumption level, coal is estimated to remain for 200 years or less, while oil is about 45 years. Fortunately, natural gas remains for around 160 years owing to the effect of recent achievement of shale gas revolution (USA) [2]. Of course these numbers are figured out based on the most pessimistic point of view. However, they cannot be doubled if we were optimistic. Therefore, a breakaway from the social structure that was thoroughly dependent on the fossil fuels is desired earnestly before the drying up of all the natural resources. As a policy of that purpose, both the fundamental studies on the utilization of the natural energy and the applied studies for the practical use of the results of fundamental researches are

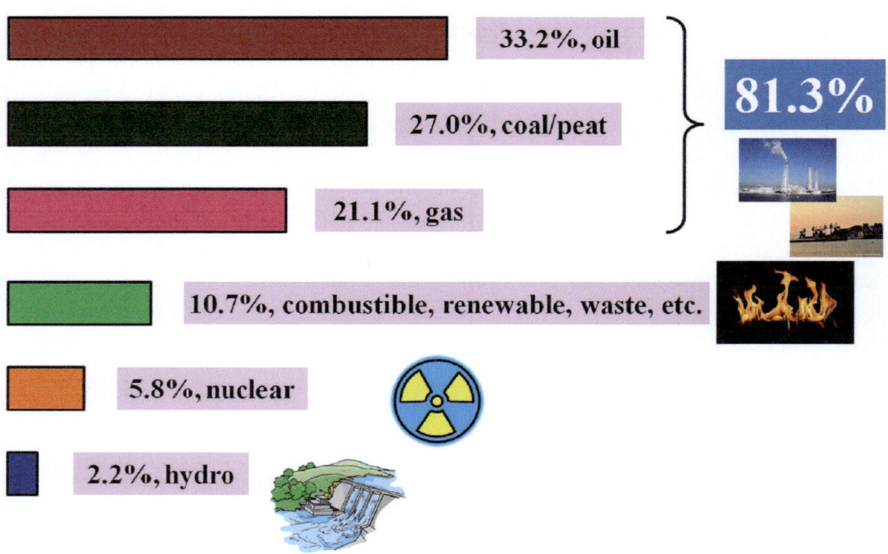

Fig. 1 The graph that shows the ratio of the primary energy supply of the world. (Statistics taken by International Energy Agency 2008)

now accelerating. Particularly, interest in utilization of the solar energy increases. This is because solar energy that falls for 1 h on the earth surface is known to be equal to the total amount of energy to be used in a year by all humankind. Solar energy is approximately inexhaustible and clean. If utilization of the solar energy increases, it is possible to reduce the emission of carbon dioxide, which is thought to be one serious factor of the global warming, to be substantially zero. To that end, it is required to correctly understand the mechanisms of photosynthesis, which are developed in the process of the evolution over 3.8 billion years, and to imitate or modify them in the form that is easy to use for the humans [3, 4].

2 From Natural Photosynthesis to Artificial Photosynthesis

Figure 2 shows the schematic drawing of light energy conversion processes carried out in photosynthetic membranes of oxygenic prototroph [5]. Photosynthesis of the plants is a reaction to produce bioenergy and carbohydrates (fuels) from water and carbon dioxide using light energy from the sun. During these processes, molecular oxygen is exhausted as a needless by-product. In other words, photosynthesis produces all the food that we eat as the origin of the food chain. It also produces all the

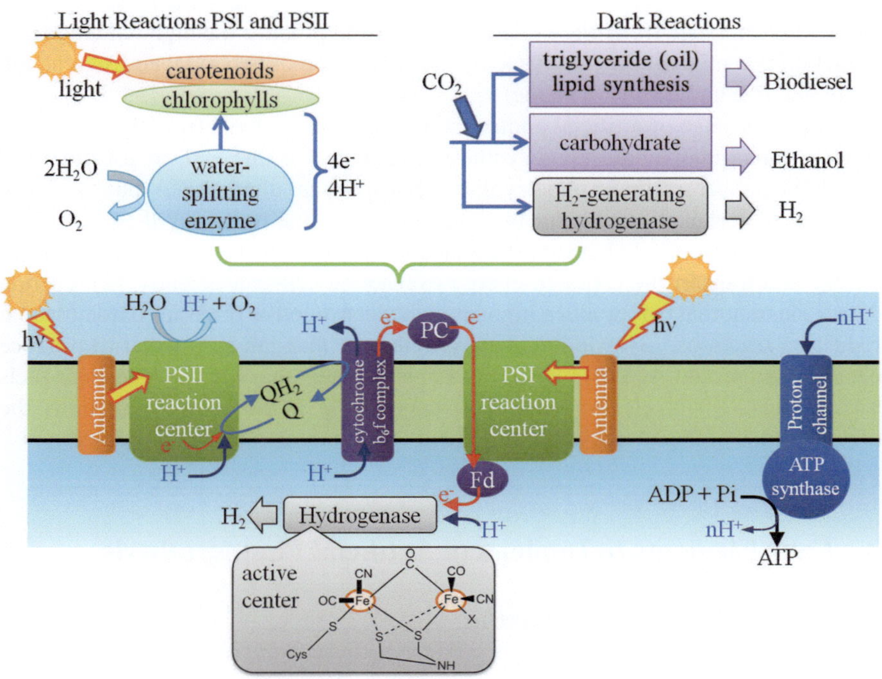

Fig. 2 A schematic description of light energy conversion processes by oxygenic photosynthesis organisms in the photosynthetic membrane. (Rearranged from [5] with permission)

oxygen that we breathe. A corpse of the phototroph originally deposited and the fossil fuels that we use were produced in the process of the earth evolution. Therefore, photosynthesis becomes the source of an all-living thing inhabiting on the earth.

Light-harvesting pigment-protein complexes called photosynthetic antennas absorb light energy from the sun to start the abovementioned photosynthesis reaction. Thus harvested solar energy is transferred in the form of excitation energy to two reaction center complexes of photosystem II (PSII) and photosystem I (PSI). In these reaction center complexes, an electric charge separation reaction takes place and electromotive force is accumulated. Biocatalyst reactions to oxidatively split water as well as to accumulate a reduction power and to synthesize NADPH (reduced form of nicotinamide adenine dinucleotide phosphate) progress using this electromotive force. PSII holds an oxidation catalyst, and PSI holds a reduction catalyst. PSII has the quinone-type reaction center that is similar to purple photosynthetic bacteria, and PSI has the iron-sulfur cluster-type reaction center that is similar to green sulfur bacteria. A remarkable point of PSII is that it demands an electronic source from "water" to fulfill a positive hole produced as a result of the electric charge separation. To this end, the biocatalyst called the Mn_4O_5Ca cluster functions. The precise structure of this Mn_4O_5Ca cluster has been determined quite recently and attracts keen attention [6]. This is because we were able to obtain the design plan of the catalyst that can take out an electron and a proton (H^+ ion) from "water," which is the most stable material on the earth. Another splendid point of the oxygen-evolving-type photosynthesis is that it allows to utilize the light of a broad spectral range from ultraviolet to infrared and produce the high oxidation and reduction potentials at the same time by the connection of two independent reaction centers of PSI and PSII (Z-scheme).

A big challenge of artificial photosynthesis is to make fuels from light from the sun by imitating splendid photosynthesis reactions. To this end, as illustrated in Fig. 3, the abovementioned photosynthesis reactions are divided into four elements, and then the production of solar fuels becomes possible by letting them link. The four elements are (1) light-harvesting antenna element that can capture the solar energy and transfer the captured energy to reaction center efficiently, (2) reaction center element that can produce high enough electromotive force for oxidation and reduction reactions, (3) oxidation catalyst element that can split water to produce electron and proton, and (4) reduction catalyst element that can produce solar fuels by reacting CO_2 with electron and proton concertedly. In the following sections, the present status of the advanced studies about these four elements is discussed.

3 Four Elements to Implement Artificial Photosynthesis

3.1 Light-Harvesting Antenna

The gross amount of the light from the sun that pours on the earth is vast. However, it does not become a big numerical value when we convert it into the number of photon flux density in a unit time. This becomes clear when the exact calculation is

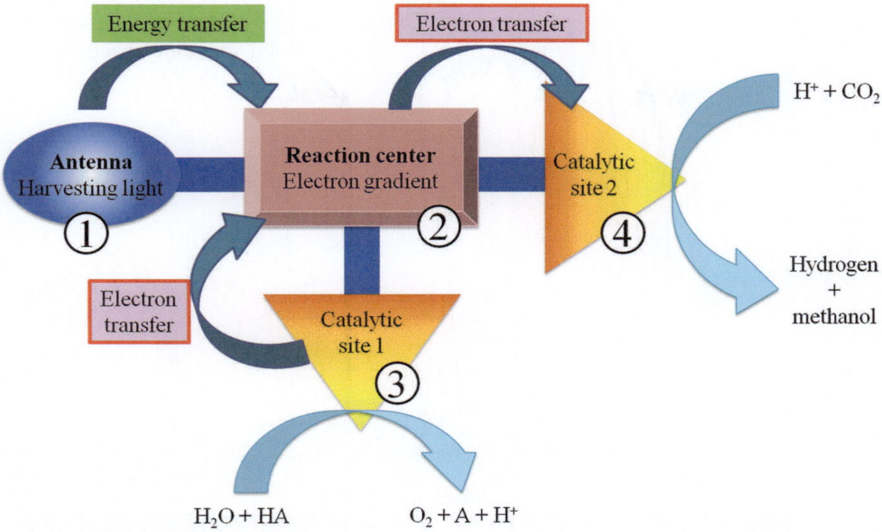

Fig. 3 A schematic description that shows the principles of artificial photosynthesis (artificial leaf)

performed. The quantity of the solar energy that hit the earth surface is prescribed with a solar constant and is approximately 1.37 kW/m^2. When exchanging this amount of energy to the number of photons, it becomes the degree of more than 6×10^{21}/m^2 s. The number of photons of 400–800 nm spectral region (visible light domain) is 1.6×10^{21}/m^2 s. This is equivalent to only a 1/400 of Avogadro's number for 1 s per 1 square meter. If we can utilize all of these photons with 100 % efficiency, only 11 mg of water can be split because four photons are necessary to split one molecule of water. In other words, the gross amount of solar energy is vast, but radiation density has a certain limit. Therefore, accumulation of the light (antenna function) becomes important.

The photosynthesis of plants has been developing unique light-harvesting antenna systems to efficiently collect light energy from the sun to overcome this problem. Various phototrophs such as higher plants, algae, and photosynthetic bacteria inhabit the earth. As illustrated in Fig. 4, when the absorption spectra of all of the antenna systems from these phototrophs are overlaid, they show a good coverage with the solar radiation spectrum. In other words, phototrophs on the earth have evolved so that they can share the solar radiation by developing their own unique antenna systems with characteristic colors. Learning this wisdom of nature itself holds the key to make feasible the breakthrough for the development of the highly efficient light-harvesting system that is the bottleneck of the full utilization of solar energy.

The phototrophs have various colors. Carotenoid is the main factor that gives the change of the color. Therefore, it is not too much to say that correct understandings of the functions of carotenoids directly connect to the understanding of the essence of the photosynthetic light harvesting. In the process of evolution, the phototroph

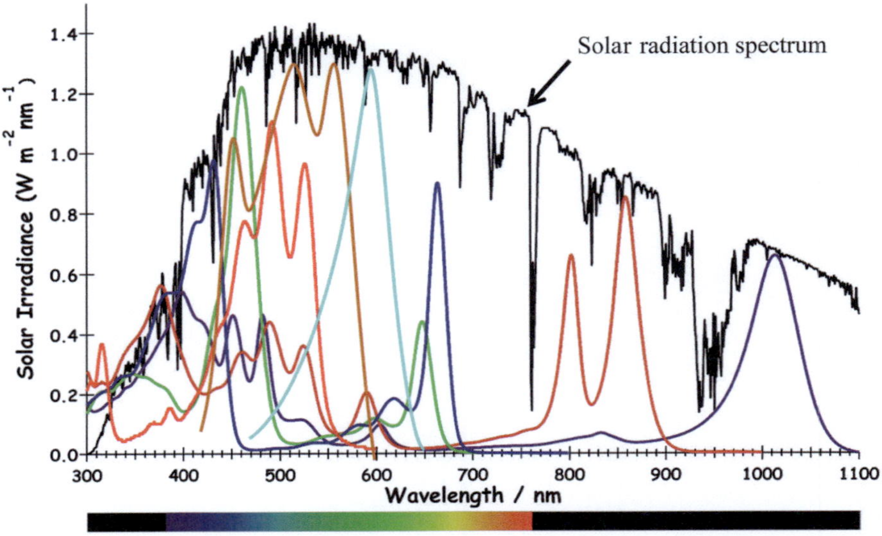

Fig. 4 The spectrum of solar radiation and the absorption spectra of light-harvesting antenna pigment-protein complexes from various photosynthetic organisms. The photosynthetic antennas that nature had created cover up perfectly all the wavelengths' range of the solar radiation. (Courtesy of Assoc. Prof. Ritsuko Fujii of Osaka City University)

got away from the ocean and lake with thin solar light radiation and developed a place of life in land with rich solar radiation. During this process, after having fully achieved the best of the solar energy conversion, the phototrophs developed a photoprotection mechanism (the defense mechanism to protect organisms against photodestruction caused by the excess amount of light exposure) that can dissipate excess amount of light energy. In a primary process of photosynthesis, carotenoids transfer their captured solar energy to chlorophylls, and then a series of photosynthesis reactions start. Photosynthesis is a multistep reaction with more than 40 steps. Surprisingly each process proceeds with the efficiency of nearly 100 %. However, the efficiency of energy transfer from carotenoid to chlorophyll varies from 30 % to nearly 100 % depending on the structure of carotenoids. Therefore, understanding the functions of carotenoids holds the key for the effective utilization of solar energy.

3.2 *Reaction Center*

Figure 5 shows the structure of the reaction center (RC) isolated from a purple photosynthetic bacterium *Rhodobacter sphaeroides*, which has been decided using X-ray crystallography [7]. RC is composed of three different subunits (proteins), H, M, and L, with different molecular weights, and photosynthetic pigments are

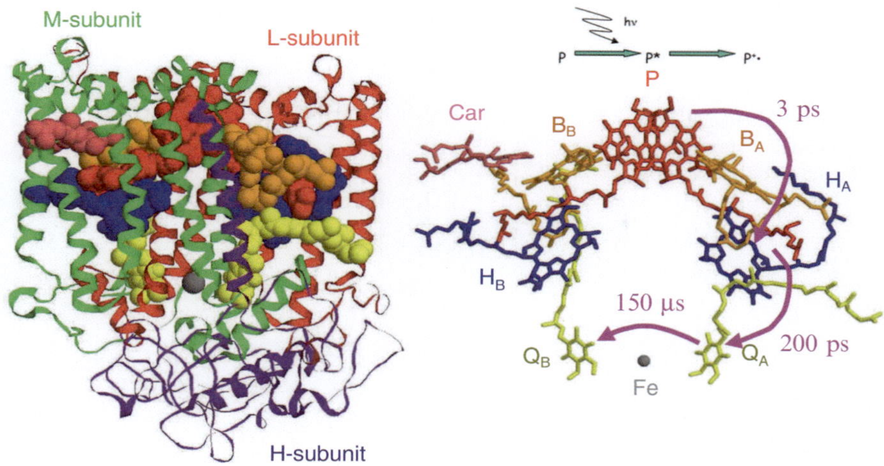

Fig. 5 The structure of the reaction center from the purple photosynthetic bacterium *Rhodobacter sphaeroides*

bound to them. The left-hand side of Fig. 5 shows the whole structure of the pigment-protein complex of RC, and the right-hand side shows arrangement of the photosynthetic pigments in RC. The ribbons of the left-hand figure are proteins (only the main chain part). The photosynthetic pigments are aligned with pseudo twofold symmetry as illustrated in the right-hand side. When dimeric bacteriochlorophylls (called special pair bacteriochlorophyll) as designated with P are excited, electric charge separation takes place, and a produced electron is transferred to quinone (Q_A) via monomeric bacteriochlorophyll (B_A) and bacteriopheophytin (H_A). The electron is transferred to Q_A with a surprisingly fast speed of 200 ps (200×10^{-12} s) after P was excited. It is interesting to note here that the pigment arrangement in the RC has pseudo twofold symmetry; the electron transport takes place only along the path denoted with the subscript A (A-branch) in purple photosynthetic bacteria.

Very interestingly, in the case of the RC of PSI and PSII, the arrangement of the photosynthesis pigments in these RCs is similar to that of purple photosynthetic bacteria. However, in these RCs, the electron transport takes place in both A- and B-branch sides and not restricted to a single pathway like purple photosynthetic bacteria. It is noteworthy to point out that the pigment arrangement in the RC is well preserved across the wall of living organisms with different categories. This means that there is an inevitable limitation for the method to make possible the electron transportation in the RC proteins that are essentially insulators. As a result of repeated trial and error in the process of evolution, the photosynthetic organisms could find out a good and unique solution for this issue. In other words, it may be able to say that the design plan to produce the electromotive source in the RC proteins has already been well established.

Fig. 6 The oxidation-reduction potential for oxidative splitting of water, hydrogen evolution, and CO_2 reduction. (Calculated against the potential of standard hydrogen electrode at pH = 7)

- $2H_2O \quad O_2 + 4H^+ + 4e^-$ +0.82 V
- $2H^+ + 2e^- \quad H_2$ -0.41 V
- $CO_2 + 8H^+ + 8e^- \quad CH_4 + 2H_2O$ -0.24 V
- $CO_2 + 6H^+ + 6e^- \quad CH_3OH + H_2O$ -0.38 V
- $CO_2 + 4H^+ + 4e^- \quad HCHO + H_2O$ -0.48 V
- $CO_2 + 2H^+ + 2e^- \quad CO + H_2O$ -0.53 V

(Aqueous solution, pH 7 versus NHE.)

3.3 Oxidation Catalyst

One of the breakthrough technologies that are desired to make possible the production of solar fuels by artificial photosynthesis is to oxidatively split water and to generate both electrons and protons from it as a source. As shown in Fig. 6, the standard electric potential of +0.82 V is required to obtain both electrons and protons while splitting water and performing oxygen evolution. In addition, the standard electric potential of −0.41 V is necessary to perform hydrogen evolution and thus produce protons. Therefore, the electric potential of at least 1.23 V is necessary in order to make feasible the splitting of water to get oxygen and hydrogen. In reality the electric potential of 1.6 V is required because the overpotential is necessary in addition to this electric potential to drive a catalyst to completely split water electrochemically. It is not easy to get such a high electric potential using only a single semiconductor photocatalyst without any help of electrochemistry. One of the most successful examples on this issue is the Honda-Fujishima effect that could completely split water photochemically by irradiating titanium dioxide with UV light [8–11].

Figure 7 shows the schematic description that explains the principle of the Honda-Fujishima effect. When titanium dioxide is irradiated with UV light whose energy is greater than 3 eV that is equivalent to the bandgap of this semiconductor photocatalyst (process (1) in Fig. 7), a pair of electron and positive hole is generated (process (2) in Fig. 7). The electron-hole pair can be effectively separated because of the presence of the bending of the bands that is formed in the semiconductor due to the interaction at the interface between semiconductor and solution or electrode. Along the curve of the bands in this semiconductor, the electron can move toward the inside of the semiconductor, while the positive hole can move to the surface of the semiconductor (process (3) in Fig. 7). The water is oxidized by the positive hole at the surface of the *n*-type semiconductor such as titanium dioxide and produces oxygen (process (4) in Fig. 7). The hydrogen ion can be reduced by the electron if a platinum catalyst is used, and molecular hydrogen is produced (process (4) in Fig. 7). This process looks quite simple at a glance, but the secret to completely split water using a single semiconductor photocatalyst is beautifully summarized. In this regard

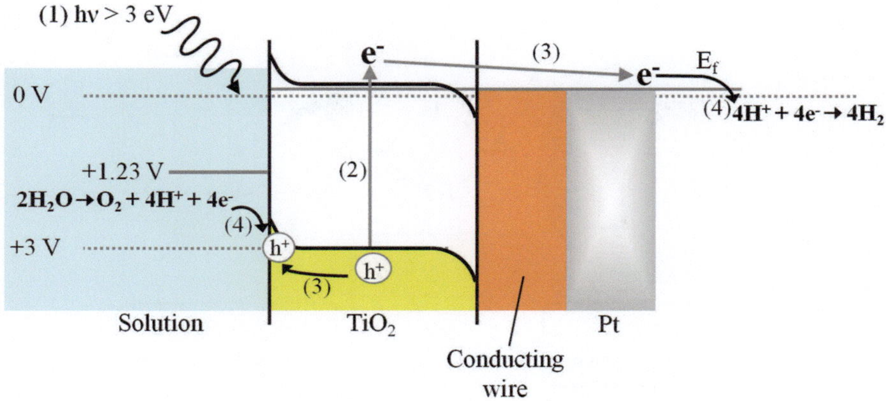

Fig. 7 A schematic description that shows the principle of photoinduced water splitting using titanium dioxide photocatalyst (Honda-Fujishima effect)

it is not too much to say that the titanium dioxide is an amazing system. According to the suggestion of Nernst, if we suppose the hydrogen standard electrode to be the base of all electrometry and define its electrode potential as 0 V in all temperature ranges, the potential equivalent to 1.23 V is necessary to split water oxidatively and produce molecular oxygen as already mentioned above. When the semiconductor photocatalyst has a catalytic ability, i.e., when the photocatalyst has some mechanisms to function, overpotential is necessary in addition to this electric potential, and hence the electric potential of around 1.6 V is required in reality. In the case of the titanium dioxide semiconductor, the top level of the valence band is prescribed by 2p orbital of oxygen. Unless the electric potential of this level exceeds 1.6 V, water cannot be split. In addition, unless the energy level of the conduction band is lower than the oxidation-reduction potential of molecular hydrogen, hydrogen production cannot be performed. Titanium dioxide has a bandgap as wide as 3 eV. Therefore this particular semiconductor can overcome the above exquisite energy balance and can show high enough potential of the valence band even considering the bend of the band in the solution/semiconductor interface.

However, on the contrary, titanium dioxide has such a wide bandgap, and light source to drive this photocatalyst is restricted in the UV region of solar radiation, where only 3 % of the light energy is included in this spectral region and hence results in a lower efficiency of conversion. The narrow-gap semiconductor that can absorb light in the visible spectral region exists. However, it is difficult to earn the high enough valence potential that can split water by a single photocatalyst. To conquer this issue, Domen et al. combined two different kinds of semiconductor photocatalysts that can absorb the visible region of light and construct Z-scheme-type arrays of these photocatalysts and were successful to split water with visible light [12]. Figure 8 shows its example [13]. Water was split oxidatively to produce oxygen using the WO_3 photocatalyst in which the cocatalyst was doped. Hydrogen was produced by the reduction of protons using ZrO_3/TaON photocatalyst in which

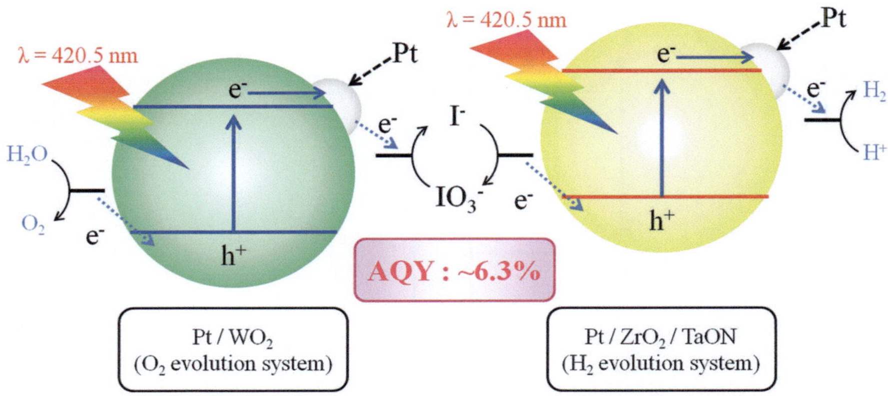

Fig. 8 An example of complete water splitting using the combination of two different kinds of visible light-responding semiconductor photocatalysts. (Reproduced from [7] with permission)

platinum was doped as a cocatalyst. Both of these photocatalysts have good response to the visible light. By connecting these photocatalysts in a Z-scheme-type manner with iodine as a redox mediator, an apparent quantum yield of 6.3 % has been achieved by visible light irradiation at 420.5 nm to split water. Because Domen's research project is very progressive as one of the huge national projects in Japan right now, it is highly expected that the success of these studies will bring us a high impact on the development of the systems that will drastically improve conversion efficiency in the very near future.

The difficult point of the complete water splitting is that we must find out good ways that both oxidation and reduction catalysts can work satisfactorily at the same time. If we can devote to only a partial reaction to produce oxygen from water, this problem is somewhat evaded, and many good results have been reported [14]. In the system using the ruthenium complex, water splitting activity at the same level as PSII has been reported and attracts attention recently [15].

3.4 Reduction Catalyst

One of the powerful charms of artificial photosynthesis is the point that it is going to generate hydrocarbon fuel with CO_2 as raw materials using solar energy. Electrochemistry shows that one electron reduction of CO_2 to generate CO_2^- needs electric potential of −2.38 V and overpotential (0.1–0.6 V). In contrast, interestingly, CO_2 can be reduced with much lower oxidation-reduction potential by reacting a proton and an electron with it concertedly as shown in Fig. 6. Therefore, proton-coupled electron transfer (PCET) becomes the important process to realize the fuel generation by the reduction of CO_2 [16]. The point that should be emphasized here is that multiple-electron reduction should always have an even number of electrons. In other words, methane generation from CO_2 has an 8-electron reduction, methanol generation 6-electron reduction, formaldehyde generation 4-electron reduction, and carbon monoxide generation 2-electron reduction (cf. Fig. 6). The development of such reduction catalysts has a long-standing history, and the development of the system using the rhenium complex advances since the first discovery by Lehn and coworkers [17]. Ishitani and coworkers reported the 59 % quantum efficiency of CO_2 to CO conversion [18]. It is necessary to understand the exact mechanism of the CO_2 conversion for the further improvement of the efficiency. 82 % of quantum yields have been achieved recently as the results of various kinds of examination [19–22].

With regard to CO_2 reduction, experimental results using the semiconductor photocatalyst system have been reported recently and they attract keen attention. Morikawa and coworkers were successful in the simultaneous oxidation of water and the reduction of CO_2 and then in the production of formic acid based on a Z-scheme by connecting titanium dioxide and InP/MCp [23]. Solar energy conversion efficiency is as low as 0.03–0.04 %, but this is the right successful example of artificial photosynthesis. Because of the later improvements, the conversion efficiency is now increased to 0.14 % [24].

Conclusion

The understandings of the key reactions of photosynthesis are highly advanced at the molecular level in recent years. As a result, "biological blueprints" which begin to make devices that mimic natural photosynthesis have been postulated. We are now approaching the stage of beginning to use this "biological blueprint" practically. This is a magnificent scientific challenge in our times. For this achievement, the development of the "artificial photosynthesis system" will become the one important step. In Europe and the USA, a number of leading researchers in the research field of photosynthesis gathered together and made very significant reports (white papers), which summarize basic science to change the dependence on fossil fuels into the use of solar energy. For those who want to know further details, the authors strongly recommend to refer to these excellent white papers. In the USA, a huge amount of research funds were cast to establish the "Joint Center for Artificial Photosynthesis (JCAP)" in order to realize the contents of the white papers. Movement that aims at the construction of a CO_2-free or CO_2-neutral society that does not have heavy load on the environment is accelerated after the serious accident at the Fukushima Nuclear Power Plant caused by the tsunami of the East Japan great earthquake disaster that happened on March 11, 2011. The authors desire it earnestly to accelerate the practical use of the next-generation fuel with the "artificial photosynthesis system" by all means.

References

1. International Energy Agency (2012) Key world energy statistics. http://www.iea.org/publications/freepublications/publication/kwes.pdf.
2. U.S. Energy Information Administration (2011) World shale gas resources: An initial assessment of 14 regions outside the United States. http://www.adv-res.com/pdf/ARI EIA Intl Gas Shale APR 2011.pdf.
3. U.S. Department of Energy (2005) Basic research needs for solar energy utilization. http://www.sc.doe.gov/bes/reports/files/SEU_rpt.pdf.
4. European Science Foundation (2005) Harnessing solar energy for the production of clean fuels. http://ssnmr.leidenuniv.nl/files/ssnmr/CleanSolarFuels.pdf.
5. Centi G, Perathoner S (2010) Towards solar fuels from water and CO_2. ChemSusChem 3: 195-208. doi:10.1002/cssc.200900289
6. Umena Y, Kawakami K, Shen J-R, Kamiya N (2011) Crystal structure of oxygen-evolving photosystem II at a resolution of 1.9 Å. Nature 473:55-60. doi:10.1038/nature09913
7. Roszak AW, McKendrick K, Gardiner AT, Mitchell IA, Isaacs NW, Cogdell RJ, Hashimoto H, Frank HA (2004) Protein regulation of carotenoid binding: Gatekeeper and locking amino acid residues in reaction centers of *Rhodobacter sphaeroides*. Structure 12 (5):765-773. doi:10.1016/j.str.2004.02.037
8. Fujishima A, Iketani H, Honda K (1970) Relation between the size of the electrode and rotation velocity at rotating ring disk electrode. Bull Chem Soc Jpn 43 (12):3949-3951. doi:10.1246/bcsj.43.3949
9. Fujishima A, Honda K (1971) Electrochemical evidence for the mechanism of the primary stage of photosynthesis. Bull Chem Soc Jpn 44 (4):1148-1150. doi:10.1246/bcsj.44.1148
10. Fujishima A, Sugiyama E, Honda K (1971) Photosensitized electrolytic oxidation of iodide ions on cadmium sulfide single crystal electrode. Bull Chem Soc Jpn 44 (1):304. doi:10.1246/bcsj.44.304
11. Fujishima A, Honda K (1972) Electrochemical photolysis of water at a semiconductor electrode. Nature 238 (5358):37-38. doi:10.1038/238037a0
12. Maeda K, Domen K (2012) Photocatalytic water splitting: Recent progress and future challenges. J Phys Chem Lett 1 (2655-2661). doi:10.1021/jz1007966
13. Maeda K, Higashi M, Lu DL, Abe R, Domen K (2010) Efficient nonsacrificial water splitting through two-step photoexcitation by visible light using a modified oxynitride as a hydrogen evolution photocatalyst. J Am Chem Soc 132 (16):5858-5868. doi:10.1021/ja1009025
14. Tanaka K, Isobe H, Yamanaka S, Yamaguchi K (2012) Similarities of artificial photosystems by ruthenium oxo complexes and native water splitting systems. Proc Nat Acad Sci USA 109 (39):15600-15605. doi:10.1073/pnas.1120705109
15. Duan L, Bozoglian F, Mandal S, Stewart B, Privalov T, Llobet A, Sun L (2012) A molecular ruthenium catalyst with water-oxidation activity comparable to that of photosystem II. Nature Chemistry 4:418-423. doi:10.1038/nchem.1301
16. Morris AJ, Meyer GJ, Fujita E (2009) Molecular approaches to the photocatalytic reduction of carbon dioxide for solar fuels. Acc Chem Res 42 (12):1983-1994. doi:10.1021/ar9001679
17. Hawecker J, Lehn JM, Ziessel R (1986) Photochemical and electrochemical reduction of carbon-dioxide to carbon-monoxide mediated by (2,2′-bipyridine)tricarbonylchlororhenium(I) and related complexes as homogeneous catalysts. Helv Chim Acta 69 (8):1990-2012. doi:10.1002/hlca.19860690824
18. Takeda H, Koike K, Inoue H, Ishitani O (2008) Development of an efficient photocatalytic system for CO_2 reduction using rhenium(l) complexes based on mechanistic studies. J Am Chem Soc 130 (6):2023-2031. doi:10.1021/ja077752e
19. Takeda H, Ishitani O (2010) Development of efficient photocatalytic systems for CO_2 reduction using mononuclear and multinuclear metal complexes based on mechanistic studies. Coord Chem Rev 254 (3–4):346-354. doi:10.1016/j.ccr.2009.09.030

20. Takeda H, Koike K, Morimoto T, Inumaru H, Ishitani O (2011) Photochemistry and photocatalysis of rhenium(I) diimine complexes. In: VanEldik R, Stochel G (eds) Advances in Inorganic Chemistry, Vol 63: Inorganic Photochemistry, vol 63. Advances in Inorganic Chemistry. pp 137-186. doi:10.1016/b978-0-12-385904-4.00007-x
21. Tamaki Y, Watanabe K, Koike K, Inoue H, Morimoto T, Ishitani O (2012) Development of highly efficient supramolecular CO_2 reduction photocatalysts with high turnover frequency and durability. Faraday Discussions 155:115-127. doi:10.1039/c1fd00091h
22. Sato S, Morikawa T, Kajino T, Ishitani O (2013) A highly efficient mononuclear iridium complex photocatalyst for CO_2 reduction under visible light. Angew Chem Int Ed 52 (3):988-992. doi:10.1002/anie.201206137
23. Sato S, Arai T, Morikawa T, Uemura K, Suzuki M, Tanaka H, Kajino T (2011) Selective CO_2 conversion to formate conjugated with H_2O oxidation utilizing semiconductor/complex hybrid photocatalysts. J Am Chem Soc 133 (39):15240-15243. doi:10.1021/ja204881d
24. Arai T, Sato S, Kajino T, Morikawa T (2013) Solar CO_2 reduction using H_2O by a semiconductor/metal-complex hybrid photocatalyst: enhanced efficiency and demonstration of a wireless system using $SrTiO_3$ photoanodes. Energy & Environmental Science 6 (4):1274-1282. doi:10.1039/c3ee24317f

Artificial Photosynthesis: From Molecular to Hybrid Nanoconstructs

Manuel J. Llansola-Portoles, Rodrigo E. Palacios, Devens Gust, Thomas A. Moore, and Ana L. Moore

Abstract Mounting environmental concerns emphasize the urgent need to transition to sustainable energy production and use. Biology offers examples of the efficient conversion of chemical energy to other forms of energy. Photosynthesis is an example of solar energy conversion to chemical potential, and many conversion steps of photosynthesis are quite efficient. Taking inspiration from the efficient processes associated with photosynthesis and the subsequent conversion of chemical potential to myriad forms of biomass, artificial photosynthesis (AP) can serve to inspire the design of efficient energy-conversion systems for human use. Herein, we point out the imperative of *efficient* solar energy conversion and describe some of the steps in the evolution of AP strategies, from early designs of artificial reaction centers (ARCs) comprising only organic chromophores to some of the most recent examples in which chromophores are attached to semiconductor nanoparticles and undergo photoinduced electron transfer processes and proton-coupled electron transfer.

Keywords Hybrid nanoconstructs • Photoinduced electron transfer • Proton-Coupled Electron Transfer • Hybrid reaction centers • Solar energy

1 Introduction

The beginning of the Anthropocene is marked by a sharp increase in population and a concomitant increase in energy demand, which are projected to continue to grow for the foreseeable future [1]. Today, the demand for energy is primarily met by burning fossil fuels, and projections extending to 2040 are that ca. 80 % of energy demand will be met with fossil fuels [2], a process which transfers carbon

M.J. Llansola-Portoles (✉) • D. Gust • T.A. Moore • A.L. Moore
Department of Chemistry and Biochemistry, Center for Bioenergy and Photosynthesis, Arizona State University, Tempe, AZ 85287-1604, USA
e-mail: mjllansola@gmail.com

R.E. Palacios
Departamento de Química, Facultad de Ciencias Exactas Físico-Químicas y Naturales, Universidad Nacional de Río Cuarto, Río Cuarto, Córdoba 5800, Argentina

E.A. Rozhkova, K. Ariga (eds.), *From Molecules to Materials*,
DOI 10.1007/978-3-319-13800-8_3

from the "slow" carbon cycle to the "fast" cycle [3]. One consequence of this is that global-scale photosynthetic activity is not able to control atmospheric CO_2 levels to within boundaries established over the last few glacial/interglacial periods and for all of the Holocene [4]. In other words, photosynthesis no longer controls the fast carbon cycle. Indeed, atmospheric CO_2 levels are rising and are higher now than ever experienced by human societies [5, 6]. Even though near-term energy demand can be met by the use of fossil fuels, the associated issues of environmental damage and carbon pollution reinforce the need for sustainable energy sources [7]. The use of solar radiation to satisfy the world's present and future energy demand is deceptively appealing considering the extraordinary rate of solar energy impinging upon the surface of the Earth (~122,500 TW) [8] and its virtual inexhaustibility [9]. The appeal of employing solar energy is somewhat deceptive because, viewed from a global perspective, it could be argued that there is not a surfeit of solar energy available for human use today. As understood from the principles of Darwinian evolution, the interconnected web of life supported by Earth systems (e.g., biodiversity, carbon, nitrogen, and phosphorous cycles, etc.) and solar energy via photosynthesis and heat extended over land and sea to occupy even the most extreme ecological niches [10, 11]. In other words, Earth was teeming with life, interwoven with geology [12], which expanded on the geological time scale to utilize available solar energy. Thus, increasing human use of solar energy will necessarily impinge upon these ecosystems. Today, to meet food, fuel, and fiber needs, it is estimated that humans appropriate globally greater than 20 % of the products of terrestrial photosynthesis [13]. The consequence of this impacts the land, nutrients, biodiversity, and solar energy supporting the appropriated photosynthesis because it largely commits those resources to, effectively, human use. This change in land use, resources, and concomitant solar energy distribution is occurring over a few hundred years, but it comes out of the resource budget of ecosystems that make up the biosphere that evolved over geological time scales. The peril to Earth *as we know it* arises because there is virtually no chance that the full range of biodiversity upon which all life depends can evolve on the 100-year time scale to adjust to new conditions. Indeed, the extraordinary extinction rate and loss of biodiversity recently reported [14, 15] have been attributed to the impact of human activity, which can be traced in part to the commandeering of photosynthesis to meet human needs. We conclude that in order to avoid the consequences to the biosphere of further resource allocation by humans, it is imperative that solar energy conversion efficiency be optimized so that minimum land and other resource use will be required [16]. It is with this concern in mind that we envision high-efficiency artificial photosynthesis (AP) supplementing photosynthesis to meet human needs. In other words, resource-efficient AP can take some of the burden of supporting human activity off of photosynthesis and thereby reduce the negative impact of human appropriation, so that the plethora of natural photosynthesis can continue to support the biosphere and lithosphere as it has for billions of years [17].

1.1 Natural Photosynthesis as Inspiration for Artificial Photosynthesis

Photosynthesis provides the design principles and essential steps and functions as starting points for bio-inspired AP. These include absorption of sunlight by antenna arrays, energy transfer, the regulation of energy flow, protection from photodamage, the conversion of excited states into redox equivalents, catalysis, and ultimately self-repair and replication [16, 18–25]. Although AP includes the design and synthesis of molecular and/or hybrid systems inspired by natural photosynthesis, the goal is not to simply replicate the natural process. Instead, those working in AP seek to learn from nature's basic principles and use this knowledge to design and build improved devices, which could be molecular based or material based or hybrids that could include living components in which the benchmark is the highly efficient photodriven production of energy-rich fuels and the net drawdown of atmospheric CO_2.

Due to the complexity of natural photosynthesis, the development of AP systems is a multidisciplinary enterprise with contributions ranging from physics through chemistry and materials science to biology and synthetic biology. In one view, the overall structures of AP constructs would resemble to some extent the photosynthetic membranes where antenna pigments convert solar photons to excited states, regulate the flow of excitation energy to reaction centers, and suppress the production of reactive oxygen species. Upon excitation, reaction centers located in the membranes generate a transmembrane redox potential, which is used for both net redox chemistry and conversion to proton motive force. However, AP is not constrained to this paradigm; it is anticipated that devices using features of our technology such as electricity, semiconductors, and materials unknown to biology will likely be an important part of a wide variety of AP systems. In natural photosynthesis, the net source of electrons is water oxidation, and the carbon for the synthesis of biomass comes from CO_2. In view of potential availability, the necessity of regaining control of the global carbon cycle, and the importance of the role of CO_2 in controlling the Earth's climate [26], it is likely that water and CO_2 will be targeted as reactants for AP systems. The reduction of protons to hydrogen, which is a vital but not an extensive process in nature, is anticipated to be an important aspect of AP systems. In addition to being a carbon-free fuel, hydrogen finds myriad uses in the global chemical industry where its production by steam reforming methane is a significant source of CO_2. This view of water oxidation, CO_2, and proton reduction as the targets of AP is predicated on the necessity of lowering the atmospheric levels of CO_2 and on the energy drivers of the global economies in place today. Hopefully, AP and other as yet unimagined manifestations of AP will become available and will change the trajectory of the Anthropocene.

As important as it is, aspects of natural photosynthesis are poorly understood and many remain controversial. Hence, further study of natural systems is imperative because the findings could play a role in the design of more efficient AP systems. For example, it is important to determine the role of quantum coherence in energy harvesting [27] where it could provide a mechanism for controlling energy flow in multi-chromophoric assemblies [28, 29] by routing energy flow to appropriately energetic reaction centers to optimize the match of the solar spectrum to the redox work of fuel synthesis and proton pumping [23]. There is also a need for a deeper understanding of the specific arrangement of the energy levels and the coupling of the cofactors in the multistep charge separation and accumulation by photosystem I (PSI) and photosystem II (PSII) [30]. Charge accumulation in the water-oxidizing catalytic center is another critical issue in photosynthesis and in the design of AP. Because each absorbed photon leads to a single electron/hole charge separation event, and catalytic water splitting and either proton or CO_2 reduction involves a multi-electron process, controlling the accumulation of electrons and holes on the appropriate catalysts is a crucial design requirement. This accumulation and the molecular steps along the reaction coordinates are poorly understood. For example, it is certain that the accumulation of oxidizing and reducing equivalents by the catalysts must involve redox leveling because the oxidant, $P680^{\bullet+}$, provides a constant ~1.3 V (vs. SHE) potential for each of the four steps. In addition to managing the thermodynamics, nature also manages the dynamics of electron/hole transfer between $P680^{\bullet+}$ and the water-oxidizing catalyst. Experiments demonstrate that the oxidizing equivalents in the water-oxidizing catalyst are stable for seconds to minutes. How does nature accomplish this?. One hint to the mechanism of redox leveling and the long lifetime of highly oxidizing species comes from studies of proton-coupled electron transfer (PCET) [31–33]. It has been shown that controlling the protonation state of a redox-active phenol hydrogen bonded to an imidazole derivative (an artificial construct that mimics a redox-active tyrosine H-bonded to a histidine in photosynthesis) controls the midpoint potential and lifetime of the phenolic radical [34–37].

2 Evolution of Molecular Reaction Centers

2.1 Early Molecular Reaction Centers

One of many approaches to the assembly of AP model compounds is to use synthetic chromophores that are related to the natural pigments but bring into play covalent bonds in place of the proteins that are used in natural systems to hold and organize the pigments. In the pigment–protein complexes of natural photosynthetic membranes, the electronic interactions, for example, between carotenoids and chlorophylls that give rise to energy and electron transfer, are a consequence of the proximity of their π-electron systems, which is a function of the distance and geometry imposed on the pigments by the secondary, tertiary, and quaternary protein structure. In general, this electronic communication among chromophores is considered to be a consequence of through-space orbital interactions. Nonetheless, in most of the artificial reaction centers constructed to date, the electronic coupling is a function of covalent bonds that hold the chromophores together. Regardless of

D1

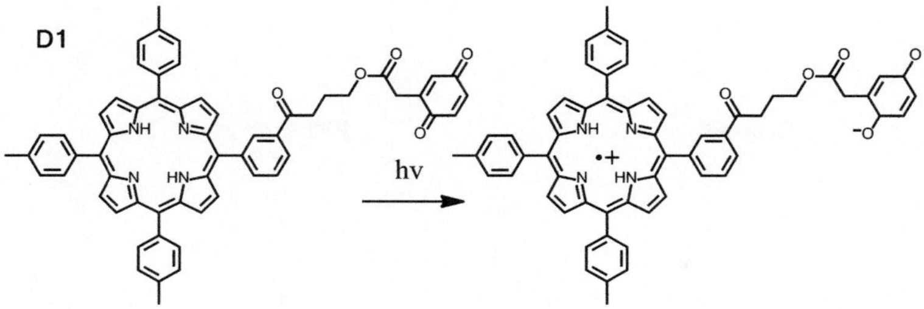

Fig. 1 Charge separation in a P–Q type dyad (**D1**) upon porphyrin excitation [42]

whether the coupling is a consequence of through-space or through-bond interactions, in a wide variety of photophysical and photochemical measurements, many photosynthetic models have demonstrated the essential features of electronic coupling required for the energy and electron transfer functions observed in the natural systems.

Several approaches to the design of artificial reaction centers have been reported since the late 1970s [38–40]. The initial approaches consisted of linking two molecular analogs of the cofactors involved in the natural photosynthetic reaction center (e.g., porphyrins, carotenoids, chlorophylls, pheophytins, phthalocyanines, quinones) with covalent bonds. These dyad compounds are the simplest artificial reaction centers and became the starting point for more sophisticated designs. Quinones are electron acceptors ubiquitous in nature because they display a series of useful properties; they have suitable redox potentials, undergo stepwise conversion into stable reduction products (hydroquinones) via semiquinones, form hydrogen bonds, undergo protonation/deprotonation reactions with ease, and are relatively small mobile molecules that can shuttle redox equivalents to the quinone pool [41]. Because of these characteristics, early work in the AP field was focused on creating porphyrin–quinone (P–Q) reaction center dyads of the type shown in Fig. 1. Upon porphyrin excitation, these compounds undergo electron transfer to form a $P^{•+}$–$Q^{•-}$ state [43, 44]. Although this charge-separated state can be produced efficiently, under optimal conditions, fast (sub-nanosecond) charge recombination precludes the direct coupling of its photogenerated redox potential to slower catalytic steps necessary for fuel synthesis in AP systems.

In the 1980s, the Gust, Moore, and Moore group (GMM) designed and synthesized a series of carotenoid–porphyrin (C–P) dyads, [45, 46] some of which were shown to undergo hole transfer from the oxidized porphyrin ($P^{•+}$) to the carotenoid to yield the $C^{•+}$–P species [47]. The idea of combining this results with the photoinduced formation of $P^{•+}$–$Q^{•-}$ led to the synthesis of a carotenoid–porphyrin–quinone (C–P–Q) triad system (Fig. 2; **T1**, **T2,** and **T3**) [48–50]. Transient spectroscopic studies of C–P–Q proved that the photogenerated excited singlet state of the porphyrin in the triad (C–^{1}P–Q) can generate the C–$P^{•+}$–$Q^{•-}$ charge-separated state by electron transfer. Subsequently, a second electron transfer step from the carotenoid to the porphyrin radical cation produces the final charge-separated state ($C^{•+}$–P–$Q^{•-}$), which has a relatively long lifetime, 170 ns in dichloromethane and 2.5 μs in

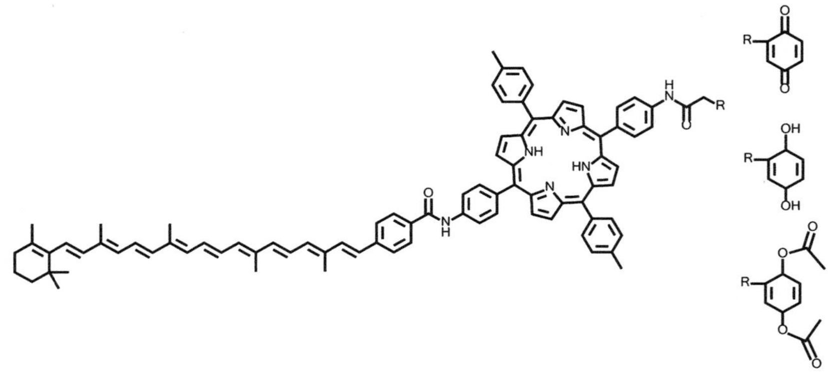

Fig. 2 Molecular structure of a carotenoid–porphyrin–benzoquinone triad and related compounds (**T1**, **T2,** and **T3**) [48]

electrolyte-saturated solvent for the best in the series [48]. Thus, these AP triad systems experimentally demonstrate one of the fundamental concepts of natural photosynthesis; i.e., a series of short-range, fast, and efficient electron transfer reactions can result in the formation of long-lived charge-separated states. The theoretical foundation for this strategy was provided by Marcus and others. The short-range, fast electron transfer steps are near the maximum ($-\Delta G^0=\lambda$) of the Marcus curve, whereas charge recombination is located deeply in the inverted region ($-\Delta G^0>\lambda$) and is therefore comparatively slow, resulting in high quantum yields of long-lived charge-separated states [51]. The concepts demonstrated by the original triad proved to be universal and have been extended to more complex multiple-component systems since then [19, 21, 52–55]. Consequently, the triad concept established a platform for engineering efficient artificial reaction centers. These constructs were later used in more complex systems to study other photosynthesis-related processes, such as photoinduced proton (and calcium ion) gradient formation across lipid membranes and its use for driving enzymatic ATP formation [56, 57].

2.2 PCET in Organic Systems

Photosystem II (PSII), the enzyme which uses solar energy to split water into molecular oxygen and reducing equivalents, contains two redox-active tyrosine residues: tyrosine Z (Tyr_Z) and tyrosine D (Tyr_D). Oxidation of the tyrosines during PSII activity is coupled to the transfer of their phenolic protons to their respective hydrogen-bonded histidine partners by a process called proton-coupled electron transfer (PCET). It has been postulated that the PCET within Tyr_z and histidine 190 plays a role in the high quantum efficiency of water oxidation. Designing synthetic models able to mimic such process is a challenging and worthwhile step in the development of AP.

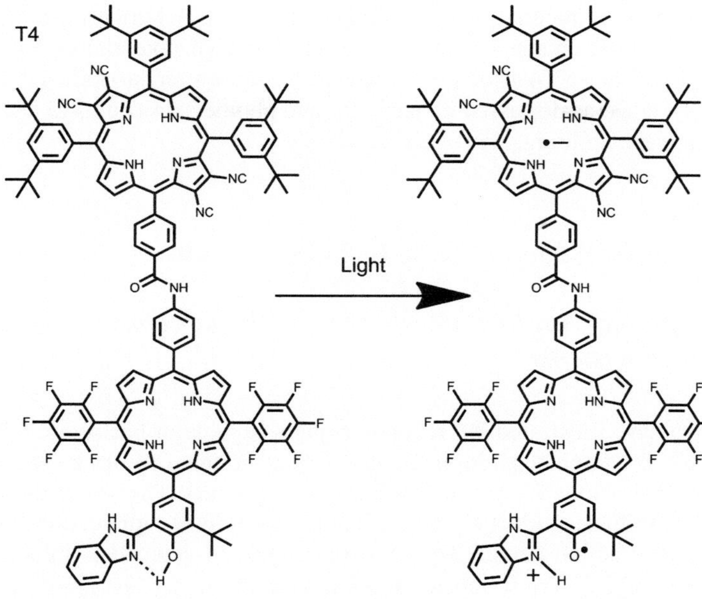

Fig. 3 Molecular structure of triad **T4**, which is composed of three covalently linked subunits that mimic the redox processes in PSII. Upon irradiation, sequential electron transfer (ET) and PCET reactions yield the final charge-separated state $BiH^+–PhO^•–PF_{10}–TCNP^{•-}$

A triad capable of PCET that bears considerable similarity to the Tyr_Z–His190–P680 portion of the electron transport chain in PSII was synthesized and studied by the GMM group [34, 36]. The fully organic triad (**T4**, Fig. 3) consisted of a high-potential porphyrin with two pentafluorophenyl groups (PF_{10}), a tetracyanoporphyrin primary electron acceptor (TCNP) and a benzimidazole–phenol secondary electron donor (Bi–PhOH) [36]. Time-resolved spectroscopic investigation of **T4** in benzonitrile reveals that with excitation of the TCNP moiety, the system undergoes hole transfer from ^1TCNP to PF_{10} ($k_{HT}=1.2\times10^9$ s^{-1}, $\tau=0.8$ ns), resulting in Bi–PhOH–$PF_{10}^{•+}$–$TCNP^{•-}$. Competing with charge recombination, a second, rapid electron transfer ($k_{PCET}>8\times10^{10}$ s^{-1}, $\tau=12.5$ ps) then occurs from Bi–PhOH to $PF_{10}^{•+}$, presumably resulting in $BiH^+–PhO^•–PF_{10}–TCNP^{•-}$. The proposed PCET reaction involves both electron transfer from the phenol to $PF_{10}^{•+}$ and transfer of the phenolic proton to the appended benzimidazole group. Indirect evidence for PCET comes from reduction potentials of model compounds [35] which indicate that $PF_{10}^{•+}$ would not generate sufficient driving force for the formation of the Bi–$PhOH^{•+}$–PF_{10}–$TCNP^{•-}$ and thus implies the formation of the BiH^+– $PhO^•–PF_{10}–TCNP^{•-}$ state. This final charge-separated state forms with a quantum yield of 0.52 and displays a relatively long lifetime ($k_{CR}=2.6\times10^5$ s^{-1}, $\tau=3.85$ μs), demonstrating that **T4** is an excellent example of the successful design of AP systems based on the combined application of two fundamental strategies used in natural photosynthetic systems, namely, vectorial–sequential electron transfer and PCET. Furthermore, the long lifetime of the final charge-separated state and the high potential of BiH^+–$PhO^•$ (1.04 V vs. SCE) make **T4** a good candidate for incorporation into model dye–semiconductor systems (as discussed in Sect. 3.5) and photoelectrochemical devices for light-driven water splitting.

3 Organic Dye: Metal Oxide Hybrid Systems

3.1 Design and Application of Dye–Semiconductor Hybrid Systems

Thus far, we have illustrated AP reaction centers capable of performing photoinduced vectorial charge separation (in some cases coupled to proton transfer) to generate relatively long-lived energetic redox equivalents. The use of these redox equivalents for fuel production and other photocatalytic applications requires "wiring" such reaction centers to semiconductors and catalysts. Unlike natural systems, most of these constructs use modified nanostructured inorganic semiconductors to carry out the photoinduced charge separation. The simplest example of this approach is a system composed of a dye attached to a semiconductor nanoparticle (SNP) [58], which is schematically shown in Fig. 4 with SNP of titanium dioxide (TiO_2). In these types of constructs, the dye excited state (D*) generated upon photoexcitation injects an electron into the conduction band (CB) of the semiconductor (Fig. 4a, path **1**) to yield an initial charge-separated state of interfacial character ($D^{+•}$-$TiO_2(e-)$). This process is usually called photosensitization. Ensuring a good dye–semiconductor electronic coupling through appropriate dye binding is a critical step for achieving efficient photosensitization. After sensitization, the oxidized dye ($D^{+•}$) can be reduced by an electron donor in solution ($S/S^{•-}$) (the I_3^-/I^- couple in this case) (Fig. 4a, path **2**). This secondary electron transfer process regenerates the dye's ground state (D) and produces a freely diffusing oxidized donor, thus effectively increasing the distance between the initially photogenerated charges in an analogous way to the previously described molecular constructs. The initially photoinjected electron in the semiconductor can follow different paths (described in detail below) in which its reducing power can advantageously be exploited. Aside from these productive processes, the injected electron can also follow undesirable energy-wasting pathways such as back electron transfer from the semiconductor to $D^{•+}$ (Fig. 4a, path **5**) or to the redox couple in solution ($S/S^{•-}$) (path **6**) and electron trapping at surface defects (path **4**). Proper design of dye-sensitized systems aims at minimizing the rates at which these unproductive processes take place. Also, dye-sensitized colloidal SNP can function as photocatalytic structures where the reducing power of the photoinjected electron is used to perform chemical transformations on dissolved substrates. Systems of this type have been designed for environmental remediation

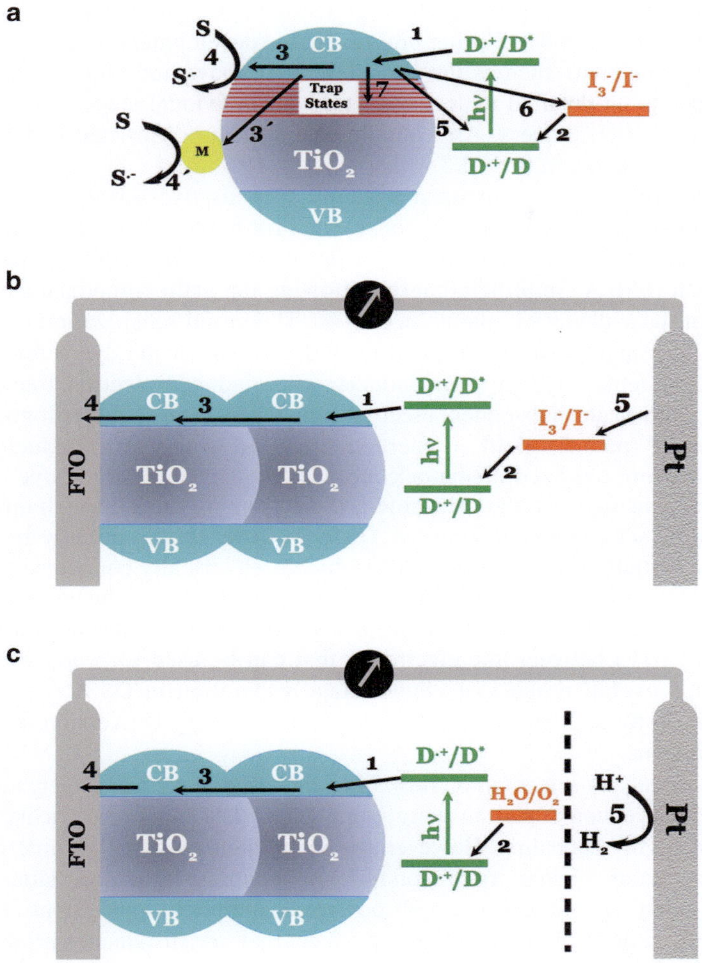

Fig. 4 Most relevant electron transfer pathways in (**a**) nanoparticulate dye–semiconductor systems, (**b**) DSSCs, and (**c**) bioinspired PECs. Dye–photon absorption is indicated by **hν**. The numbers indicate selected electron transfer pathways. See text for details

applications such as pollutant degradation [59, 60]. The photogenerated electrons are used to degrade organic pollutants by directly reducing them (Fig. 4a, paths **3** and **4**) or by reductively generating reactive oxygen species such as superoxide and hydroxyl radical which eventually degrade the contaminants. Another application that has been demonstrated for these dye-sensitized SNP colloidal systems is the direct photocatalytic production of reduced fuels such as H_2, methane, methanol, formic acid, etc [61]. Usually for these applications, the SNP bears noble metal aggregates or particles in good electrical contact, which act both as secondary electron acceptors (path **3′**) and as catalytic sites for specific chemical transformations (path **4′**). Further progress in the development of nanoparticulate-sensitized constructs such as the ones described above and more advanced systems depends on detailed understanding of the individual steps involved in the energy-conversion process. An example of a system constructed and studied to this end is discussed in Sect. 3.5.

Closely related dye-sensitized systems such as dye-sensitized solar cells (DSSCs) have been designed and constructed for photovoltaic applications [62–64], and currently they have achieved up to 13 % efficiency of solar energy conversion [65]. A simplified diagram showing the main components and functioning of these devices is shown in Fig. 4b. The initial sensitization mechanism in DSSCs is analogous to that previously described for the dye semiconductor colloidal systems. Dye to semiconductor-photoinduced electron transfer (P_{eT}) (Fig. 4b, path **1**) and subsequent electron transfer from an electron donor in solution to dye$^{•+}$ (path **2**) yield an energetic electron in the semiconductor CB, a freely diffusing oxidized mediator couple (I_3^-) and a regenerated dye molecule (D). After sensitization, the photoinjected electron migrates through the CB of a mesoporous semiconductor network (path **3**) to reach an optically transparent electrode (usually ITO or FTO, path **4**) which is electrically connected to a metal cathode electrode (usually platinum) where dissolved I_3^- is reduced to regenerate I^- (path **5**). The flow of electrons in this electrical circuit produces a potential difference (ΔV) between the electrodes that can be used to produce electrical work. This overall process of photovoltage generation in DSSCs is similar in many ways to the photon to redox-potential energy-conversion process in photosynthesis.

Since electricity is a common form of energy used by human society nowadays, the direct conversion of photon energy supplied by the sun into electrical energy by DSSCs is highly appealing. However, the daily intensity variation, intermittency, and diffuse nature of solar irradiation make photovoltaic technology alone impractical for many on-demand and high power consumption applications. For these reasons, the use of high energy–density fuels that are sustainably produced and which have low environmental impact is highly desirable. Bio-inspired photoelectrochemical cells (PECs) should be able to produce such fuels. A PEC is a device with an architecture analogous to that of the DSSC (see Fig. 4c) but designed to use the photon energy for the production of chemical fuels though electro-reductive synthesis [20]. Figure 4c shows the main processes which occur in PECs. Paths **1**, **3**, and **4** are identical to those previously described for DSSCs, path **5** involves the reduction of two protons to produce H_2 (a fuel), and path **2** indicates the regeneration of the sensitizing dye by oxidation of water. The development of these types of devices represents a huge leap forward toward the production of clean and renewable energy since they produce O_2 and H_2 from water and light. The redox-driven chemical transformations in PECs can be catalyzed by organic or inorganic catalysts, macromolecular natural catalysts (i.e., enzymes), or a combination of both approaches. As a result PECs emulate photosynthesis by using the photogenerated redox potential to perform catalytically assisted chemical transformations, ultimately storing part of the solar energy as chemical energy. Two examples of PECs developed in the GMM group based on a hybrid organic–inorganic system are described in Sect. 3.6.

3.2 Anchoring Groups and Bridges

As mentioned before, the attachment of dyes to the metal oxide surface is a key step in the construction of hybrid dye–semiconductor systems [66]. Creating suitable bonds between the sensitizer dye and the metal oxide surface offers several advantages: increasing the surface coverage, improving electronic coupling between dye-occupied orbitals and the semiconductor CB, and enhancing stability, all of which could result in increased performance [67]. Suitable anchor groups for dyes used in water-splitting photoanodes must provide strong chemical bonding between the molecular species and the semiconductor surface functionalities, should be stable under both aqueous and oxidative conditions, and should provide electronic coupling between the dye and semiconductor that enhances charge injection and retards charge recombination.

One of the moieties most frequently used to attach organic molecules to TiO_2 is the carboxylic acid group [34, 37, 68–71]. This group can bind to metal oxide surfaces by forming a single bond through one oxygen atom (Fig. 5a) or via a resonant bidentate or bridging structure as shown in Fig. 5b, c. The carboxylic acid functionality can be incorporated as such in the dye structure or as precursors capable of forming carboxylic moieties under certain conditions. An example of this latter case is the incorporation of cyclic anhydrides in perylenes. The hydrolysis of these cyclic structures yields two carboxylic groups that act as efficient anchors to TiO_2 nanoparticles [72]. Although carboxylic groups attach efficiently to TiO_2 in dry organic solvents, the binding is labile in presence of water, especially at acidic pH [70]. Another extensively used anchor group is the phosphonic acid group which forms a more stable attachment in aqueous and acidic conditions as compared to the carboxylic acid group (Fig. 5f–h). However, the weaker dye–TiO_2 electronic coupling and the low surface coverage obtained with phosphonic acid-functionalized dyes as compared to carboxylic acid-functionalized dyes result in DSSC devices that perform less well. In addition to carboxylic acid [34, 37, 68–71] and phosphonic acid groups [68, 71, 73, 74], other moieties have proven capable of dye–semiconductor binding. Among them, we find nitro groups [75], pyridine rings [76], salicylic acid [77], catechol [78–81], siloxyl [71], and hydroxamic acid groups [82, 83]. However, in the literature there is no general systematic study of all these anchors to directly compare their stability and performance. Thus, it remains controversial as to what group is the most appropriate for general use.

On the other hand, the process of the electron injection into the CB of the metal oxide is not only influenced by the chromophore and the anchoring group. The

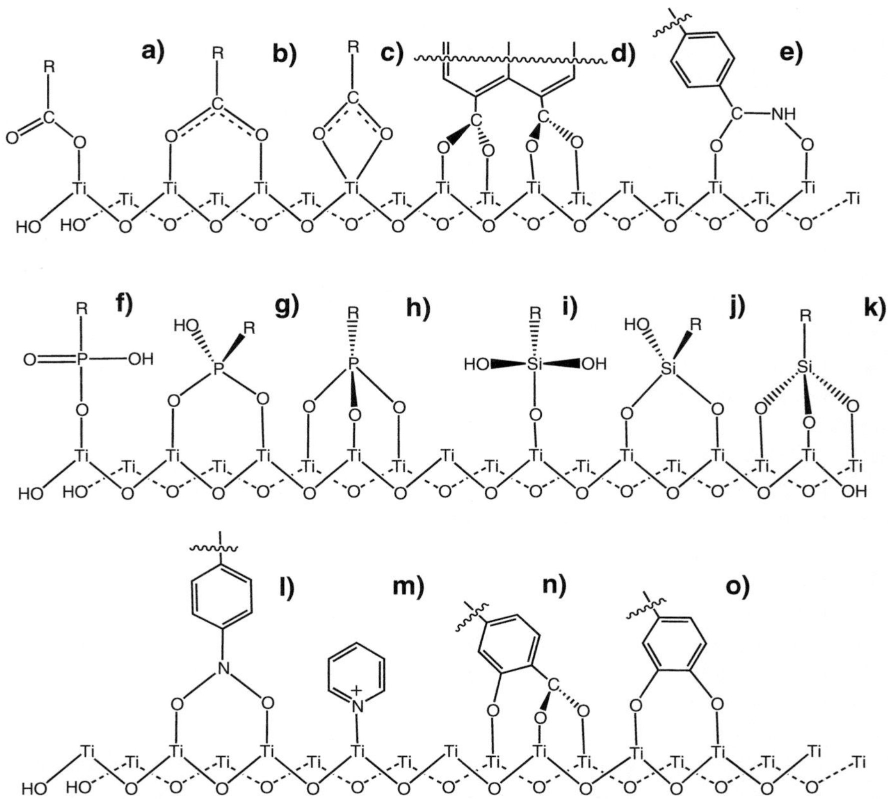

Fig. 5 Anchoring groups with their more widely accepted anchoring modes carboxylate monodentate (**a**), carboxylate bidentate in a bridge (**b**), carboxylate bidentate (**c**), dicarboxylate (**d**), hydroxamate (**e**), phosphonate monodentate (**f**), phosphonate bidentate (**g**), phosphonate tridentate (**h**), siloxyl monodentate (**i**), siloxyl bidentate (**j**), siloxyl tridentate (**k**), nitro (**l**), pyridine ring (**m**), salicylate (**n**), and catechol (**o**)

bridge between the chromophore and the binding group also plays an important role in determining the electronic coupling between molecular excited states and the semiconductor. This interaction can be modified through the use of different bridges. Carefully designed linkers can be useful for fixing the distance of the dye from the semiconductor surface, tuning the properties of the dye, preventing aggregation of the chromophores, and preparing models for interfacial electron transfer studies. These bridges can be of a variety of types and based on alkyl chains (flexible or semirigid bridges) [84], rigid rods (rigid linkers), tripod-shaped molecules (tripodal linkers) [85], or molecular containers (cyclodextrins, cucurbiturils, Cram-type hosts, and calixarenes) [86]. Due to the increased control over the distance and orientation of the chromophore relative to the TiO_2, rigid linkers of varying complexities, with large footprints and multiple anchoring groups, are rapidly replacing simple alkyl chains. Recent studies have demonstrated that rigid rod and tripod linkers are able to fix the position of the sensitizers on the surface, are useful models to investigate interfacial electronic processes, and can tune the properties of the sensitizer [85].

3.3 Dye–Semiconductor Nanoparticle (Dye–SNP) Systems

The study of electron and energy transfer processes in simple and well-defined dye–SNP colloidal systems is important not only for direct photocatalytic applications but also for understanding the key processes that control the efficiency of devices such as DSSC and PEC. Due to their simplicity, dye–SNP systems allow detailed studies of the parameters that control photosensitization such as dye binding (and its correlation with dye structure) and the relative energies of the states involved in the photoinduced electron transfer (P_{eT}) process. In contrast, the study of these parameters in actual devices is usually complicated by several factors including dye aggregation at the semiconductor surface and the poorly defined nature of the dye–semiconductor attachment, which are partially a consequence of the high degree of chemical and structural heterogeneity in electrode–semiconductor surface-binding sites [87]. Particularly, the study of nanoparticulate dye–TiO_2 systems suspended in organic solvents is of special interest due to its relevance for the design of DSSC and PEC. However these experiments are usually also complicated because colloidal TiO_2 is highly unstable and tends to aggregate and precipitate in organic solvents.

In order to prevent aggregation, particle encapsulation methods involving the use of inverse micelles have been developed and applied to a number of dye–TiO_2 nanoparticles systems [88, 89]. Using this strategy, simple dye–TiO_2 systems consisting of perylene derivatives and TiO_2 nanoparticles are encapsulated inside AOT reverse micelles (perylene–TiO_2@micelles) [72]. These systems were originally designed to study the effect of semiconductor Fermi level change on the distribution of single dye–TiO_2 sensitization efficiencies by applying single-molecule fluorescence microscopy techniques. Information with this level of detail is especially important for characterizing intrinsically heterogeneous systems, and it cannot be obtained directly from ensemble measurements using conventional (bulk) techniques. Although these experiments are currently in progress, a thorough study of the perylene–TiO_2@micelles systems has been carried out using conventional (bulk) spectroscopy techniques [72]. The results of these experiments are useful to illustrate some of the key factors that control the photosensitization process and that need to be considered when designing optimized dye–semiconductor-based devices. Two of these factors are the shape and size of heterogeneous SNPs. In this regard encapsulation by micelles not only helps to solve solubility problems but also provides a convenient nano-environment for controlling SNP synthesis. Thus, in the

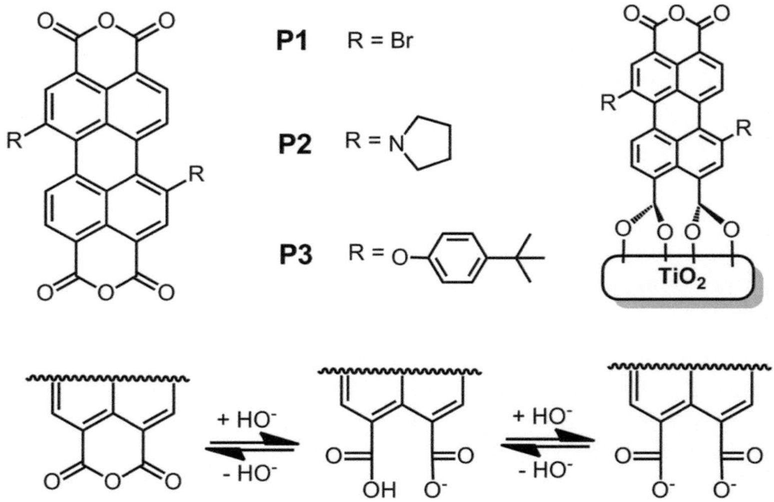

Fig. 6 Structures of perylene derivatives (**P1**, **P2**, and **P3**) used to attach to TiO_2 through carboxylate groups

preparation of perylene–TiO_2@micelle systems, the water pool inside the inverse micelles is used as a nanoreactor where the hydrolysis of semiconductor precursors and SNP nucleation and growth take place, producing well-defined nearly spherical SNPs with controlled size (average diameter ~ 3.5 nm) and extremely narrow size distribution (sd ≅ 0.7 nm). As mentioned before, dye binding to TiO_2 is crucial for photosensitization. The binding strategy chosen for the perylene–TiO_2@micelle systems involves the opening of a cyclic anhydride group by hydrolysis to generate two carboxylic acid groups that can attach to the semiconductor surface (see Fig. 6).

Because the anhydride ring is part of the perylene π-conjugated chromophoric structure, binding of the dye is conveniently followed by monitoring absorption and emission spectral changes generated upon hydrolysis. For the series of analogous perylene derivatives **P1**, **P2**, and **P3** shown in Fig. 6, it was shown that dye to TiO_2 binding efficiency depended among other factors on the dye molecular structure. In particular **P1** and **P2** showed efficient binding, whereas **P3** did not bind, presumably due to steric effects induced by the bulky *t*-butyl groups. It was also observed that the micelle structure plays a role in the binding process; fluorescence anisotropy measurements indicate that **P1** and **P2** are preferentially solubilized in the micellar structure creating a relatively large local concentration that presumably favors the attachment of the dye to the TiO_2 surface. Photoinduced charge injection from **P1** and **P2** to the TiO_2 CB was indicated by emission quenching of the TiO_2-bound form of the dyes and confirmed by transient absorption measurements of the radical cation of the dyes (**P1•+** and **P1•+**) and free carriers (injected electrons) in the TiO_2 semiconductor ($TiO_2(e^-)$) [72]. These results are consistent with the energy diagram shown in Fig. 7 which indicates that both **P1** and **P2** are energetically capable of

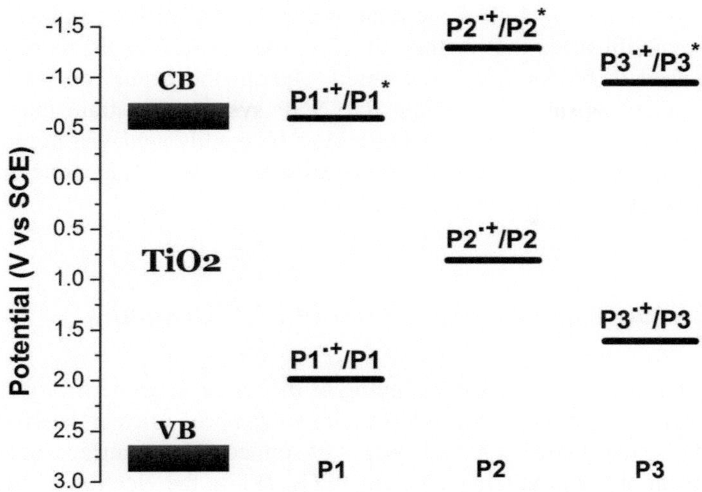

Fig. 7 Diagram depicting the potentials versus SCE for the first oxidation and for the oxidation of the photoexcited states of **P1**, **P2**, and **P3** and the TiO_2 CB and VB

photoinjecting an electron into the conduction band of TiO_2. The **P1**–TiO_2@micelle and **P2**–TiO_2@micelle systems undergo fast photoinduced electron transfer (k_{PeT} < 210 fs) to form the dye$^{•+}$–$TiO_2(e^-)$@micelle charge-separated state, whereas charge recombination occurs on a comparatively very long time scale (k_{rec}(**P1**) > 14 μs and k_{rec}(**P2**) = 30 ms). Thus, these systems are good examples of hybrid dye–SNP systems capable of producing long-lived energetic charge-separated states.

Simpler model systems constructed with the same perylene dyes but attached to "naked" TiO_2 NP (dye–TiO_2) suspended in an organic solvent were prepared and studied [90]. In these systems the SNPs were synthesized by a method that produces surfactant-free SNPs with a narrow size distribution (radius: 4 ± 1 nm) [74] which are readily suspended in an organic solvent such as tetrahydrofuran (THF). As was mentioned before, the use of well-defined (in size, shape, and composition) SNPs is important for the construction of model systems that are amenable to detailed photoinduced electron transfer studies. In particular, by eliminating the need of micelle encapsulation, this approach produces simpler colloidal systems more closely related to the structures encountered in functioning dye–semiconductor devices. The "naked" systems showed similar results to the encapsulated systems with respect to dye binding and sensitization. However, the distinct solvent environment sensed in each case (THF and micelle–water pool for the "naked" and encapsulated systems, respectively) creates a considerably different energetic situation for the two systems. In particular the CB edge in the "naked" system (E_{CB}(naked)) is more negative than that of the micelle encapsulated system (E_{CB}(micelle)) [90]. Consequently, the driving force for P_{eT} in "naked" systems is smaller than that in the micelle system, and the rates for photoinduced charge separation (k_{PeT}) follow the relationship k_{PeT}(naked) < k_{PeT}(micelle). Thus, these systems illustrate that factors affecting the position of the semiconductor E_{CB} also need to be carefully considered. Overall the results obtained for the "naked" and encapsulated colloidal dye–SNP systems illustrate that efficient formation of long-lived redox equivalents can be readily achieved in model constructs by careful consideration of parameters such as dye–semiconductor binding and the energetics of the system.

3.4 Relevance of TiO_2 Trap States in PEC Dynamics

In the design of PECs, the understanding of the intrinsic properties of the semiconductor is as important as the understanding of the properties of the dye attached to it. In particular, knowledge and control of semiconductor surface and energetics are essential for producing efficient PECs. While the electronic structure of bulk TiO_2 is now well understood [59, 91], the structure and properties of the surface remain controversial and not completely clear [91, 92]. The TiO_2 surface consists of high-energy configurations of bonds in which the terminal Ti and O atoms rearrange so that the symmetry and electronic structure of the resulting surface is different from that of the bulk [93]. This creates new energetic states that would be considered defects in the bulk material such as under-coordinated Ti^{4+}sites due to surface curvature and oxygen vacancies resulting in Ti^{3+}sites [94]. These surface defects influence the interfacial charge transfer kinetics [95, 96] and modify the reactivity in catalytic processes [97]. The network defects in the TiO_2 surface act as both anchoring sites and charge injection/recombination sites and thereby play a crucial role in the electron injection and recombination dynamics of the material [94, 98, 99]. It may be possible to harness synergistic effects between surface sites, chromophores, and bulk TiO_2 by precisely engineering these sites [100].

In order to determine the role and behavior of defect states in TiO_2 for P_{eT} processes, two different TiO_2 nanoparticles with low and high density of surface defects (1-TiO_2 and 2-TiO_2, respectively) have been designed [74]. Then a perylene dye bearing a phosphonic acid group as a binding moiety (**P4** in Fig. 8) was attached to the surface of both kinds of the nanoparticles. Figure 8 depicts the relative energetics for the CB, VB, and defect state of TiO_2, the potential for the first oxidation ($D^{\bullet+}/D$), and the potential for the oxidation of the excited state ($D^{\bullet+}/D^*$) of the **P4** dye. As shown in Fig. 8, the excited singlet state of the dye is thermodynamically capable of photoinjecting an electron into the conduction band and/or defect states of TiO_2, and theoretical work affirms that defect states may act as electron acceptors and thereby increase the driving force for the P_{eT} process [101, 102]. It is expected that for TiO_2 nanoparticles with low defect surface density (1-TiO_2), the most probable electron acceptor is the CB and electron transfer would follow the pathway depicted by the blue arrows in Fig. 8: injection from the excited **P4** into the CB and recombination from the CB to the oxidized **P4**. On the other hand, for

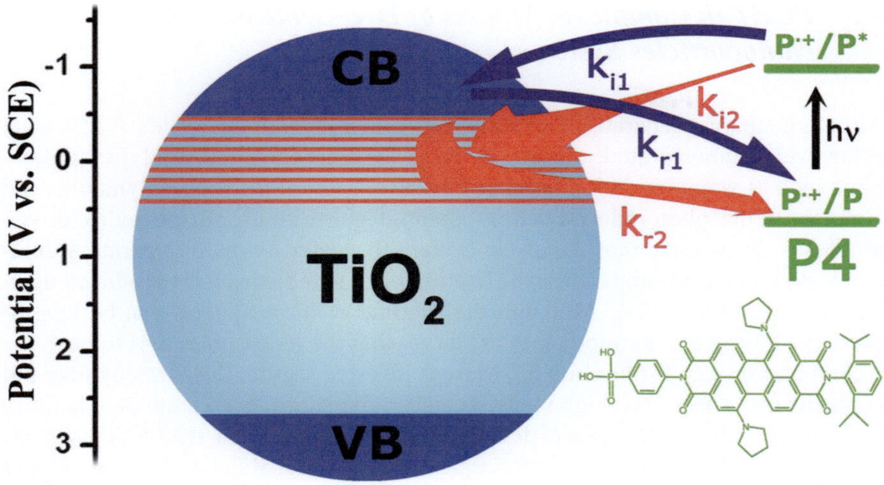

Fig. 8 Diagram depicting the potentials for the ground and the lowest excited singlet states of the perylene dye (**P4**), TiO_2 conduction band energy (CB), TiO_2 valence band energy (VB), and TiO_2 energy levels due to defect states (*red lines*)

TiO_2 nanoparticles with a high surface density of defects (2-TiO_2), injections into both CB states (indicated by the blue arrow) and defect states (indicated by the red arrow) are possible.

The role of defect states in the kinetics of the P_{eT} reactions in **P4**–1-TiO_2 and **P4**–2-TiO_2 systems in acetonitrile solution was studied using femtosecond transient absorption spectroscopy. The results showed clear evidence that both systems undergo photoinduced electron transfer to yield the **P4**$^{\bullet+}$–$TiO_2(e^-)$ charge-separated state. The apparent forward (k_{i1}) and reverse (k_{r1}) rate constants for the formation and decay of the **P4**$^{\bullet+}$–1-$TiO_2(e^-)$ state are $k_{i1} \sim 5 \times 10^{12}$ s^{-1} (1/200 fs) and $k_{r1} \sim 2 \times 10^{11}$ s^{-1} (1/5 ps), respectively. The apparent forward rate constant for the formation of the **P4**$^{\bullet+}$–2-$TiO_2(e^-)$ state is $k_{i2} \sim 1 \times 10^{13}$ s^{-1} (1/100 fs). The decay (by charge recombination) of this state is composed of two fast main components $k_{r2} \sim 9.1 \times 10^{10}$ s^{-1} (1/11 ps (50 %)) and $k_{r2} \sim 4.4 \times 10^{9}$ s^{-1} (1/228 ps (46 %)) and a slow minor (4 %) component associated with deeply trapped electrons. The faster injection is likely due to the stronger electronic coupling at defect sites and the increased driving force from P4* to TiO_2 defect states. Once **P4**$^{\bullet+}$–2-$TiO_2(e^-)$ is formed, charge recombination is relatively slow (as compared to that in **P4**$^{\bullet+}$–1-$TiO_2(e^-)$) and multi-exponential, indicating the presence of multiple relaxation pathways. These pathways likely include electron migration to CB states and/or to other surface defect states with lower energy where the electron might be "trapped" for long time periods before charge recombination. The electrons in the CB for **P4**–1-TiO_2 have higher mobility (since the CB edge electronic states are nearly degenerate and mostly unoccupied) and are more readily available for recombination.

3.5 PCET in Complexes Models of Dye–Semiconductor Nanoparticles Systems

Although charge separation in simple dye–semiconductor systems has been extensively demonstrated, the use of these systems for photocatalytic applications such as water oxidation requires careful control of the thermodynamics and kinetics of the photoinduced redox-potential generation and the redox-driven processes. In practice most catalytic centers for water oxidation operate at relatively slow rates as compared to the lifetimes of the redox species produced upon initial charge separation. Although the lifetime of these species can be lengthened by altering the thermodynamics (increasing the redox potential) to push the recombination rate further into the Marcus inverted region, this strategy may not be sufficient to result in a high yield of catalyst oxidation. In an attempt to address these issues, the GMM group developed a series of model dye–SNP systems based on the bioinspired proton-coupled electron transfer strategy previously described for the molecular triad **T4** (Bi–PhOH–PF_{10}–TCNP; see Fig. 3). These constructs consist of molecular dyads composed of Bi–PhOH moieties covalently attached to a PF_{10} porphyrin which bear a carboxylic acid as an anchoring group to attach to SNP (see Fig. 9, systems **H1**, **H2,** and **H3**). In these hybrid systems,

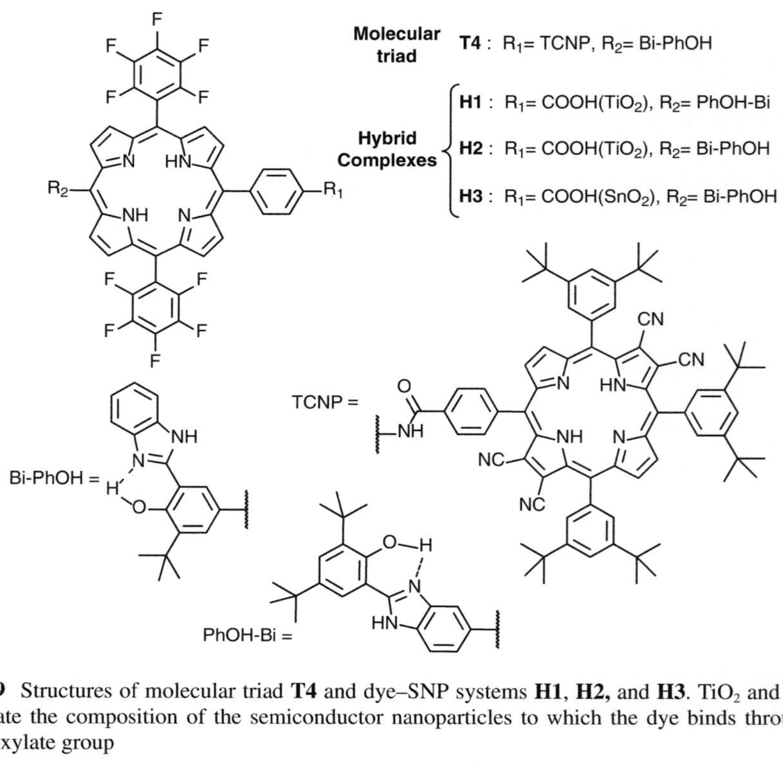

Fig. 9 Structures of molecular triad **T4** and dye–SNP systems **H1**, **H2,** and **H3**. TiO_2 and SnO_2 indicate the composition of the semiconductor nanoparticles to which the dye binds through a carboxylate group

each SNP can bind more than one molecular dyad on the surface. Nonetheless they can be studied as constructs effectively functioning as triads. For all these dye–SNP constructs, the high-potential porphyrin (PF_{10}) acts as the primary photoinduced electron donor, the SNP acts as the primary electron acceptor (replacing the organic TCNP moiety of triad **T4**), and either the benzimidazole–phenol (Bi–PhOH) or the phenol–benzimidazole (PhOH–Bi) acts as a secondary electron donor (see Fig. 9). As was explained for triad **T4,** the phenolic hydrogen is capable of forming a hydrogen bond with the lone pair electrons of the imidazole nitrogen. During the oxidation of the phenol by the $PF_{10}^{\bullet+}$, the phenol transfers its proton to the benzimidazole group by a PCET mechanism to produce a neutral phenoxyl radical ($E = 1.06$ V vs. SCE in the case of **H1**) that is thermodynamically capable of water oxidation ($E^0 = +0.58$ V vs. SCE, pH 7).

As shown in Fig. 9, the dye–SNP system **H1** consists of a benzimidazole–phenol (PhOH–Bi) moiety covalently bound to PF_{10}, and the primary electron acceptor is a TiO_2 nanoparticle. In this construct the presence of a final charge-separated state that involves a phenoxyl radical and injected electrons in TiO_2 was demonstrated by X-band and D-band EPR spectroscopies at low temperature [34]. The high spectral resolution of high-frequency (D-band) EPR measurements allowed for detailed identification of the molecular structures involved in the electron transfer. Results show that only 48 % of the holes localized on PhOH–Bi and 52 % on PF_{10} upon illumination at 4.2 K. However, when experiments are performed at 80 K, 95 % of the holes are localized on the PhOH–Bi moiety. This observed temperature dependence of the charge shift is attributed to restricted nuclear motion at low temperature.

The **H2** system is similar in construction to **H1**, but it bears a benzimidazole–phenol group (Bi–PhOH) in which the phenol moiety (instead of the benzimidazole) is covalently attached to PF_{10} (see Fig. 9) [37]. In **H2** a stronger internal hydrogen bond than in **H1** is observed. For **H2**, the detection of the phenoxyl radicals was performed by high-field EPR. It was shown that a higher-energy state formed at 13 K, while a relaxed state was observed after raising the temperature up to 100 K and then reducing it back to 13 K for the measurement. Theoretical DFT calculations were performed that indicated that solvent molecules that solvate the initial site O–H···N undergo a reorganization to solvate the newly formed O···H–N site. However, this reorganization is not possible at 13 K, and then the system results in a higher-energy state with a g_x value of 2.0056. According to DFT calculations, when the system is warmed up to 100 K, the solvent molecules around the new O···H–N site reorganize. This solvent reorganization and some minor structural rearrangements around the oxidized benzimidazole–phenol result in a relaxed final structure with a g_x value of 2.0061. The PCET process in the case of **H1** and **H2** is probably occurring by a concerted mechanism [36]. However, the thermal relaxation observed by high-field EPR of **H2** is not a general phenomenon. As mention before, construct **H1** with a weaker intramolecular hydrogen bond and the benzimidazole group as part of the linkage between the porphyrin and phenol moieties does not show such relaxation [34]. Thus, the simplicity of these dye–SNP systems and the ability of **H2** to reproduce the EPR thermal relaxation observed in the Tyr_Z–His190–P680 portion of the PSII electron transport chain helped to show that the strong intramolecular hydrogen bond in the Bi–PhOH group of **H2** is the main structural component responsible for its biomimetic behavior [37].

The system **H3** uses SNP small (2.5 nm diameter) SnO_2 nanoparticles instead of TiO_2. The SnO_2 acts as the primary electron acceptor, and again PF_{10} is the photoinduced electron donor, and a Bi–PhOH group in which the phenol moiety is covalently attached to PF_{10} acts as a secondary electron donor (see Fig. 9) [103]. In construct **H3**, the generation of the final charge-separated state BiH^+–$PhO^•$–PF_{10}–$SnO_2(e^-)$ was proven using time-resolved transient absorption measurements in aqueous solution at pH 7 at room temperature. Global analysis of the kinetics at different wavelengths yielded an evolution-associated difference spectrum (EADS) of 93 μs that was associated with decay of BiH^+–$PhO^•$–PF_{10}–$SnO_2(e^-)$ based on the lack of PF_{10} ground-state bleaching, and the observation of $PhO^•$ and $SnO_2(e^-)$ induced absorption around 400 nm and 600–1,000 nm, respectively. This long-lived charge-separated species (BiH^+–$PhO^•$–PF_{10}–$SnO_2(e^-)$) demonstrates the key role played by the Bi–PhOH in stabilizing the charge-separated state even when semiconductors with high electron mobility, such as SnO_2, are involved.

3.6 PCEs for Water Splitting

The knowledge gained in the construction and study of hybrid model dye–SNP systems has been applied to the design of functional photocatalytic devices. In this section we describe two approaches to the design of PCEs for water splitting developed by the GMM group in collaboration with the Mallouk group at Pennsylvania State University. The first approach consisted of a PEC containing a substituted trisbipyridylruthenium(II) $Ru(bpy)_3^{2+}$ complex as the sensitizing dye bound to a hydrated iridium oxide nanoparticle ($IrOx{\cdot}nH_2O$) which works as the water oxidation catalyst (Fig. 10) [104]. In this PEC the reductive synthesis of H_2 is catalyzed by a platinum cathode in aqueous solution. The system resembles a triad-type reaction center with $Ru(bpy)_3^{2+}$ acting both as light absorber and primary electron donor, a TiO_2 nanostructured electrode being the primary acceptor, and $IrOx{\cdot}nH_2O$ functioning as the secondary electron donor. Ideally, photoinduced electron transfer from the ruthenium dye to the TiO_2 electrode results in current flow to the platinum counter electrode to produce H_2, and transport of four holes to the $IrOx{\cdot}nH_2O$ nanoparticle results in the oxidation of water to molecular oxygen. Transient absorption experiments in this PCE system indicate that recombination of the initial photoinduced charge-separated state by migration of an electron back to the oxidized ruthenium complex is rapid ($\tau = 0.37$ ms) compared to the rate of electron transfer from the $IrOx{\cdot}nH_2O$ nanoparticle to the oxidized chromophore ($\tau = 2.2$ ms).

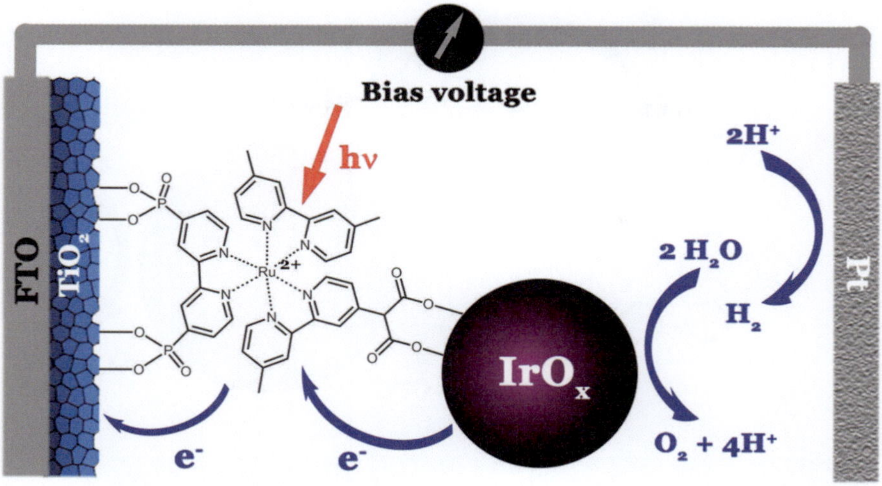

Fig. 10 Scheme of bio-inspired dye-sensitized water-splitting cell. The sequence of events begins upon light absorption by the ruthenium complex; then there is an electron injection from the ruthenium complex into TiO_2; and subsequently a secondary electron injection from $IrOx{\cdot}nH_2O$ into the oxidized ruthenium complex

This occurs because a significant fraction of photoinjected electrons are trapped in states below the TiO_2 CB edge that do not have sufficient potential to generate hydrogen (Fig. 4a, path 7). These electrons rapidly recombine with the oxidized dye (Fig. 4a, path 5). To overcome this kinetic impediment and achieve overall water splitting in these PCE systems, it was necessary to apply a small external bias (~200 mV).

The fast back electron transfer relative to the rate of water oxidation results in low quantum yields for water splitting in these systems. Hence, the photoanode of the previously described water-splitting PEC was redesigned. A bioinspired PCET strategy analogous to that illustrated for the model dye–SNP systems in Sect. 3.5 was employed. The same PhOH–Bi group of system **H1** was used as a redox mediator in the new PEC. This PhOH–Bi moiety was modified with a dicarboxylate functionality to produce colloidal $IrOx{\cdot}nH_2O$ particles decorated with both the PhOH–Bi mediator and 2-dicarboxyethylphosphonic acid (CEPA) [105]. When the $Ru(bpy)_3^{2+}$ dye and the PhOH–Bi/CEPA–$IrOx{\cdot}nH_2O$ nanoparticles are co-immobilized in the FTO-TiO_2 photoanode of the new PCE, the PhOH–Bi mediator is found to improve the performance of the system (Fig. 11). As compared to a cell without the mediator (using only CEPA-capped $IrOx{\cdot}nH_2O$), higher photocurrents and greater overall efficiency (by about a factor of three) are observed. Transient absorption experiments indicate that the improved performance results from faster regeneration of the ground-state dye from the oxidized state in the presence of PhOH–Bi in the system, confirming the validity of the bioinspired PCET strategy [105].

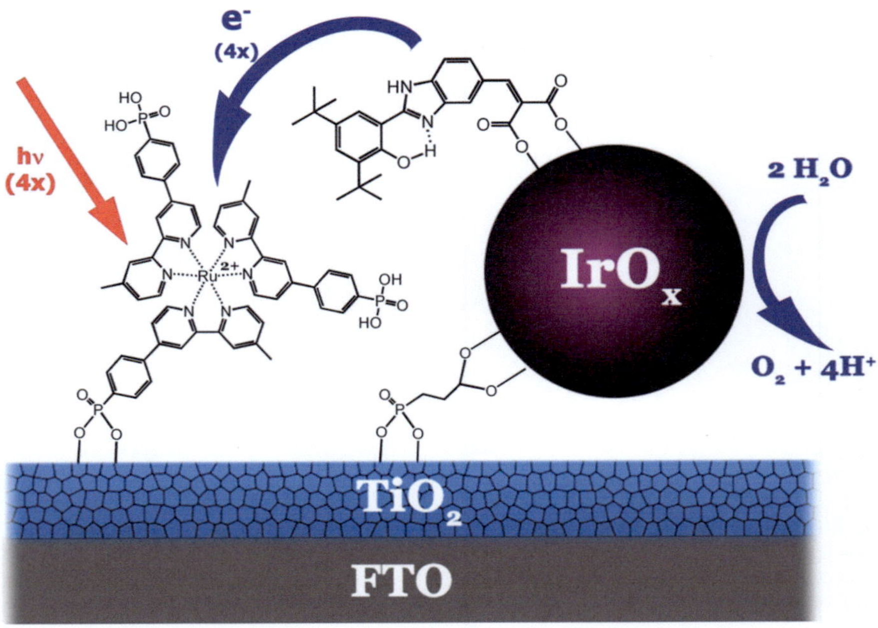

Fig. 11 Electron transfer reactions in the mediator-based, dye-sensitized TiO_2 photoanode showing molecular structures of the PhOH–Bi mediator, $Ru(bpy)_3^{2+}$, and CEPA linking spacer

Conclusions

In conclusion, this chapter shows how it is possible to employ the knowledge obtained from natural photosynthesis to design and build artificial photosynthetic systems that use sunlight to generate fuels. From the examples illustrated, it becomes clear that the coordinated effort of several disciplines such as chemistry, biochemistry, and materials science are essential to accomplish this goal. The trajectory from early simple molecular reaction center constructs up to the development of more complex systems for the conversion of solar energy to a fuel was summarized, and the main breakthroughs are highlighted. Future efforts in the field should be aimed at improving the efficiency of water-splitting PCEs by fine-tuning the application of previously demonstrated bioinspired concepts and by designing and testing new AP systems. Figure 12 shows a direct structural comparison between the reaction center of PSII [106], the structure of the fully organic **T4** reaction center, and the hybrid reaction center **H2**. There is a clear functional similarity of the artificial constructs and the cofactors of PSII. It is expected that artificial systems should be able to split water following steps analogous to those of the natural counterpart after linking the water oxidation catalyst (IrO*x* or other heterogeneous or molecular catalyst) to the PhOH–Bi redox relay. Ultimately, useful AP systems for fuel production will need

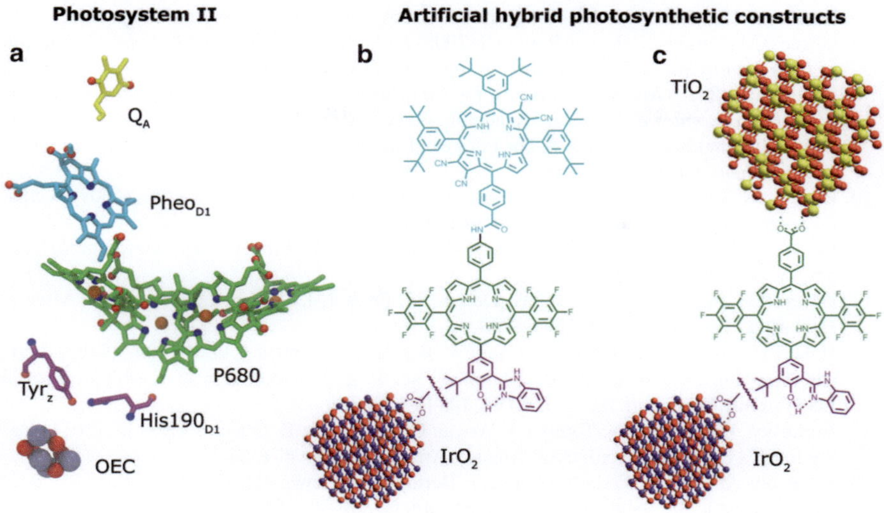

Fig. 12 Reaction center of PSII [106] (**a**) compared to structures of reaction centers in artificial photosynthetic constructs (**b**) and (**c**). The attachment of the catalyst IrO_2 has not yet been accomplished

to employ catalysts and chromophores based on abundant, inexpensive elements rather than the relatively rare and expensive iridium, ruthenium, and platinum.

While the AP concept is inspired by natural photosynthesis, the design of specific systems to carry out the reactions is based upon fundamental principles taken not only from biology but also from physics, chemistry, and materials science. Importantly, much of the progress to date comes from synergies and interactions at the intersections of these disciplines. Furthermore, it is anticipated that some features of AP will be used in hybrid photosynthetic systems so that the features of biology including self-assembly, replication, and repair can be implemented in energy-converting constructs to meet human needs.

References

1. U.S. Department of Energy (2014) *U.S. Energy Information Administration.*
2. ExxonMobil Corporation (2014) The Outlook for Energy: A View to 2040.
3. Olah GA, Prakash GKS, and Goeppert A (2011) Anthropogenic Chemical Carbon Cycle for a Sustainable Future. *J. Am. Chem. Soc.* **133**, 12881-12898.
4. Stocker TF, Qin D, Plattner G-K, Tignor M, Allen SK, Boschung J et al. (2013) Fifth Assessment Report (AR5). Climate Change 2013: The Physical Science Basis. In http://www.climatechange2013.org.
5. Sabine CL, Feely RA, Gruber N, Key RM, Lee K, Bullister JL et al. (2004) The Oceanic Sink for Anthropogenic CO_2. *Science* **305**, 367-371.

6. Keeling R (2014) Scripps CO_2 Program. http://scrippsco2.ucsd.edu/home/index.php
7. Grimes CA, Varghese OK, and Ranjan S (2008) *Light, Water, Hydrogen. The Solar Generation of Hydrogen by Water Photoelectrolysis*. pp. 546, Springer, New York.
8. Sorensen B (2010) *Renewable Energy, Fourth Edition: Physics, Engineering, Environmental Impacts, Economics & Planning*. Academic Press, Oxford.
9. Armaroli N and Balzani V (2007) The Future of Energy Supply: Challenges and Opportunities. *Angew. Chem. Int. Ed.* **46**, 52-66.
10. Rockstrom J, Steffen W, Noone K, Persson A, Chapin FS, Lambin EF et al. (2009) A Safe Operating Space for Humanity. *Nature* **461**, 472-475.
11. Lemonick MD (2014) The Hunt for Life Beyond Earth. In *Astrobiology*. www.nationalgeographic.com/astrobiology.
12. Hazen RM, Papineau D, Bleeker W, Downs RT, Ferry JM, McCoy TJ et al. (2008) Mineral Evolution. *Am. Mineral.* **93**, 1693-1720.
13. Haberl H, Erb KH, Krausmann F, Gaube V, Bondeau A, Plutzar C et al. (2007) Quantifying and Mapping the Human Appropriation of Net Primary Production in Earth's Terrestrial Ecosystems. *Proc. Natl. Acad. Sci.* **104**, 12942-12947.
14. Barnosky AD, Matzke N, Tomiya S, Wogan GOU, Swartz B, Quental TB et al. (2011) Has the Earth's Sixth Mass Extinction Already Arrived? *Nature* **471**, 51-57.
15. Barnosky AD, Hadly EA, Bascompte J, Berlow EL, Brown JH, Fortelius M et al. (2012) Approaching a State Shift in Earth's Biosphere. *Nature* **486**, 52-58.
16. Sherman BD, Vaughn MD, Bergkamp JJ, Gust D, Moore AL, and Moore TA (2014) Evolution of Reaction Center Mimics to Systems Capable of Generating Solar Fuel. *Photosynth. Res.* **120**, 59-70.
17. Morton O (2009) *Eating the Sun: How Plants Power the Planet*. pp. 480, Harper Perennial, New York.
18. Gust D, Moore TA, and Moore AL (1993) Molecular Mimicry of Photosynthetic Energy and Electron Transfer. *Acc. Chem. Res.* **26**, 198-205.
19. Gust D, Moore TA, and Moore AL (2000) Mimicking Photosynthetic Solar Energy Transduction. *Acc. Chem. Res.* **34**, 40-48.
20. Grätzel M (2001) Photoelectrochemical Cells. *Nature* **414**, 338-344.
21. Gust D, Moore TA, and Moore AL (2009) Solar Fuels via Artificial Photosynthesis. *Acc. Chem. Res.* **42**, 1890-1898.
22. Hambourger M, Moore GF, Kramer DM, Gust D, Moore AL, and Moore TA (2009) Biology and Technology for Photochemical Fuel Production. *Chem. Soc. Rev.* **38**, 25-35.
23. Blankenship RE, Tiede DM, Barber J, Brudvig GW, Fleming G, Ghirardi M et al. (2011) Comparing Photosynthetic and Photovoltaic Efficiencies and Recognizing the Potential for Improvement. *Science* **332**, 805-809.
24. Concepcion JJ, House RL, Papanikolas JM, and Meyer TJ (2012) Chemical Approaches to Artificial Photosynthesis. *Proc. Natl. Acad. Sci.* **109**, 15560-15564.
25. Rybtchinski B and Wasielewski MR (2012) Artificial Photosynthesis for Solar Energy Conversion. In *Fundamentals of Materials for Energy and Sustainability*. Ginley D and Cahen D (ed.), pp. 349-364. Cambridge University Press,
26. Lacis AA, Schmidt GA, Rind D, and Ruedy RA (2010) Atmospheric CO_2: Principal Control Knob Governing Earth's Temperature. *Science* **330**, 356-359.
27. Scholes GD, Fleming GR, Olaya-Castro A, and van Grondelle R (2011) Lessons from Nature About Solar Light Harvesting. *Nature Chem.* **3**, 763-774.
28. Olaya-Castro A, Lee CF, Olsen FF, and Johnson NF (2008) Efficiency of Energy Transfer in a Light-Harvesting System under Quantum Coherence. *Phys. Rev. B* **78**, 085115.
29. Fassioli F, Nazir A, and Olaya-Castro A (2010) Quantum State Tuning of Energy Transfer in a Correlated Environment. *The Journal of Physical Chemistry Letters* **1**, 2139-2143.
30. Rutherford AW, Osyczka A, and Rappaport F (2012) Back-Reactions, Short-Circuits, Leaks and Other Energy Wasteful Reactions in Biological Electron Transfer: Redox Tuning to Survive Life in O_2. *FEBS Lett.* **586**, 603-616.

31. Hammes-Schiffer S (2009) Theory of Proton-Coupled Electron Transfer in Energy Conversion Processes. *Acc. Chem. Res.* **42**, 1881-1889.
32. Hammarstrom L and Styring S (2011) Proton-Coupled Electron Transfer of Tyrosines in Photosystem Ii and Model Systems for Artificial Photosynthesis: The Role of a Redox-Active Link between Catalyst and Photosensitizer. *Energy & Environmental Science* **4**, 2379-2388.
33. Warren J and Mayer J (2013) Proton-Coupled Electron Transfer. In *Encyclopedia of Biophysics*. Roberts GK (ed.), pp. 2112-2114. Springer Berlin Heidelberg.
34. Moore GF, Hambourger M, Gervaldo M, Poluektov OG, Rajh T, Gust D et al. (2008) A Bioinspired Construct That Mimics the Proton Coupled Electron Transfer between P680•+and the TyrZ-His190 Pair of Photosystem II. *J. Am. Chem. Soc.* **130**, 10466-10467.
35. Moore GF, Hambourger M, Kodis G, Michl W, Gust D, Moore TA et al. (2010) Effects of Protonation State on a Tyrosine–Histidine Bioinspired Redox Mediator. *J. Phys. Chem. B* **114**, 14450-14457.
36. Megiatto JD, Antoniuk-Pablant A, Sherman BD, Kodis G, Gervaldo M, Moore TA et al. (2012) Mimicking the Electron Transfer Chain in Photosystem Ii with a Molecular Triad Thermodynamically Capable of Water Oxidation. *Proc. Natl. Acad. Sci.* **109**, 15578-15583.
37. Megiatto JD, Méndez-Hernández DD, Tejeda-Ferrari ME, Teillout A-L, Llansola-Portoles MJ, Kodis G et al. (2014) A Bioinspired Redox Relay That Mimics Radical Interactions of the Tyr-His Pairs of Photosystem II. *Nature Chem.* **6**, 423–428.
38. Gust D and Moore TA (1987) Electron Transfer in Model Systems for Photosynthesis. In *Photoinduced Charge Separation and Energy Migration in Supramolecular Species*. Balzani V (ed.), Vol. pp. 267-282. D. Reidel Press, Boston.
39. Moore TA and Gust D (1987) Charge Separation in Model Compounds for Photosynthesis. In *Proceedings, Protein Structure: Molecular and Electronic Reactivity Conference*. Austin R, Buhks E, Chance B, DeVault D, Dutton PL, Fravenfelder H and Gol'damskii VI (ed.), Vol. pp. 389-398. Springer-Verlag, New York.
40. Gust D, Moore TA, and Moore AL (2012) Realizing Artificial Photosynthesis. *Faraday Discuss.* **155**, 9-26.
41. Kurreck H and Huber M (1995) Model Reactions for Photosynthesis—Photoinduced Charge and Energy Transfer between Covalently Linked Porphyrin and Quinone Units. *Angew. Chem. Int. Ed.* **34**, 849-866.
42. Kong JLY and Loach PA (1980) Syntheses of Covalently-Linked Porphyrin-Quinone Complexes. *J. Heterocycl. Chem.* **17**, 737-744.
43. Kong JLY and Loach PA (1978) *Frontieres of Biological Energetics*. pp. 73, Academic Press, New York.
44. Tabushi I, Koga N, and Yanagita M (1979) Efficient Intramolecular Quenching and Electron Transfer in Tetraphenylporphyrin Attached with Benzoquinone or Hydroquinone as a Photosystem Model. *Tetrahedron Lett.* **20**, 257-260.
45. Bensasson RV, Land EJ, Moore AL, Crouch RL, Dirks G, Moore TA et al. (1981) Mimicry of Antenna and Photo-Protective Carotenoid Functions by a Synthetic Carotenoporphyrin. *Nature* **290**, 329-332.
46. Moore AL, Joy A, Tom R, Gust D, Moore TA, Bensasson RV et al. (1982) Photoprotection by Carotenoids During Photosynthesis: Motional Dependence of Intramolecular Energy Transfer. *Science* **216**, 982-984.
47. Land EJ, Lexa D, Bensasson RV, Gust D, Moore TA, Moore AL et al. (1987) Pulse Radiolytic and Electrochemical Investigations of Intramolecular Electron Transfer in Carotenoporphyrins and Carotenoporphyrin-Quinone Triads. *J. Phys. Chem.* **91**, 4831-4835.
48. Moore TA, Gust D, Mathis P, Mialocq J-C, Chachaty C, Bensasson RV et al. (1984) Photodriven Charge Separation in a Carotenoporphyrin-Quinone Triad. *Nature* **307**, 630-632.
49. Gust D, Mathis P, Moore AL, Liddell PA, Nemeth GA, Lehman WR et al. (1983) Energy Transfer and Charge Separation in Carotenoporphyrins. *Photochem. Photobiol.* **37**, S46.
50. Gust D and Moore TA (1989) Mimicking Photosynthesis. *Science* **244**, 35-41.

51. Marcus RA (1993) Electron Transfer Reactions in Chemistry: Theory and Experiment (Nobel Lecture). *Angew. Chem. Int. Ed.* **32**, 1111-1121.
52. Gust D, Moore TA, Moore AL, Barrett D, Harding LO, Makings LR et al. (1988) Photoinitiated Charge Separation in a Carotenoid-Porphyrin-Diquinone Tetrad: Enhanced Quantum Yields Via Multistep Electron Transfers. *J. Am. Chem. Soc.* **110**, 321-323.
53. Gust D, Moore TA, Moore AL, Lee SJ, Bittersmann E, Luttrull DK et al. (1990) Efficient Multistep Photoinitiated Electron Transfer in a Molecular Pentad. *Science (New York, N.Y.)* **248**, 199-201.
54. Kodis G, Liddell PA, de la Garza L, Clausen PC, Lindsey JS, Moore AL et al. (2002) Efficient Energy Transfer and Electron Transfer in an Artificial Photosynthetic Antenna–Reaction Center Complex. *J. Phys. Chem. A* **106**, 2036-2048.
55. Kodis G, Liddell PA, Moore AL, Moore TA, and Gust D (2004) Synthesis and Photochemistry of a Carotene–Porphyrin–Fullerene Model Photosynthetic Reaction Center. *J. Phys. Org. Chem.* **17**, 724-734.
56. Steinberg-Yfrach G, Liddell PA, Hung S-C, Moore AL, Gust D, and Moore TA (1997) Conversion of Light Energy to Proton Potential in Liposomes by Artificial Photosynthetic Reaction Centres. *Nature* **385**, 239-241.
57. Steinberg-Yfrach G, Rigaud J-L, Durantini EN, Moore AL, Gust D, and Moore TA (1998) Light-Driven Production of Atp Catalysed by F0f1-Atp Synthase in an Artificial Photosynthetic Membrane. *Nature* **392**, 479-482.
58. Anderson NA and Lian T (2005) Ultrafast Electron Transfer at the Molecule-Semiconductor Nanoparticle Interface. *Annu. Rev. Phys. Chem.* **56**, 491-519.
59. Hoffmann MR, Martin ST, Choi W, and Bahnemann DW (1995) Environmental Applications of Semiconductor Photocatalysis. *Chem. Rev. (Washington, DC, U. S.)* **95**, 69-96.
60. Park H, Park Y, Kim W, and Choi W (2013) Surface Modification of TiO_2 Photocatalyst for Environmental Applications. *J. Photochem. Photobiol. C* **15**, 1-20.
61. Fox MA and Dulay MT (1993) Heterogeneous Photocatalysis. *Chem. Rev. (Washington, DC, U. S.)* **93**, 341-357.
62. O'Regan B and Grätzel M (1991) A Low-Cost, High-Efficiency Solar Cell Based on Dye-Sensitized Colloidal TiO_2 films. *Nature* **353**, 737-740.
63. Shah A, Torres P, Tscharner R, Wyrsch N, and Keppner H (1999) Photovoltaic Technology: The Case for Thin-Film Solar Cells. *Science* **285**, 692-698.
64. Grätzel M (2004) Conversion of Sunlight to Electric Power by Nanocrystalline Dye-Sensitized Solar Cells. *Journal of Photochemistry and Photobiology A: Chemistry* **164**, 3-14.
65. Mathew S, Yella A, Gao P, Humphry-Baker R, CurchodBasile FE, Ashari-Astani N et al. (2014) Dye-Sensitized Solar Cells with 13 % Efficiency Achieved through the Molecular Engineering of Porphyrin Sensitizers. *Nature Chem.* **6**, 242–247.
66. Kathiravan A and Renganathan R (2009) Effect of Anchoring Group on the Photosensitization of Colloidal TiO_2 Nanoparticles with Porphyrins. *J. Colloid Interface Sci.* **331**, 401-407.
67. Young KJ, Martini LA, Milot RL, Snoeberger Iii RC, Batista VS, Schmuttenmaer CA et al. (2012) Light-Driven Water Oxidation for Solar Fuels. *Coord. Chem. Rev.* **256**, 2503-2520.
68. Bae E, Choi W, Park J, Shin HS, Kim SB, and Lee JS (2004) Effects of Surface Anchoring Groups (Carboxylate vs Phosphonate) in Ruthenium-Complex-Sensitized TiO_2 on Visible Light Reactivity in Aqueous Suspensions. *J. Phys. Chem. B* **108**, 14093-14101.
69. Kilså K, Mayo EI, Brunschwig BS, Gray HB, Lewis NS, and Winkler JR (2004) Anchoring Group and Auxiliary Ligand Effects on the Binding of Ruthenium Complexes to Nanocrystalline TiO_2 Photoelectrodes. *J. Phys. Chem. B* **108**, 15640-15651.
70. Park H, Bae E, Lee J-J, Park J, and Choi W (2006) Effect of the Anchoring Group in Ru–Bipyridyl Sensitizers on the Photoelectrochemical Behavior of Dye-Sensitized TiO_2 Electrodes: Carboxylate versus Phosphonate Linkages. *J. Phys. Chem. B* **110**, 8740-8749.
71. Brennan BJ, Llansola-Portoles MJ, Liddell PA, Moore TA, Moore AL, and Gust D (2013) Comparison of Silatrane, Phosphonic Acid, and Carboxylic Acid Functional Groups for Attachment of Porphyrin Sensitizers to TiO_2 in Photoelectrochemical Cells. *Phys. Chem. Chem. Phys.* **15**, 16605-16614.

72. Hernandez LI, Godin RP, Bergkamp JJ, Llansola-Portoles MJ, Sherman BD, Tomlin J et al. (2013) Spectral Characteristics and Photosensitization of TiO_2 Nanoparticles in Reverse Micelles by Perylenes. *J. Phys. Chem. B* **117**, 4568-4581.
73. Guerrero G, Mutin PH, and Vioux A (2001) Anchoring of Phosphonate and Phosphinate Coupling Molecules on Titania Particles. *Chem. Mater.* **13**, 4367-4373.
74. Llansola-Portoles MJ, Bergkamp JJ, Finkelstein-Shapiro D, Sherman BD, Kodis G, Dimitrijevic NM et al. (2014) Controlling Surface Defects and Photophysics in TiO_2 Nanoparticles. *J. Phys. Chem. A* DOI: 10.1021/jp506284q
75. Cong J, Yang X, Liu J, Zhao J, Hao Y, Wang Y et al. (2012) Nitro Group as a New Anchoring Group for Organic Dyes in Dye-Sensitized Solar Cells. *Chem. Commun. (Cambridge, U. K.)* **48**, 6663-6665.
76. Cui J, Lu J, Xu X, Cao K, Wang Z, Alemu G et al. (2014) Organic Sensitizers with Pyridine Ring Anchoring Group for p-Type Dye-Sensitized Solar Cells. *J. Phys. Chem. C*
77. Gou F, Jiang X, Li B, Jing H, and Zhu Z (2013) Salicylic Acid As a Tridentate Anchoring Group for Azo-Bridged Zinc Porphyrin in Dye-Sensitized Solar Cells. *ACS Appl. Mater. Interfaces* **5**, 12631-12637.
78. Rice CR, Ward MD, Nazeeruddin MK, and Gratzel M (2000) Catechol as an Efficient Anchoring Group for Attachment of Ruthenium-Polypyridine Photosensitisers to Solar Cells Based on Nanocrystalline TiO_2 Films. *New J. Chem.* **24**, 651-652.
79. Abuabara SG, Cady CW, Baxter JB, Schmuttenmaer CA, Crabtree RH, Brudvig GW et al. (2007) Ultrafast Photooxidation of Mn(Ii)–Terpyridine Complexes Covalently Attached to TiO_2 nanoparticles. *J. Phys. Chem. C* **111**, 11982-11990.
80. Li S-C, Chu L-N, Gong X-Q, and Diebold U (2010) Hydrogen Bonding Controls the Dynamics of Catechol Adsorbed on a TiO_2(110) Surface. *Science* **328**, 882-884.
81. Liu L-M, Li S-C, Cheng H, Diebold U, and Selloni A (2011) Growth and Organization of an Organic Molecular Monolayer on TiO_2: Catechol on Anatase (101). *J. Am. Chem. Soc.* **133**, 7816-7823.
82. McNamara WR, Milot RL, Song H-e, Snoeberger Iii RC, Batista VS, Schmuttenmaer CA et al. (2010) Water-Stable, Hydroxamate Anchors for Functionalization of Tio2 Surfaces with Ultrafast Interfacial Electron Transfer. *Energy & Environmental Science* **3**, 917-923.
83. Brewster TP, Konezny SJ, Sheehan SW, Martini LA, Schmuttenmaer CA, Batista VS et al. (2013) Hydroxamate Anchors for Improved Photoconversion in Dye-Sensitized Solar Cells. *Inorg. Chem.* **52**, 6752-6764.
84. Johansson PG, Kopecky A, Galoppini E, and Meyer GJ (2013) Distance Dependent Electron Transfer at TiO_2 Interfaces Sensitized with Phenylene Ethynylene Bridged RuII–Isothiocyanate Compounds. *J. Am. Chem. Soc.* **135**, 8331-8341.
85. Galoppini E (2004) Linkers for Anchoring Sensitizers to Semiconductor Nanoparticles. *Coord. Chem. Rev.* **248**, 1283-1297.
86. Freitag M and Galoppini E (2011) Molecular Host-Guest Complexes: Shielding of Guests on Semiconductor Surfaces. *Energy & Environmental Science* **4**, 2482-2494.
87. Stergiopoulos T, Bernard M-C, Hugot-Le Goff A, and Falaras P (2004) Resonance Micro-Raman Spectrophotoelectrochemistry on Nanocrystalline Tio2 Thin Film Electrodes Sensitized by Ru(II) Complexes. *Coord. Chem. Rev.* **248**, 1407-1420.
88. Joselevich E and Willner I (1994) Photosensitization of Quantum-Size TiO_2 Particles in Water-in-Oil Microemulsions. *J. Phys. Chem.* **98**, 7628-7635.
89. Zang L, Liu C-Y, and Ren X-M (1995) Photochemistry of Semiconductor Particles 5. Location of Dyes in Reverse Micelles Containing TiO_2 Nanoparticles and Effects on Photoinduced Interfacial Electron Transfer. *J. Photochem. Photobiol. A* **88**, 47-51.
90. Llansola-Portoles MJ, Bergkamp JJ, Tomlin J, Moore TA, Kodis G, Moore AL et al. (2013) Photoinduced Electron Transfer in Perylene-TiO2 Nanoassemblies. *Photochem. Photobiol.* **89**, 1375-1382.
91. Diebold U (2003) The Surface Science of Titanium Dioxide. *Surf. Sci. Rep.* **48**, 53-229.
92. Thompson TL and Yates JT (2006) Surface Science Studies of the Photoactivation of TiO_2 - New Photochemical Processes. *Chem. Rev. (Washington, DC, U. S.)* **106**, 4428-4453.

93. Diebold U (2010) Oxide Surfaces: Surface Science Goes Inorganic. *Nat. Mater.* **9**, 185-187.
94. Vijayan B, Dimitrijevic NM, Rajh T, and Gray K (2010) Effect of Calcination Temperature on the Photocatalytic Reduction and Oxidation Processes of Hydrothermally Synthesized Titania Nanotubes. *J. Phys. Chem. C* **114**, 12994-13002.
95. Gregg BA, Chen S-G, and Ferrere S (2003) Enhanced Dye-Sensitized Photoconversion Efficiency via Reversible Production of UV-Induced Surface States in Nanoporous TiO_2. *J. Phys. Chem. B* **107**, 3019-3029.
96. Green ANM, Palomares E, Haque SA, Kroon JM, and Durrant JR (2005) Charge Transport versus Recombination in Dye-Sensitized Solar Cells Employing Nanocrystalline TiO_2 and SnO2 Films. *J. Phys. Chem. B* **109**, 12525-12533.
97. Gong X-Q, Selloni A, Batzill M, and Diebold U (2006) Steps on Anatase TiO_2(101). *Nat. Mater.* **5**, 665-670.
98. Linsebigler AL, Lu G, and Yates JT (1995) Photocatalysis on TiO_2 Surfaces: Principles, Mechanisms, and Selected Results. *Chem. Rev. (Washington, DC, U. S.)* **95**, 735-758.
99. Hurum DC, Agrios AG, Gray KA, Rajh T, and Thurnauer MC (2003) Explaining the Enhanced Photocatalytic Activity of Degussa P25 Mixed-Phase TiO_2 Using EPR. *J. Phys. Chem. B* **107**, 4545-4549.
100. Finkelstein-Shapiro D, Petrosko SH, Dimitrijevic NM, Gosztola D, Gray KA, Rajh T et al. (2013) CO_2 Preactivation in Photoinduced Reduction via Surface Functionalization of TiO_2 Nanoparticles. *The Journal of Physical Chemistry Letters* **4**, 475-479.
101. Jakubikova E, Snoeberger Iii RC, Batista VS, Martin RL, and Batista ER (2009) Interfacial Electron Transfer in TiO_2 Surfaces Sensitized with Ru(II)–Polypyridine Complexes. *J. Phys. Chem. A* **113**, 12532-12540.
102. Martsinovich N and Troisi A (2012) How TiO_2 Crystallographic Surfaces Influence Charge Injection Rates from a Chemisorbed Dye Sensitiser. *Phys. Chem. Chem. Phys.* **14**, 13392-13401.
103. Llansola-Portoles MJ, Palacios RE, Kodis G, Jackson D. Megiatto J, Moore AL, Moore TA et al. (2013) One Approach to Artificial Photosynthesis. *European Photochemistry Association Newsletters* **84**, 98-105.
104. Youngblood WJ, Lee S-HA, Kobayashi Y, Hernandez-Pagan EA, Hoertz PG, Moore TA et al. (2009) Photoassisted Overall Water Splitting in a Visible Light-Absorbing Dye-Sensitized Photoelectrochemical Cell. *J. Am. Chem. Soc.* **131**, 926-927.
105. Zhao Y, Swierk JR, Megiatto JD, Sherman B, Youngblood WJ, Qin D et al. (2012) Improving the efficiency of water splitting in dye-sensitized solar cells by using a biomimetic electron transfer mediator. *Proc. Natl. Acad. Sci.* **109**, 15612–15616.
106. Umena Y, Kawakami K, Shen J-R, and Kamiya N (2011) Crystal Structure of Oxygen-Evolving Photosystem II at a Resolution of 1.9 A. *Nature* **473**, 55-60.

Enzymes as Exploratory Catalysts in Artificial Photosynthesis

Andreas Bachmeier, Bhavin Siritanaratkul, and Fraser A. Armstrong

Abstract Water and sunlight are the most abundant resources on Earth: sunlight has more than enough energy to photolyse the oceans, producing H_2 and O_2, but this does not happen unassisted—light harvesters and catalysts are required. This chapter discusses how enzymes, which have evolved to be highly efficient catalysts for biological photosynthetic and energy conserving reactions, have an important role in designing and developing artificial photosynthesis systems at the model level. Enzymes facilitate the study of integrated model systems because they remove the burden of poor electrocatalytic rates and efficiencies that are characteristic of most systems (exception being platinum) and complicate the overall picture. In contrast to most artificial catalysts, enzymes operate efficiently at neutral pH and can hence be used to mimic conditions under which future technologies will have to operate.

Keywords Enzymes • Artificial photosynthesis • Hydrogenases • Semiconductors • Integrated model systems

1 Introduction

Enzymes make possible the design and construction of bench-scale test systems that are instructive because the chemical reactions of solar-fuel production are rendered so fast and efficient that they do not limit the performance. The active sites of enzymes are superb catalysts for H_2 evolution, CO_2 reduction and O_2 evolution; thus, if attached to nanoparticles so that interfacial electron transfers are fast, they serve as smooth, proficient depolarisers to allow electron flow. Exploitation of enzymes in this way makes possible the evaluation of semiconducting materials for artificial photosynthesis (AP) development in cases where those materials may be highly suited for light harvesting but very poor catalysts.

There is a close connection between redox catalysis that occurs on the surface of a photoexcited semiconductor and electrocatalysis that occurs at electrodes [1].

A. Bachmeier • B. Siritanaratkul • F.A. Armstrong (✉)
Department of Chemistry, University of Oxford, South Parks Road, Oxford, UK
e-mail: fraser.armstrong@chem.ox.ac.uk

E.A. Rozhkova, K. Ariga (eds.), *From Molecules to Materials*,
DOI 10.1007/978-3-319-13800-8_4

From extensive investigations with a suite of techniques called protein film electrochemistry (PFE), it is now established that many redox enzymes that catalyse AP chemical reactions are extremely fast and efficient, despite their giant size [2, 3]. This information is implicit in numerous observations; namely, that fuel-forming electrocatalytic reactions—H_2 evolution and CO_2 reduction—are reversible when catalysed by the appropriate enzyme and require just the minimal overpotential to operate in either direction; hence, they are efficient. Given the crucial importance, for scaling up AP, of having catalysts based upon inexpensive, abundant materials, it is highly significant that these investigations have proved that solar-fuel production *does not* require precious metal catalysts like Pt, Ru or Ir. Indeed, experiments with enzymes show instead that given the right atomic environments (we know the structures of many enzymes to high resolution), the 3d elements Mn, Fe, Ni and Cu perform excellently. Importantly also, these inexpensive, 'dressed-up' catalysts perform under ambient, neutral conditions.

Studies with enzymes provide hints at other advantages, one being the ability of an enzyme to trap and accumulate all the charge carriers (electrons or holes) required for the reaction (usually two, sometimes four). In AP, a margin of irreversibility is desirable to allow the chemical reactions to compete against wasteful recombination processes, but this margin can disappear if the potential needed to drive the chemical reaction is close to the relevant band potentials of the semiconductor [4]. In this chapter we explore concepts that are helpful in understanding relationships between the properties of the semiconducting material (*p* vs. *n* materials, bandgaps, conduction band potentials, surface charges and stability) and the actual catalytic reactions, using two classes of fuel-forming enzymes that have been extensively characterised. Hence, in obtaining design principles for artificial photosynthesis, we ask 'what can we learn from biology?'

2 A Picture of the Concepts

The simplest way to depict how enzymes are integrated into AP research is shown in Fig. 1. The enzyme is attached to the surface of a semiconductor nanoparticle and receives electrons that have been promoted to the conduction band (CB) by bandgap excitation (as indicated) or by injection from a dye. The holes (in valence band or oxidised dye) are filled by electrons that are ideally derived from water (using an O_2-evolving catalyst, OEC) although practicalities usually favour the use of cheap and reliable systems based on small molecules, oxidation of which is initiated by a fast, one-electron transfer. The opposite situation arises for studying an OEC. Strangely, these agents are often called 'sacrificial reagents' as if water were not one.

The idea behind using enzymes as the catalysts to investigate potential AP systems stems from their extraordinary performance in H_2 production, CO_2 reduction and O_2 evolution. The active sites of enzymes have evolved to optimise selectivity and reaction rates by tuning reduction potentials, coupling proton and electron transfers so they are synchronous and stabilising intermediates by just the right degree (avoiding high-energy states). Most attempts to mimic enzymes have focused

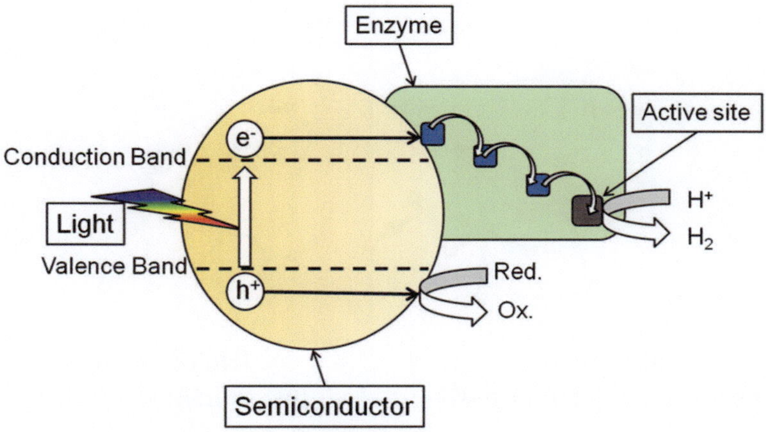

Fig. 1 *Cartoon* showing the special role for enzymes in AP, in this case as fuel-forming catalysts

on the active site, but it is important to consider how and why the inner coordination complex is held in a special environment (the outer shell). Away from the active site (which is usually buried), electron relay centres have evolved to have small reorganisation energies, optimal reduction potentials and separation distances just short enough to allow fast and efficient electron transfer (ET). The complete enzyme has the capacity to store all the charge carriers (electrons or holes) that are required to perform a complete half-cell reaction—a property that contrasts with most small metal-complex catalysts. The charge carriers may be 'trapped' within the enzyme, thus impeding recombination within the semiconductor. How important then is the outer shell, with its special environment—a factor distinguishing redox enzymes from small-molecule electrocatalysts?

3 Enzymes and Their Proficiency

3.1 The Importance of the Environment Beyond the Catalyst Centre

Structures of the active sites of four important enzymes are shown in Fig. 2, in each case including some of the amino acid side chains that lie in the outer shell.

An excellent example of the special nature of the active sites of enzymes is provided by a recent discovery made during investigations of the biosynthesis of microbial enzymes known as hydrogenases (H_2ases) that catalyse H_2 evolution and oxidation. The active site of the so-called [FeFe]-hydrogenases, known as the 'H-cluster', consists of two components: the part responsible for dealing directly with H_2 and atomic/ionic H-species is a binuclear (Fe) complex containing CO, CN^- and a bridging azadithiolate; this unit ($2Fe_H$) is linked by a cysteine-S ligand to a [4Fe–4S] cluster that acts as an electron storage site/conduit. The structure is shown in Fig. 2.

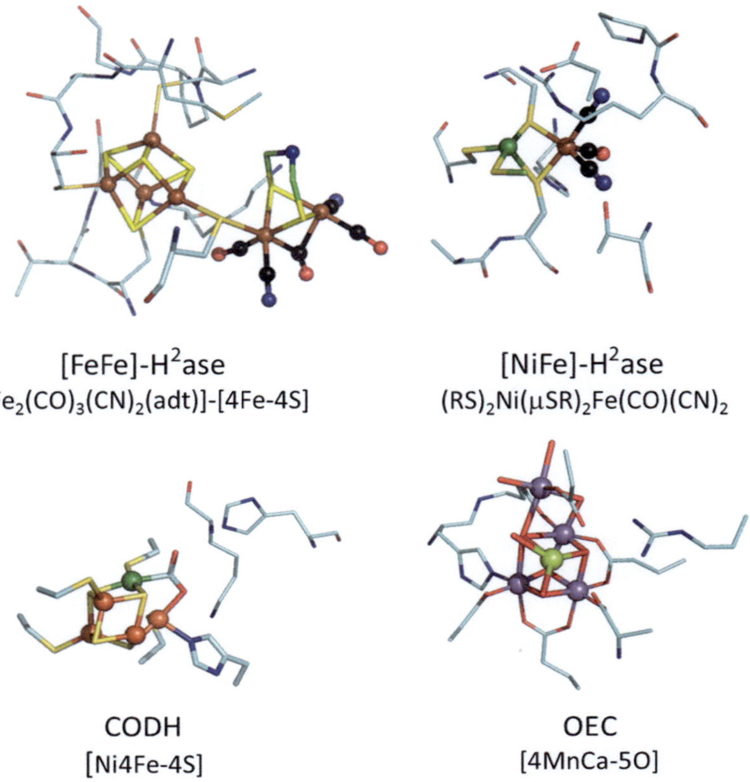

Fig. 2 Active sites of [FeFe]-H_2ase, [NiFe]-H_2ase, [CODH] and OEC of photosystem II. Generated in PyMOL using the Protein Data Bank structures, 3C8Y, 3USC, 3B52 and 3ARC, respectively

In 2013, two papers appeared showing how the Fe-carbonyl precursor of $2Fe_H$ (which shows little catalytic activity) becomes extremely active when inserted into the $2Fe_H$ binding cavity of the preformed 'apoenzyme' already equipped with the [4Fe–4S] cluster [5, 6]. The conversion is shown pictorially in Fig. 3a. All that occurs in terms of the inner-sphere coordination chemistry is that one CO is replaced by the cysteine-S ligand to link the two components together. The structures of the $2Fe_H$ precursor, the apoenzyme and the active site of the assembled H-cluster within the binding cavity are depicted in Fig. 3b.

One important possibility is that the complete H-cluster is able to delocalise, or accommodate more easily, a second electron. A recent experiment by Rauchfuss et al. showed that attaching a ferrocene unit to an Fe-carbonyl analogue of $2Fe_H$ conferred catalytic activity for H_2 oxidation that was previously absent [7]. However, the activity remained orders of magnitude lower than for the enzyme.

This discovery establishes the importance of providing the right environment (i.e. second- and outer-shell atoms), not just the first coordination shell. Despite the

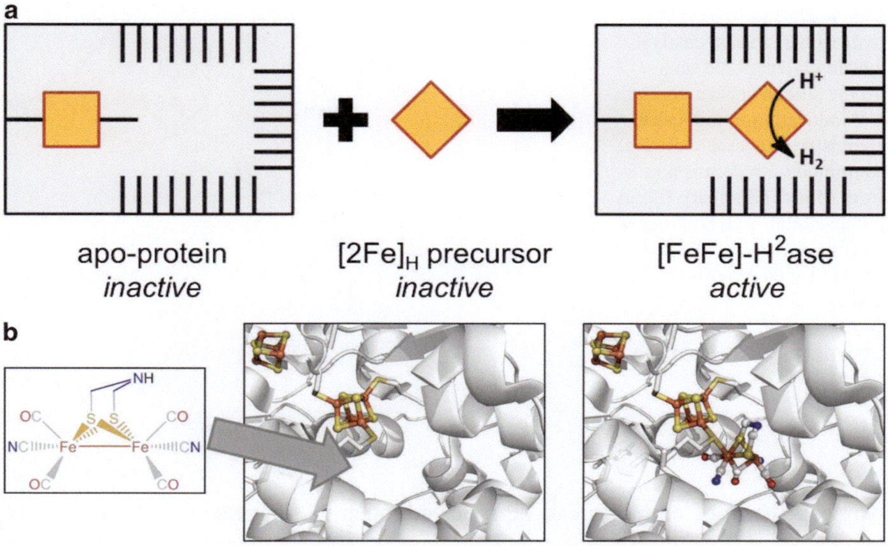

Fig. 3 (**a**) *Cartoon* showing the insertion of the $2Fe_H$ precursor into the H_2ase apoprotein to give active enzyme. Binding of $2Fe_H$ to the [4Fe–4S] cluster via a cysteine-S ligand occurs after loss of one CO ligand. The *grey box* represents the binding cavity, with amino acids surrounding the active site depicted as *black lines*. The [4Fe–4S] fragment is shown as *orange square*; $2Fe_H$ is depicted as *orange rhombus*. (**b**) Structural representation of the insertion reaction. Depictions of the active site and the surrounding cavity were generated in PyMOL using the Protein Data Bank structure 3C8Y. Some amino acids are omitted to provide an unobstructed view of the active site and the surrounding environment. Colour coding: iron, *orange*; sulphur, *yellow*; carbon, *grey*; nitrogen, *blue*; oxygen, *red*

much longer distance that is required to transfer electrons to and from the buried active site of the enzyme compared to the small complex, the enzyme is the clear winner. A buried active site is not a problem if electron relay centres are present, which trap and store electrons (or 'holes') within the macromolecule in addition to providing a fast conduit.

Once charge transfer into the electrode has occurred, fast intramolecular ET to the active site should compete very favourably with destructive *back* ET into the semiconductor valence band (VB). Furthermore, as the active site of the enzyme is deeply buried within the protein, the catalytic redox chemistry of the enzyme is very likely to be independent of the electrode environment, in contrast to molecular catalysts. Unlike a small molecular catalyst surrounded by solvent, the active site of an enzyme is surrounded by specific groups that can be identified and located precisely at long range, thanks to the power of modern crystallography.

The high performance of enzymes comes at a cost: their activities are short lived—a consequence of the fact that they are so highly engineered. Biology has no difficulty in compensating for this problem because enzymes are constructed, in just a few rapid stages, by a polymerisation process that follows a blueprint. Even

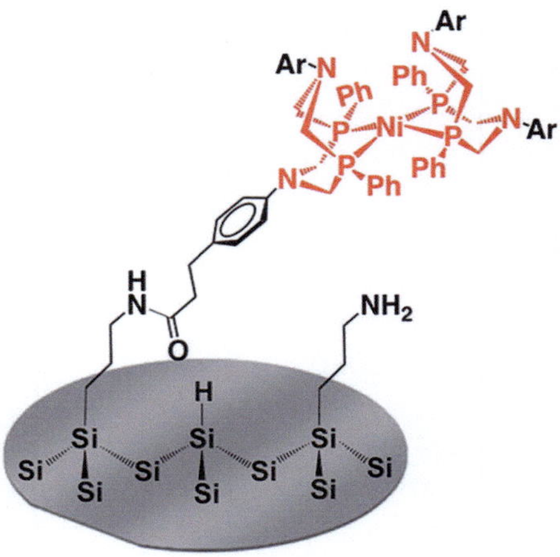

Fig. 4 DuBois-type [10] hydrogen evolution catalyst covalently anchored on a p-type silicon surface through UV-induced immobilisation chemistry of alkene moieties. Reprinted with permission from Moore GF, Sharp ID.; A Noble-Metal-Free Hydrogen Evolution Catalyst Grafted to Visible Light-Absorbing Semiconductors. *J. Phys. Chem. Lett. 4*, 568–572. Copyright (2013) American Chemical Society [12]

worn-out parts of enzymes can be replaced, such as the oxidised D1 subunit of photosystem II, responsible for housing the O_2-evolving catalyst [8, 9]. The clear lesson here is that artificial photosynthesis cannot ultimately use enzymes, which are best left to biology. A Formula 1 car has outstanding performance, but unacceptable service requirements and maintenance costs! Our interest in enzymes is (a) to establish what is possible using abundant elements, (b) to assess the importance of the catalyst environment (the outer shell that includes electron relays) and (c) to see how their fast and efficient catalytic rates allow electronic and charge transfer properties of the light-harvesting semiconductor to be revealed in steady-flow catalytic conditions.

3.2 Hydrogenases

Hydrogenases are exceptional catalysts of H_2 production and oxidation, and their active sites have been compared favourably with Pt. We mentioned above how the inner-shell coordination complex $2Fe_H$ is inactive until it has docked with the [4Fe–4S] cluster to give the intact H-cluster and is embedded in its outer-shell environment. Such H_2ases are known as [FeFe]-hydrogenases, reflecting the fact that the $2Fe_H$ unit is the actual site of H_2 interaction (see Fig. 2). The mechanism probably involves the interaction between a reactive H^- ligand formed at the distal Fe (Fe_d) and a H^+ bound to the bridgehead N atom of the azadithiolate. This type of mechanism is favoured for functional analogues synthesised and investigated by the DuBois group [10]. These molecular benchmark catalysts have recently been interfaced with dye-sensitised TiO_2 nanoparticles [11] and p-type GaP and Si photocathodes. Figure 4 shows such a catalyst, effectively a functional mimic of a hydrogenase, attached to the Si surface: such an arrangement is central to the theme of this chapter [12].

In the enzyme, the efficient two-electron interconversion of a metal-bound hydride to a proton is assisted by its ability to use the [4Fe–4S] cluster to house an electron transiently. The other main class, [NiFe]-hydrogenases, is generally less active, but many have the important property of being able to function indefinitely in the presence of O_2 [13]. The structure of the active site of a [NiFe]-hydrogenase is shown in Fig. 2. Although many more states of [NiFe]-hydrogenase have been characterised by spectroscopy, the mechanism is less clear than for the [FeFe]-enzymes.

3.3 Carbon Dioxide-Reducing Enzymes

Two-electron reduction of CO_2 in biology is catalysed by two very different types of enzymes, formate dehydrogenases and carbon monoxide dehydrogenases. Both are electroactive and set excellent benchmarks for the performance that must be possible with small molecular or surface catalysts. Carbon dioxide is a difficult molecule to activate: the HOMO is localised on the two O-atoms, and the linear molecule must bend and convert to a C-bonded carbonyl. The large reorganisational energy required to form the bent one-electron reduced radical anion CO_2^- is reflected in its thermodynamic potential of −1.9 V (vs. SHE, pH 7). Through proton-coupled electron transfer (PCET), enzymes are capable of avoiding this high-energy intermediate in favour of the two-electron reduced species CO or HCOOH. Although, ultimately, AP should produce liquid fuels such as methanol from carbon dioxide, CO and HCOOH are valuable feedstock in industrial processes, e.g. the Fischer–Tropsch process that converts carbon monoxide and hydrogen into liquid hydrocarbon fuel.

Formate dehydrogenase (FRD) was among the first enzymes to be studied in the context of artificial photosynthesis. A paper published in 1984 described how electrons from photoexcited *p*-InP photocathode are transferred via methyl viologen to the enzyme which carries out CO_2 reduction to formic acid [14]. Photogenerated electrons in the semiconductor can be produced with wavelengths of up to 900 nm. More recent work has shown that FRD is a reversible electrocatalyst when adsorbed directly on a graphite electrode [15]. However, no studies have so far been reported on FRD attached to a semiconductor.

Carbon monoxide dehydrogenases (CODHs) fall into two main types, aerobic and anaerobic, but it is the latter, containing a [Ni4Fe-4S] cluster, that have been established to be excellent electrocatalysts for CO_2 reduction to CO [16]. These enzymes operate virtually reversibly, contrasting markedly with the situation for small molecular catalysts. The 3D structure of the CODH active site is shown in Fig. 2. As with the hydrogenases, the surrounding outer-shell environment probably holds the secret to the very high activity. Evidence suggests that a bond forms between the C-atom and the Ni subsite which is very electrophilic in the reduced enzyme: an O-atom is then removed (as oxide/hydroxide) by the pendant Fe in a PCET process that is exquisitely organised by the side chains of appropriately positioned side chains.

Numerous synthetic homogeneous and heterogeneous catalysts have been developed and interfaced with light absorbers to achieve photo-driven reduction of CO_2 [17, 18]. Heterogeneous metal catalysts often suffer from high overpotential requirements and low faradaic efficiencies that result in product mixtures [19, 20]. Notable organometallic electrocatalysts include Ni(II)-cyclam systems originally created by Sauvage and co-workers [21] and Lehn's Re(bipy)$(CO)_3$X catalysts, with X being Cl^- or Br^- [22, 23]. Both classes of catalysts have been intensively studied and further developed by many research groups [24]. Rhenium and iridium, however, are among the rarest elements on Earth. A recent advance is the incorporation of an earth-abundant manganese ion as the metal centre in the bipyridyl-tricarbonyl ligand environment, which lowers the overpotential for CO_2 reduction by 0.4 V while maintaining selectivities and faradaic efficiencies that are comparable to the rhenium analogues [25]. Iron tetraphenylporphyrins developed by the Saveant group account for another class of molecular electrocatalysts for the conversion of CO_2 to CO based on earth-abundant metal ions [26]. Modification of the porphyrin framework with ortho-phenol moieties results in a high local concentration of protons built into the molecule, which significantly enhances catalysis by the transiently generated iron(0) complex. Reduction of CO_2 is catalysed with an average turnover frequency (TOF) of 3,500 s^{-1} over 4 h at an overpotential requirement of close to 0.5 V, which is, although being a significant advancement for synthetic catalysts, still far from reversible catalysis displayed by enzymes.

3.4 *The Oxygen-Evolving Enzyme*

Both H_2 production and CO_2 reduction involve electron-rich centres in which nucleophilic intermediates are produced. In contrast, for O_2 evolution, the catalyst must survive the production of very electrophilic and highly oxidising intermediates, which seems at first incompatible with organic chemistry. Biology has indeed adopted a cofactor that is 'mineral-like' but housed in a subunit that is regularly replaced to offset oxidative damage. The [4MnCa-5O] cluster of photosystem II, depicted in Fig. 2 [27], is currently the best catalyst for O_2 evolution, and apart from a few catalysts based on Ru or Ir complexes, the best contenders from the chemistry world are transition metal oxides. A bipyridine-isoquinoline ruthenium complex co-developed by Llobet and Sun exhibits a water oxidation TOF of 300 s^{-1} under acidic conditions, which is comparable to the rates of photosystem II in vivo (100–400 s^{-1}) [28]. Although the turnover number (TON) of 8,400 might indicate chemical stability, the compound degrades within a minute.

The catalytic activity and stability of several heterogeneous oxygen evolution catalysts have recently been 'benchmarked' by Jaramillo and co-workers, at a current density of 10 mA cm^{-2} that corresponds to a 10 % solar-to-fuel conversion efficiency (Fig. 5) [29]. In alkaline solution, every non-noble metal system achieved this current density at similar operating overpotentials between 0.35 and 0.43 V. However, only IrO_x is stable under oxidative conditions in acidic solutions.

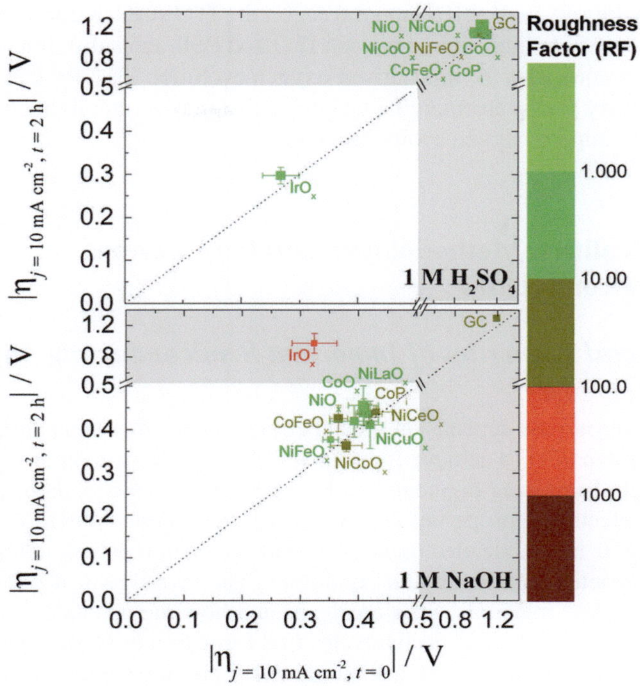

Fig. 5 Comparison of inorganic oxygen evolution catalysts. Comparison of catalytic activity, stability and electroactive surface area for heterogeneous water oxidation catalysts in acidic and basic media. The overpotential requirement for operation at 10 mA cm^{-2} of a given catalyst after $t=2$ h is plotted against the overpotential requirement at $t=0$. Hence, the *dashed line* indicates a stable catalyst. Only catalysts with activity and stability parameters falling into the non-shaded area of the plot have sufficient activity to be considered benchmarks. Reprinted with permission from McCrory CCL, Jung S, Peters JC, Jaramillo TF.; Benchmarking Heterogeneous Electrocatalysts for the Oxygen Evolution Reaction. *J Am Chem Soc. 135*, 16977–16987. Copyright (2013) American Chemical Society [29]

It is important to note that although benchmarking experiments are easiest to carry out in acidic or basic media, a future AP system will have to operate at neutral pH, ideally using waste or seawater. The energetic cost of separating protons from hydroxide anions would make any integrated device commercially unviable.

The good news is that in the context of the ability to operate efficiently at neutral pH, first-row transition elements feature well: even the isolated [4MnCa-5O] cluster is predicted to be a good O_2-evolving catalyst. To be efficient the catalyst must be able to store four oxidising equivalents at the right potential and/or stabilise intermediates just sufficiently so that they are not trapped.

Cyanobacterial photosystem II (PSII) has been shown to be electroactive on mesoporous indium tin oxide (*meso*-ITO) electrodes, and catalytic water oxidation has been observed under red light irradiation [30]. Orientation-controlled adsorption of the water oxidising enzyme on *meso*-ITO through electrostatic or covalent interaction with self-assembled monolayers on the ITO surface improved the interfacial electronic communication between ITO and PSII and substantially enhanced the turnover frequencies for light-driven oxygen evolution [31]. Protein film photoelectrochemistry [32] is an emerging technique that may further light on the reactivity of nature's superb oxygen evolution system.

4 Practicalities, Methodology: Linking Enzymes to Semiconductors

4.1 General Properties of Important Semiconducting Materials

Natural photosynthesis depends on special arrangements of organic pigments (chlorophylls, carotenoids) to achieve light harvesting and electron–hole separation. In AP, these light-harvesting pigments can be replaced by semiconductors. A guide to optical and electronic properties of some important semiconducting materials is shown in Fig. 6, alongside electrochemical data (reduction potentials) for fuel- and O_2-forming reactions under neutral conditions. The investigator needs to consider both n- and p-type materials, their bandgaps and how the conduction and valence band potentials are suited, respectively, for fuel formation or O_2 evolution. The values shown are for the bulk materials, but they still provide a good guide as to what is expected for the respective nanoparticles. The materials can be modified in such a way that light does not excite across the bandgap, but excites via an external, attached sensitiser. The surface chemistry of the material becomes important when considering strategies for attaching sensitisers and catalysts (enzymes).

Figure 6 shows the electronic band structures of selected semiconductors, with the blue and red lines for n-type and p-type, respectively. To be useful for AP, the

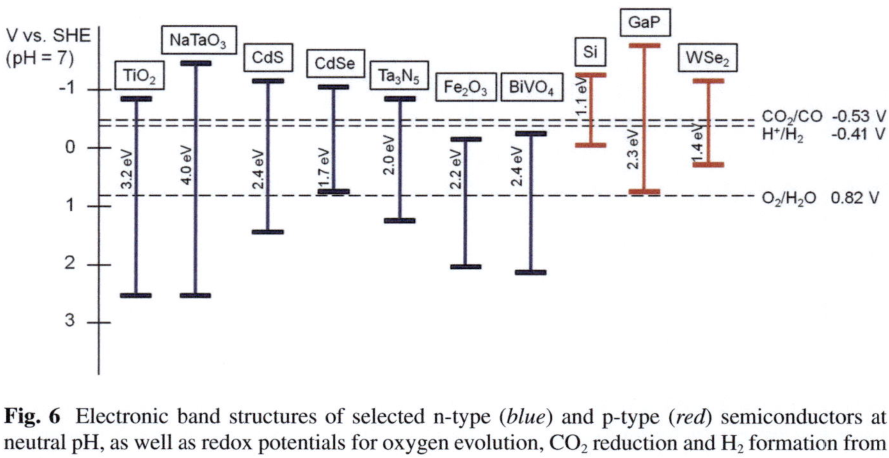

Fig. 6 Electronic band structures of selected n-type (*blue*) and p-type (*red*) semiconductors at neutral pH, as well as redox potentials for oxygen evolution, CO_2 reduction and H_2 formation from water

conduction band minimum must be more negative than the reduction potential of fuel-forming reactions (H_2 evolution or CO_2 reduction) and the valence band maximum more positive than the reduction potential of the oxidation reaction (ideally O_2 evolution from water).

Metal oxides such as TiO_2 satisfy these conditions, but at the cost of a wide bandgap, resulting in lower utilisation of the whole solar spectrum. Some non-oxide materials, e.g. chalcogenides and nitrides, can have a smaller bandgap due to the contribution of less electronegative S and N atoms to the valence band. The stability of the semiconductors in water must also be considered, as chalcogenides can undergo self-decomposition in water under irradiation.

Some materials are only suitable for half of the overall reaction, due to insufficient positioning of either the conduction band or valence band. For example, $BiVO_4$ can only conduct O_2 evolution, due to its conduction band minimum lying more positive than the H_2 evolution potential. To construct an overall water splitting system, such materials can be used with applied bias in a photoelectrochemical set-up or used in conjunction with another light-absorbing material that is able to evolve H_2.

4.2 *Photoelectrochemical vs. Photochemical Water Splitting*

Semiconducting light-harvesting materials for artificial photosynthesis can be employed in different design architectures. Generally, one has to distinguish between *photoelectrochemical* and *photochemical* water-splitting/solar-fuel production. The first proof-of-concept example of photo-induced water splitting was a photoelectrochemical cell (PEC) developed by Fujishima and Honda in 1972 [33]. In this system, the reactions are driven by UV light absorbed by a TiO_2 photoanode. The electrons excited into the conduction band are transferred through an external circuit to a 'dark' platinum cathode, where the H_2 evolution reaction takes place. Water is oxidised by high-energy holes in the TiO_2 valence band, the latter reaction being costly as it proceeds via high-energy intermediates such as hydroxyl radicals [34].

4.2.1 Photoelectrochemical Cells

Photoelectrochemical cells can be constructed from single p-type or n-type semiconductor materials to give single-bandgap devices (where the second half-reaction takes place at a dark electrode similar to the Fujishima–Honda system) or from two semiconductors connected in series, i.e. to form a dual-bandgap device [35]. In dual-bandgap photoelectrochemical cells, combinations of semiconductor materials with different bandgaps and complementary absorption characteristics are employed to use the solar spectrum to its maximum extent, particularly to harvest low-energy photons in the near-infrared region (even light with a wavelength of 1,000 nm has enough energy to split water). A dual-bandgap photoelectrochemical cell is depicted schematically in Fig. 7a.

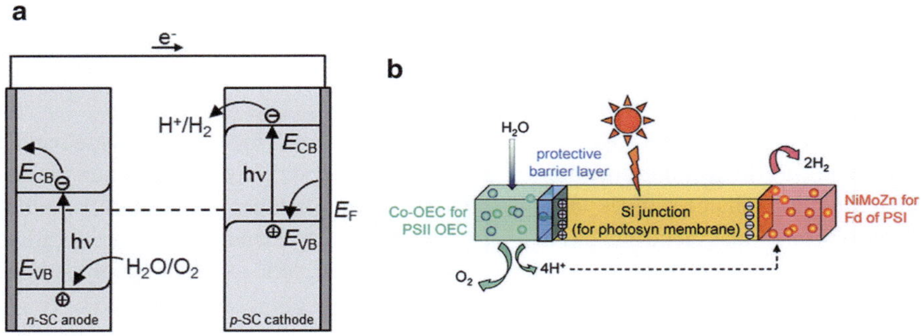

Fig. 7 (**a**) Schematic representation of a dual-bandgap PEC for water splitting with water oxidation taking place at the n-type photoanode and proton reduction at the p-type photocathode. (**b**) Wireless solar-fuel generator ('artificial leaf') developed by Nocera and co-workers. Reprinted with permission from Nocera DG. The Artificial Leaf. *Acc Chem Res. 45*, 767–776. Copyright (2012) American Chemical Society. (**b**) taken from [36]

The minority carriers generated in the photoelectrodes carry out the electrochemical reactions, with charge separation being assisted by the electric fields that arise at the semiconductor/electrolyte (or semiconductor/catalyst) interface. In the case of an n-type semiconducting photoanode, a depletion layer prevents oxidative currents in the dark. Upon illumination, the electron–hole pair is separated by the electric field of the space-charge region, resulting in a photo-oxidation current: electrons, the majority carriers, move through the conduction band into the bulk of the material (and to the counter electrode for fuel formation). In the valence band the minority carriers, holes, move to the semiconductor/electrolyte interface where they drive the water oxidation reaction. p-Type materials show analogous behaviour, but with holes as majority carriers; they are dark anodes and photocathodes.

One of the inherent advantages of a 'wired' photoelectrochemical cell is that this design utilises fast electronic conductivities: an electronic conductor completes the circuit between the anode and the cathode by a relatively long path, but with the ionic path being as short as possible [37]. Placing the two electrodes in close proximity limits 'bottleneck' effects of the intrinsically lower ionic conductivities. In contrast, 'wireless' PEC architectures provide a direct connection between anode and cathode with the ionic conductor taking a much longer path. 'Wireless' p/n-PEC device configurations, first developed by Nozik and referred to as 'photochemical diodes' [38], are intrinsically simpler in structure and are thus more suited for a low-cost, manufactured device structure [35]. The intimate integration of the light-harvesting, charge separation and chemical bond-forming functions into one single structure is advantageous compared to a multi-component process in which electricity is a tangible intermediary: integration avoids the cost and system constraints associated with electrical contacts, wires and inverters [39]. Consequently, several 'wireless' integrated architectures for AP have been designed and evaluated (both theoretically [40, 41] and experimentally) in recent years; an example of an artificial leaf is depicted in Fig. 7b [42].

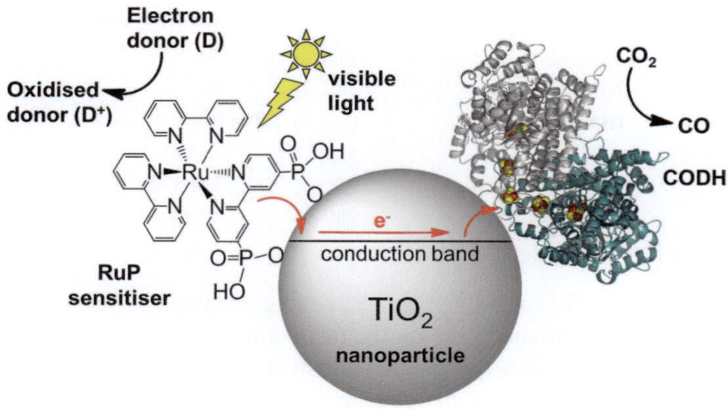

Fig. 8 *Cartoon* representation of a CO_2 photoreduction system with CODH adsorbed on dye-sensitised TiO_2 nanoparticles. Sacrificial electron donors such as triethanolamine or ascorbic acid regenerate the ground state of the oxidised sensitiser

4.2.2 Photochemical Systems

To study the mechanistic interplay between catalysts and light-absorbing components, it is advantageous to examine the water oxidation or fuel-forming 'half-systems' separately through *photochemical* experiments in (ideally homogeneous) solution: equilibria can be shifted in the desired direction and reaction rates can be controlled through the use of sacrificial reactants (in very large excess). Mechanisms can be elucidated through a variety of spectroscopic tools, most of which are easiest to perform in solution, such as steady-state UV/Vis or time-resolved optical spectroscopy.

Through rapid advances in nanosciences over the last 20 years, we have now the ability to manipulate matter on the nanometre scale at which elementary photosynthetic processes need to be controlled. It is extremely unlikely that light-driven overall water splitting, i.e. simultaneous water oxidation and fuel production, could occur efficiently at a single nanoparticle site. Reasons include the lack of spatial charge separation to suppress recombination and destructive back reactions, as well as the difficulty in dealing with two sets of highly reactive intermediates during a four-photon cycle. Nevertheless, colloidal semiconductor nanocrystals are excellent light-harvesting elements for AP model systems because they are strong light absorbers with tunable particle size, shape, electronic structure, absorption spectra and surface chemistry [43–50].

A variety of fuel-forming catalysts have been coupled with nanoparticulate, colloidal semiconductor materials, with benchmark systems for both CO_2 reduction and H_2 production relying on enzyme catalysts. An example is represented in Fig. 8. Most semiconductor/enzyme systems developed so far focus on dye-sensitised TiO_2 nanoparticles and CdS nanostructures of various geometries, mainly quantum dots and nanorods (see Table 1). Both materials offer several advantages for their use in semiconductor/enzyme hybrid model systems: their photochemistry and excited-state

Table 1 Overview of enzyme–semiconductor systems for photochemical solar-fuel generation

Semiconductor	Enzyme	Evolved gas	TOF (s^{-1})	QY (%)	Reference
RuP/TiO_2	NiFeSe	H_2	50	N/A	[68]
RuP/TiO_2	CODH	CO	0.14	0.07	[69]
CdS	CODH	CO	1.23	N/A	[70]
CdTe	HydA	H_2	25	9 at 532 nm	[71]
CdS	HydA	H_2	380–900	20 at 405 nm	[60]

kinetics are well established [51, 52] and the conduction band edges (see Fig. 6 for a guide) provide enough driving force for the fuel-forming reactions to occur.

Both TiO_2 and CdS are n-type semiconductors that normally form a depletion layer at open circuit. A positive charge is associated with the space-charge region, which is reflected in an upward bending of the band edges at the semiconductor–electrolyte interface [53]. Size becomes crucial: At the *macroscale*, excited electrons in the conduction band migrate from the surface into the interior and render the material unsuitable for light-harvesting applications for the fuel-generating half-reaction. However, at the *nanoscale*, formation of a space-charge layer is improbable due to the tiny dimensions. Semiconductor nanocrystals with very small particle diameters do not contain enough electrons to support an effective space-charge layer. In practice, this lack of space-charge effects applies to both nanoparticle suspensions and highly porous electrodes [54]. Hence, on the nanometre scale, both TiO_2 and CdS are useful light absorbers for light-driven fuel formation.

Commercially available TiO_2 nanoparticles (such as 'P25', a mixture of rutile and anatase) are widely used in self-cleaning wall paint (exploiting hydroxyl radicals generated by UV radiation). The disadvantage for AP is their transparency towards visible light, with bandgaps of 3.03 eV for rutile and 3.20 eV for anatase, respectively [55]. To overcome this issue, TiO_2 is sensitised with several molecular dyes, most commonly tris(bipyridine)ruthenium(II) derivatives [56] and high-potential porphyrins [57]. Both sensitisers have large extinction coefficients in the visible region and long-lived charge-separated (triplet) states, leading to efficient charge injection from the sensitiser to the semiconductor [51, 58]. Due to the position of its conduction band potential (Fig. 5), TiO_2 provides only a marginal driving force for fuel formation reactions, especially for CO_2 reduction. This is considered to be another disadvantage, but poses little problem for a reversible electrocatalyst. In this respect, H_2ases and CODH have the ability to catalyse H_2 formation and CO_2 reduction at minimal driving force, which renders them particularly good candidates for coupling to TiO_2 nanostructures. Owing to a combination of advantages that include low cost, TiO_2 has been widely used in research on dye-sensitised solar cells (DSSCs) and is one of the best characterised semiconductor materials.

In contrast to TiO_2, CdS which has a bandgap of 2.3 eV absorbs visible light below a wavelength of about 540 nm. The conduction band edge of bulk CdS lies at a significantly more negative potential than for TiO_2: with $E_{CB} = -0.87$ V vs. SHE (pH 6) there is a much greater driving force available for fuel-forming reactions (e.g. consider CO_2 to CO, which has a reduction potential of –0.46 V vs. SHE, at pH 6).

4.3 Linking Enzymes to Semiconductors

We consider next the problem of obtaining stable and efficient linkages between catalyst and semiconductor surface. The linkage can be a deciding factor in terms of charge transport yields and chemical robustness under reaction conditions. A large fraction of electrons or holes is typically diverted by destructive side or back reactions or trapped at defect sites resulting in low photochemical quantum yields (QY). Ideally, enzymes should be attached such that the electron entry/exit point, usually lying close to the protein surface, comes into close contact with the semiconductor surface, thus minimising the interfacial electron tunnelling barrier. This problem is far greater for enzymes than for molecular catalysts, due to their relatively large dimensions. A covalent linker improves stability but need not improve interfacial ET rates. Efficient charge transport is crucial at every stage of the catalytic cycle, but for efficient catalysts that operate close to reversibility, charge injection from the excited semiconductor to the catalyst is most critical. This step is always in direct competition with several destructive decay processes in the semiconductor such as photoemission, non-radiative decay, intersystem crossing, etc.

Several strategies for linking enzymes to semiconductor materials are in use. Electrostatic interactions can be exploited, using variations in pH and ionic composition and exploiting favourable charge distributions on the enzyme surface. The surface of TiO_2 is covered with hydroxyl groups that are in acid–base equilibrium with H_2O, and the enzyme's electrostatic surface can be computed from a crystal structure. Electrostatic interactions alone usually result in a wide distribution of enzyme orientations on the semiconductor and may limit experiments to certain narrow pH regimes in which semiconductor and the optimal contact region on the enzyme have opposite charges.

Semiconductor surfaces can be modified with small linker molecules to give more specific enzyme attachment. For example, the [FeFe]-hydrogenase (HydA) from *Clostridium acetobutylicum* has a positively charged region of surface close to the [4Fe–4S] cluster lying at end of the relay system (and hence termed 'distal') [59]. This is the site at which electrons enter or leave the enzyme and, in vivo, engages in a dynamic ET complex with its natural redox partner ferredoxin. Exploiting this knowledge, HydA is attached to CdS nanorods capped with a monolayer of negatively charged 3-mercaptopropionic acid (MPA) [43, 60].

Enzymes can be attached to carbon nanotubes resulting in excellent ET kinetics, usually measured by electrochemical methods. By attaching HydA directly to single-walled carbon nanotubes [61], King and co-workers obtained an order-of-magnitude increase in catalytic rates for both H^+ reduction and H_2 oxidation [62]. Another example has been developed in the course of electrochemical studies where the aim has been to maximise coverage and stability for a membraneless H_2/O_2 fuel cell [63]. First, a graphite electrode is pre-coated with multiwalled carbon nanotubes, producing a dense permeable conducting layer. Second, a [NiFe]-hydrogenase (Hyd-1) from *Escherichia coli* is covalently linked to 1-pyrenebutyric acid units via peptide coupling. Finally the modified H_2ase is introduced to the electrode, whereupon it becomes strongly bonded to the carbon nanotube scaffold through strong π–π interactions. This modification leads to an order-of-magnitude enhancement of an enzymatic hydrogen–air fuel cell compared to analogous cells in which the enzymes are conventionally adsorbed on graphite. The powerful π–π-stacking capability of pyrene moieties has also been exploited to attach small *molecular* fuel-forming catalysts to electrodes. Gray and co-workers attached small organometallic H_2-evolving and CO_2 reduction catalysts to a graphite electrode via pyrene-appended bipyridine ligands; even so, the resulting electrocatalytic rates were much lower than observed with enzymes as catalysts [64].

Another strategy to link [FeFe]-H_2ases covalently to graphite electrodes is via amide coupling to a graphite electrode that has been functionalised with a layer of either 4-aminophenyl or 4-carboxyphenyl moieties, depending on the surface charge of the respective enzyme [65]. In this approach, covalent bond formation is achieved through peptide coupling with *N*-hydroxysuccinimide.

Based upon the Moser–Dutton model of electron transfer in proteins [66, 67], efficient ET can occur between redox centres in proteins over a distance up to 14 Å. This estimate is based upon an exponential decay constant of 1.1 $Å^{-1}$ which is a reasonable estimate for the interior of a protein. The FeS relay centres in H_2ases and CODH are appropriately spaced, so intramolecular ET *through the enzyme* to and from the active site should be faster than the inherent rate of turnover at the active site and therefore not be rate limiting. Intramolecular ET rate constants of the order of 10^9 s^{-1} for [FeFe]-hydrogenases and 10^8 s^{-1} for CODH have been calculated according to the above-mentioned model. Taking the view that once electrons are in the enzyme, the reaction proceeds rapidly and efficiently, the restricting factor is the rate of interfacial ET, which can be particularly problematic if the enzyme is badly oriented or the linker molecule itself acts as a tunnel barrier. In the example of HydA adsorbed on MPA capped CdS nanoparticles, the closest point of contact between the enzyme and the surfactant molecules on the surface can vary up to 8 Å, with an additional tunnelling distance of ca. 5 Å through the MPA molecule. The realistic interfacial tunnelling distance is therefore at least 13 Å [43].

Ultimately, the issue of rate-limiting tunnel barriers for electron injection from the semiconductor into the enzyme could be overcome by introducing unnatural amino acids with anchoring points for covalent linkage close to the distal FeS cluster. This would lead to a directed, oriented attachment and minimise both distribution of enzyme orientations on the electrode/particle surface and the distance for intermolecular charge transfer.

5 Progress on Steady-State Rates and Stability

Gas evolution rates for enzyme–semiconductor systems are usually measured chromatographically. Typically, enzyme-modified semiconductor particles are suspended in solution with a sacrificial reagent in order to enhance the desired half-reaction; then the gas evolved under irradiation is measured by taking samples and analysis by gas chromatography (GC). To compare gas evolution across various systems, a common measure of rates must be used. One measure is the TOF, which is the number of evolved gas molecules per enzyme molecule per second. Another is the quantum yield (QY), which is the proportion of the total number of incoming photons that is converted into the product. A selection of recent measurements of various enzyme-modified semiconductors is listed in Table 1.

Direct comparison between enzymatic and non-enzymatic systems is deceptive. Semiconductors modified with noble metals such as Pt and Rh usually show a much higher H_2 evolution rate per mass of the semiconductor; however, this is due to the larger amount of loaded metal catalyst, typically upwards of 1 % by mass. In contrast, an enzyme is an extremely large molecule (often > 100 kDa), which greatly restricts the amount of active sites on a given surface.

The TOFs of enzymes on semiconductors are much lower than values obtained from chemical assays and electrochemical methods, and this general observation raises important questions. The reaction rate may be influenced by the band structure close to the semiconductor surface and by the transient nature of the availability of photogenerated electrons (including competition with recombination) which can be demonstrated by measuring the light intensity dependence.

6 Advanced Mechanistic Studies of Enzyme-Catalysed AP

6.1 Rate and Efficiency Parameters

Several spectroscopic and mechanistic tools have been employed to study the intricate mechanisms and the interactions of light-absorbing semiconductors with enzymatic fuel-forming catalysts. First, the efficiencies of the fuel-forming reactions have to be determined, usually via monitoring of product formation by GC as described above. The solar-to-hydrogen (STH) or CO_2 activation efficiencies can be calculated, and from the amount of enzyme used, the TON and TOF values on a per enzyme basis can be determined. In photoelectrochemical cells, solar energy conversion efficiencies can also be determined from catalytic current densities when the faradaic efficiencies for product formation are known. Enzymes offer much more selectivity than heterogeneous surface electrocatalysts, which often give complex mixtures of products, particularly in the case of CO_2 reduction. Therefore, in enzymes, faradaic efficiencies are close to 100 %. Enzymes are also sensitive to selective inhibitors, which can help to confirm whether or not a product or current arises from catalytic conversion. However, direct product detection is always highly desirable, where possible.

It is important to note that only STH efficiencies (or solar-to-CO_2 reduction product efficiencies, respectively) measured without the use of sacrificial reactants or applied bias can serve as benchmark values to compare against other materials/systems. 'Half-cell' efficiencies can inflate the apparent performance of a system and cannot be converted into a meaningful device performance metric [72].

Although not suitable for standardisation, the other efficiency parameters described below can still give important information about mechanistic details and are valuable for understanding and improving material and system performance.

The external quantum efficiency (EQE), equivalent to the incident photon-to-current efficiency (IPCE), is an important characteristic of a solar-fuel system. It describes the fraction of catalytic current generated per incoming photon flux ('electrons out per photons in'), taking into account the spectral variation of incident photons at each energy [72]. The IPCEs are usually measured by chronoamperometry experiments under varying monochromatic light exposure. The internal quantum efficiency (IQE) or absorbed photon-to-current efficiency (APCE) characterises the inherent performance of a material/solar-to-fuel system, since losses from reflected or transmitted photons are subtracted. The APCE can be determined by dividing the IPCE by the absorptance, $\eta_{e-/h+}$, which is defined as the fraction of electron–hole pairs generated per incident photon flux and can be estimated from Beer's law [72].

These measures are important characteristics of solar-fuel systems, but do not give detailed information about the mechanisms of light-driven fuel formation. To characterise the catalytic performances in enzyme/semiconductor systems, TOF, TON and rates of electron injection and charge transport under catalytic conditions (i.e. light exposure and presence of substrate) are the most valuable parameters.

6.2 *Case Studies*

6.2.1 Continuous Catalytic Rates

On tris(bipyridine)ruthenium(II) (RuP)-sensitised TiO_2, a benchmark turnover frequency for visible light-driven proton reduction of 50 s^{-1} (on a per catalyst basis) has been measured, using a [NiFeSe]-hydrogenase enzyme as catalyst [68]. In a study of [FeFe]-hydrogenase HydA attached to CdS, King and co-workers measured a per enzyme TOF of 380 s^{-1} [60]. These values are much higher than what has so far been achieved for molecular catalysts on surfaces. However, as mentioned above, the TOFs measured under photocatalytic conditions are significantly lower than those measured in solution assays in the dark or in protein film electrochemistry experiments. As another example, for CO_2 reduction to CO, the highest photocatalytic TOF obtained so far (on a per catalytic unit basis) stems from using CODH adsorbed on CdS nanorods [70]. The value of only 1.23 s^{-1} (three orders of magnitude slower than the theoretical maximum) emphasises the need for improvement in photo-driven CO_2 reduction and to understand why there is such a difference compared to electrocatalytic rates.

Dukovic, King and co-workers carried out an in-depth analysis of the photochemical processes that occur in the CdS:HydA system. Scheme for photocatalytic H_2 production in these assemblies is shown in Fig. 9. Light absorption by the semiconducting nanorods is followed by quenching of the VB hole by the electron donor,

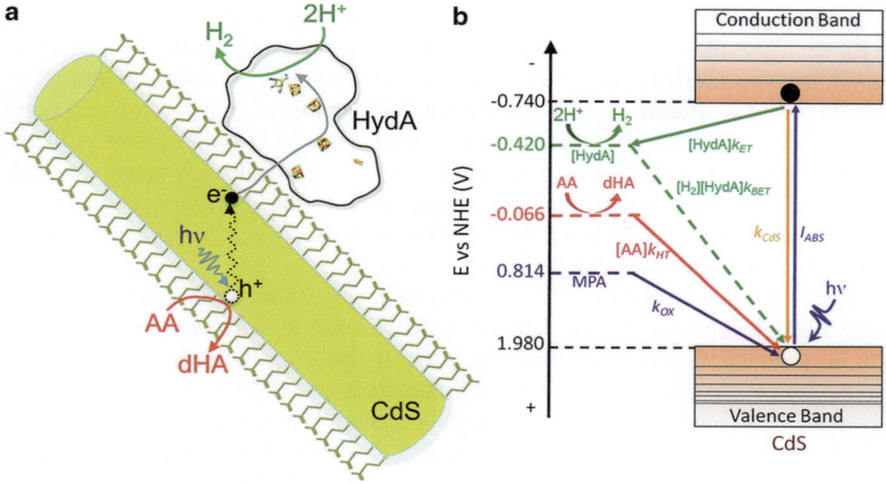

Fig. 9 Proposed scheme for photocatalytic H_2 production by CdS:HydA complexes. Reprinted with permission from Brown KA, Wilker MB, Boehm M, Dukovic G, King PW. Characterization of Photochemical Processes for H_2 Production by CdS Nanorod–[FeFe] Hydrogenase Complexes. *J. Am. Chem. Soc. 134*, 5627–5636. Copyright (2012) American Chemical Society [60]

ascorbic acid (AA). Each excited conduction band electron is injected into the distal [4Fe–4S] cluster of HydA. In mixtures of 1:0.67 CdS:HydA, a quantum yield of 20 % for H_2 production (based on two photons yielding 1 eq. H_2) has been measured using 405 nm laser light as the excitation source. The H_2 generation rate scales linearly with light intensity, indicating that the system is limited by the availability of photogenerated electrons rather than catalysis.

6.2.2 Transient Processes

Hydrogenase at CdS. In a recent, more advanced mechanistic study, Dukovic and colleagues have examined the kinetics of electron injection from CdS into HydA via time-resolved optical measurements, namely, transient absorption spectroscopy (TAS) [43]. In this technique, samples of interest are excited by a short-lived laser pulse. The processes that follow excitation, such as the formation of transients, are monitored by a second light pulse, either from a white-light-emitting flash lamp or from a laser in the case of ultrafast experiments.

The decay of the absorbance of the excited semiconductor CB electron should coincide exactly with the arrival of the electron in the enzyme. However, because HydA does not have absorptive features with intensities comparable to the strength of the CdS bandgap absorption [73], no transient features corresponding to excited electrons in the enzyme are found in transient absorption spectra recorded on CdS:HydA [43]. The broad absorption bands of FeS clusters in the relay do not give rise to sharp, distinct changes upon excitation.

For the CdS:HydA system, ET from CdS occurs with a rate constant k_{ET} of ~10^7 s^{-1}. This value is similar to the excited-state decay rate constant k_{CdS}, resulting in a quantum efficiency for ET of 42 % for 1:1 CdS:HydA complexes. Hence, forward ET and destructive recombination are in direct kinetic competition and occur with similar probabilities. Measurements with inactive enzyme, either with the H-cluster inhibited by CO or with the apoprotein that lacks the active site (see Fig. 3b), give electron transfer rates similar to injection into the active enzyme. These results demonstrate that ET to the FeS cluster relay is independent of catalysis at the active site.

Quantum *efficiencies* for ET (Q_{ET}) can be raised by increasing the number of enzyme molecules per CdS particle, which however results in lower quantum *yields* for H_2 production. Since the intramolecular ET rate is of the order of 10^9 s^{-1} (see above), increasing k_{ET} by two orders of magnitude would lead to Q_{ET} being 99 % (provided k_{CdS} remained unchanged) and the system becoming limited by catalytic turnover at the H-cluster. Electron transfer in the picosecond regime, as observed for small molecules on semiconductors, would therefore not be necessary for maximising catalytic efficiency. The ET rate from CdS to H_2ase is relatively slow compared with ET to Pt ($k_{ET} > 10^{11}$ s^{-1}) meaning the fraction of photoexcited electrons available for reduction is higher in CdS:Pt than in CdS:HydA structures. Conversely, QY values for H_2 formation are much higher in the enzyme system (20 %) than with Pt as the catalyst (1–4 %). This highlights the proficiency of the enzyme. Photo-induced H_2 generation is a two-electron process driven by two 1-photon events. Due to its FeS relay, the H_2ase can very efficiently stabilise the first reductive equivalent and separate it from the interface over a relatively large distance of 29 Å, away from the distal cluster and in towards the active site. Back electron transfer is considered to be very slow because of this large spatial separation distance. In CdS:Pt assemblies, in contrast, the catalyst is located much closer to the nanocrystal surface and therefore potentially more likely to recombine with a hole in the valence band. These results alone serve as justification for the use of enzymes in hybrid model systems for AP, as they outperform the best available synthetic catalysts.

Electron transfer at TiO2. In the case of TiO_2-based systems, charge transfer events are less well understood. Electron injection from the excited state of the ruthenium sensitiser into the TiO_2 conduction band is well established (due to the intense research efforts into DSSCs over the last decades). This first ET event is very efficient and occurs on the picosecond timescale (see above). The lifetime of the charge-separated state is also considerably long, with a value of ca. 1 ms [74]. Electron injection into an enzyme however has not been measured to date. It is known that ET from TiO_2 to a synthetic cobaloxime H^+ reduction catalyst occurs with a rate constant of 10^5 s^{-1} [74]. This gives a ratio of forward ET/back ET of 100, which is more favourable than in the CdS:HydA system. As outlined above, electron injection into an organometallic catalyst is considered to be much faster than ET to an enzyme due to the smaller tunnelling barrier. Nevertheless, even slower forward ET rates of 10^4–10^5 s^{-1} would compare favourably with back ET, mainly due to the relatively long lifetime of the charge-separated state in dye-sensitised TiO_2. The importance of catalysis in light-driven fuel formation is highlighted by

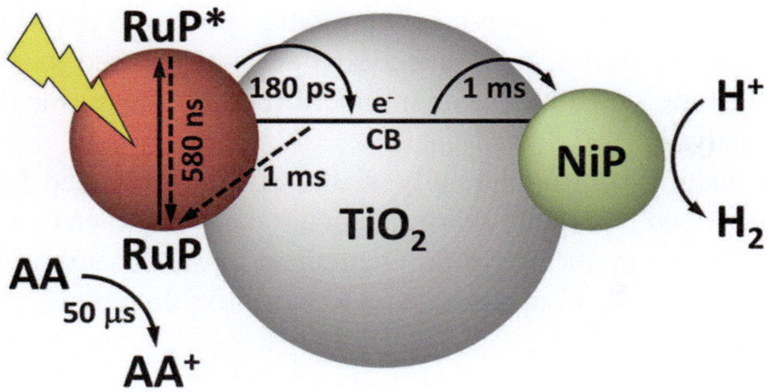

Fig. 10 Electron transfer dynamics in a RuP–TiO_2–NiP assembly for H_2 evolution [11]

the fact that the TOF for H_2 production of RuP–TiO_2–cobaloxime [74] is much lower than when (under otherwise similar conditions) a DuBois-type catalyst ('NiP', Fig. 10) is employed instead [11], even though electron injection into the former is much faster.

To date, transient absorption measurements have shed some light on the mechanism of light-driven fuel generation at hybrid enzyme–semiconductor systems. Electron injection into the enzyme seems to be the rate-limiting step. However, it is essentially irreversible because both H_2ase and CODH enzymes are very efficient in charge separation and transport, shuttling the electrons away from the interface into the enzyme.

Other spectroscopic techniques have yet to be developed for use with enzymes. Time-resolved infrared (IR) measurements could potentially be employed to monitor catalysis under steady-state conditions. The CO and CN^- ligands of the H-cluster, for instance, show very characteristic IR bands. A general advantage of IR spectroscopy over optical techniques is its structure specificity, i.e. transients can be correlated with distinct chemical groups. Another technique of potential use is time-resolved X-ray absorption spectroscopy (XAS). Transient XAS could give precise information about bond-length and oxidation state changes in the catalytic cycle while leading to less radiation damage than crystallography.

7 Where Do We Go from Here?

Although enzymes are wholly unsuited for long-term, large-scale systems, they provide valuable insight for developing integrated artificial systems where getting as close as possible to reversible catalysis is important for efficiency. Employing enzymes in integrated devices can relieve the related burdens imposed by high overpotential requirement and low TOF and allow mechanistic studies to be carried out under operational conditions. The fact that H_2ases show H_2 evolution activities comparable to Pt, while no reversible and selective CO_2 reduction catalyst other than CODH and FRD is known, is proof that AP has a future, based on highly abundant, first-row transition elements. Enzymes are relatively easy to deposit on materials (albeit not necessarily with ideal connections) and, as opposed to molecular catalysts, their active site structures remain unperturbed by the surface and therefore intact during catalysis.

Ultimately, integrated, commercially viable AP devices need to be based on robust materials and scalable, earth-abundant and inexpensive components. Therefore, one of the major scientific challenges in AP research is 'how do we modify common materials to upgrade them to work like enzymes?'

References

1. Woolerton TW, Sheard S, Chaudhary YS, Armstrong FA. Enzymes and bio-inspired electrocatalysts in solar fuel devices. Energy & Environmental Science. 2012;5:7470-90.
2. Armstrong FA, Hirst J. Reversibility and efficiency in electrocatalytic energy conversion and lessons from enzymes. Proc Natl Acad Sci U S A. 2011;108:14049-54.
3. Hexter SV, Esterle TF, Armstrong FA. A unified model for surface electrocatalysis based on observations with enzymes. PCCP. 2014.
4. Bachmeier A, Wang VCC, Woolerton TW, Bell S, Fontecilla-Camps JC, Can M, et al. How Light-Harvesting Semiconductors Can Alter the Bias of Reversible Electrocatalysts in Favor of H2 Production and CO2 Reduction. J Am Chem Soc. 2013;135(40):15026-32.
5. Berggren G, Adamska A, Lambertz C, Simmons TR, Esselborn J, Atta M, et al. Biomimetic assembly and activation of [FeFe]-hydrogenases. Nature. 2013;499(7456):66-9.
6. Esselborn J, Lambertz C, Adamska-Venkatesh A, Simmons T, Berggren G, Noth J, et al. Spontaneous activation of [FeFe]-hydrogenases by an inorganic [2Fe] active site mimic. Nat Chem Biol. 2013;9(10):607-9.
7. Camara JM, Rauchfuss TB. Combining acid-base, redox and substrate binding functionalities to give a complete model for the [FeFe]-hydrogenase. Nat Chem. 2012;4:26-30.
8. Aro E-M, Virgin I, Andersson B. Photoinhibition of Photosystem II. Inactivation, protein damage and turnover. Biochimica et Biophysica Acta (BBA) - Bioenergetics. 1993;1143(2):113-34.
9. van Wijk KJ, Nilsson LO, Styring S. Synthesis of reaction center proteins and reactivation of redox components during repair of photosystem II after light-induced inactivation. J Biol Chem. 1994;269(45):28382-92.
10. Rakowski Dubois M, DuBois DL. Development of molecular electrocatalysts for CO_2 reduction and H_2 production/oxidation. Acc Chem Res. 2009;42:1974-82.
11. Gross MA, Reynal A, Durrant JR, Reisner E. Versatile Photocatalytic Systems for H_2 Generation in Water Based on an Efficient DuBois-Type Nickel Catalyst. J Am Chem Soc. 2013;136(1):356-66.
12. Moore GF, Sharp ID. A Noble-Metal-Free Hydrogen Evolution Catalyst Grafted to Visible Light-Absorbing Semiconductors. The Journal of Physical Chemistry Letters. 2013;4(4):568-72.
13. Evans RM, Parkin A, Roessler MM, Murphy BJ, Adamson H, Lukey MJ, et al. Principles of Sustained Enzymatic Hydrogen Oxidation in the Presence of Oxygen – The Crucial Influence of High Potential Fe–S Clusters in the Electron Relay of [NiFe]-Hydrogenases. J Am Chem Soc. 2013;135(7):2694-707.
14. Parkinson BA, Weaver PF. Photoelectrochemical pumping of enzymatic CO_2 reduction. Nature. 1984;309(5964):148-9.
15. Reda T, Plugge CM, Abram NJ, Hirst J. Reversible interconversion of carbon dioxide and formate by an electroactive enzyme. Proc Natl Acad Sci U S A. 2008;105(31):10654-8.

16. Can M, Armstrong FA, Ragsdale SW. Structure, Function, and Mechanism of the Nickel Metalloenzymes, CO Dehydrogenase, and Acetyl-CoA Synthase. Chem Rev. 2014.
17. Kumar B, Llorente M, Froehlich J, Dang T, Sathrum A, Kubiak CP. Photochemical and Photoelectrochemical Reduction of CO_2. Annu Rev Phys Chem. 2012;63(1):541-69.
18. Appel AM, Bercaw JE, Bocarsly AB, Dobbek H, DuBois DL, Dupuis M, et al. Frontiers, Opportunities, and Challenges in Biochemical and Chemical Catalysis of CO_2 Fixation. Chem Rev. 2013;113(8):6621-58.
19. Wenzhen L. Electrocatalytic Reduction of CO_2 to Small Organic Molecule Fuels on Metal Catalysts. Advances in CO_2 Conversion and Utilization. ACS Symposium Series. 1056: American Chemical Society; 2010. p. 55-76.
20. Jhong H-RM, Ma S, Kenis PJA. Electrochemical conversion of CO_2 to useful chemicals: current status, remaining challenges, and future opportunities. Current Opinion in Chemical Engineering. 2013;2(2):191-9.
21. Beley M, Collin J-P, Ruppert R, Sauvage J-P. Nickel(II)-cyclam: an extremely selective electrocatalyst for reduction of CO_2 in water. J Chem Soc, Chem Commun. 1984(19):1315-6.
22. Hawecker J, Lehn J-M, Ziessel R. Efficient photochemical reduction of CO_2 to CO by visible light irradiation of systems containing $Re(bipy)(CO)_3X$ or $Ru(bipy)_3^{2+}$-Co^{2+} combinations as homogeneous catalysts. J Chem Soc, Chem Commun. 1983(9):536-8.
23. Hawecker J, Lehn J-M, Ziessel R. Photochemical and Electrochemical Reduction of Carbon Dioxide to Carbon Monoxide Mediated by (2,2′-Bipyridine)tricarbonylchlororhenium(I) and Related Complexes as Homogeneous Catalysts. Helv Chim Acta. 1986;69(8):1990-2012.
24. Benson EE, Kubiak CP, Sathrum AJ, Smieja JM. Electrocatalytic and homogeneous approaches to conversion of CO_2 to liquid fuels. Chem Soc Rev. 2009;38:89-99.
25. Bourrez M, Molton F, Chardon-Noblat S, Deronzier A. [Mn(bipyridyl)$(CO)_3$Br]: An Abundant Metal Carbonyl Complex as Efficient Electrocatalyst for CO_2 Reduction. Angew Chem Int Ed. 2011;50(42):9903-6.
26. Costentin C, Drouet S, Robert M, Savéant J-M. A Local Proton Source Enhances CO_2 Electroreduction to CO by a Molecular Fe Catalyst. Science. 2012;338(6103):90-4.
27. Umena Y, Kawakami K, Shen J-R, Kamiya N. Crystal structure of oxygen-evolving photosystem II at a resolution of 1.9 Å. Nature. 2011;473(7345):55-60.
28. Duan L, Bozoglian F, Mandal S, Stewart B, Privalov T, Llobet A, et al. A molecular ruthenium catalyst with water-oxidation activity comparable to that of photosystem II. Nat Chem. 2012;4(5):418-23.
29. McCrory CCL, Jung S, Peters JC, Jaramillo TF. Benchmarking Heterogeneous Electrocatalysts for the Oxygen Evolution Reaction. J Am Chem Soc. 2013;135(45):16977-87.
30. Kato M, Cardona T, Rutherford AW, Reisner E. Photoelectrochemical Water Oxidation with Photosystem II Integrated in a Mesoporous Indium–Tin Oxide Electrode. J Am Chem Soc. 2012;134(20):8332-5.
31. Kato M, Cardona T, Rutherford AW, Reisner E. Covalent Immobilization of Oriented Photosystem II on a Nanostructured Electrode for Solar Water Oxidation. J Am Chem Soc. 2013;135(29):10610-3.
32. Kato M, Zhang JZ, Paul N, Reisner E. Protein film photoelectrochemistry of the water oxidation enzyme photosystem II. Chem Soc Rev. 2014.
33. Fujishima A, Honda K. Electrochemical photolysis of water at a semiconductor electrode. Nature. 1972;238(5358):37-8.
34. Linsebigler AL, Lu G, Yates JT. Photocatalysis on TiO_2 Surfaces: Principles, Mechanisms, and Selected Results. Chem Rev. 1995;95(3):735-58.
35. Walter MG, Warren EL, McKone JR, Boettcher SW, Mi Q, Santori EA, et al. Solar Water Splitting Cells. Chem Rev. 2010;110:6446-73.
36. Nocera DG. The Artificial Leaf. Acc Chem Res. 2012;45(5):767-76.
37. Newman J. Scaling with Ohm's Law; Wired vs. Wireless Photoelectrochemical Cells. J Electrochem Soc. 2013;160(3):F309-F11.
38. Nozik AJ. Photochemical diodes. Appl Phys Lett. 1977;30(11):567-9.

39. Lewis NS, Nocera DG. Powering the Planet. Proc Natl Acad Sci U S A. 2006;103(43): 15729–35.
40. Haussener S, Xiang C, Spurgeon JM, Ardo S, Lewis NS, Weber AZ. Modeling, simulation, and design criteria for photoelectrochemical water-splitting systems. Energy & Environmental Science. 2012;5(12):9922-35.
41. Haussener S, Hu S, Xiang C, Weber AZ, Lewis NS. Simulations of the irradiation and temperature dependence of the efficiency of tandem photoelectrochemical water-splitting systems. Energy & Environmental Science. 2013;6(12):3605-18.
42. Kanan MW, Nocera DG. In Situ Formation of an Oxygen-Evolving Catalyst in Neutral Water Containing Phosphate and Co2+. Science. 2008;321(5892):1072-5.
43. Wilker MB, Shinopoulos KE, Brown KA, Mulder DW, King PW, Dukovic G. Electron Transfer Kinetics in CdS Nanorod–[FeFe]-Hydrogenase Complexes and Implications for Photochemical H_2 Generation. J Am Chem Soc. 2014;136(11):4316-24.
44. Yin Y, Alivisatos AP. Colloidal nanocrystal synthesis and the organic-inorganic interface. Nature. 2005;437(7059):664-70.
45. Burda C, Chen X, Narayanan R, El-Sayed MA. Chemistry and Properties of Nanocrystals of Different Shapes. Chem Rev. 2005;105(4):1025-102.
46. Bawendi MG, Steigerwald ML, Brus LE. The Quantum Mechanics of Larger Semiconductor Clusters ("Quantum Dots"). Annu Rev Phys Chem. 1990;41(1):477-96.
47. Cozzoli PD, Pellegrino T, Manna L. Synthesis, properties and perspectives of hybrid nanocrystal structures. Chem Soc Rev. 2006;35(11):1195-208.
48. Brus L. Electronic wave functions in semiconductor clusters: experiment and theory. The Journal of Physical Chemistry. 1986;90(12):2555-60.
49. Talapin DV, Lee J-S, Kovalenko MV, Shevchenko EV. Prospects of Colloidal Nanocrystals for Electronic and Optoelectronic Applications. Chem Rev. 2009;110(1):389-458.
50. Wilker MB, Schnitzenbaumer KJ, Dukovic G. Recent Progress in Photocatalysis Mediated by Colloidal II-VI Nanocrystals. Isr J Chem. 2012;52(11–12):1002-15.
51. Tachibana Y, Moser JE, Grätzel M, Klug DR, Durrant JR. Subpicosecond Interfacial Charge Separation in Dye-Sensitized Nanocrystalline Titanium Dioxide Films. The Journal of Physical Chemistry. 1996;100(51):20056-62.
52. Tseng H-W, Wilker MB, Damrauer NH, Dukovic G. Charge Transfer Dynamics between Photoexcited CdS Nanorods and Mononuclear Ru Water-Oxidation Catalysts. J Am Chem Soc. 2013;135(9):3383-6.
53. Bott A. Electrochemistry of semiconductors. Curr Sep. 1998;3:87-91.
54. Beranek R. (Photo)electrochemical Methods for the Determination of the Band Edge Positions of TiO_2-Based Nanomaterials. Advances in Physical Chemistry. 2011:786759.
55. Scanlon DO, Dunnill CW, Buckeridge J, Shevlin SA, Logsdail AJ, Woodley SM, et al. Band alignment of rutile and anatase TiO_2. Nat Mater. 2013;12(9):798-801.
56. O'Regan B, Gratzel M. A low-cost, high-efficiency solar cell based on dye-sensitized colloidal TiO_2 films. Nature. 1991;353(6346):737-40.
57. Moore GF, Konezny SJ, Song H-e, Milot RL, Blakemore JD, Lee ML, et al. Bioinspired High-Potential Porphyrin Photoanodes. The Journal of Physical Chemistry C. 2012;116(7): 4892-902.
58. Milot RL, Moore GF, Crabtree RH, Brudvig GW, Schmuttenmaer CA. Electron Injection Dynamics from Photoexcited Porphyrin Dyes into SnO_2 and TiO_2 Nanoparticles. The Journal of Physical Chemistry C. 2013;117(42):21662-70.
59. King PW. Designing interfaces of hydrogenase–nanomaterial hybrids for efficient solar conversion. Biochimica et Biophysica Acta (BBA) - Bioenergetics. 2013;1827(8–9):949-57.
60. Brown KA, Wilker MB, Boehm M, Dukovic G, King PW. Characterization of Photochemical Processes for H_2 Production by CdS Nanorod–[FeFe] Hydrogenase Complexes. J Am Chem Soc. 2012;134(12):5627–36.
61. McDonald TJ, Svedruzic D, Kim Y-H, Blackburn JL, Zhang SB, King PW, et al. Wiring-Up Hydrogenase with Single-Walled Carbon Nanotubes. Nano Lett. 2007;7(11):3528-34.

62. Svedružić D, Blackburn JL, Tenent RC, Rocha J-DR, Vinzant TB, Heben MJ, et al. High-Performance Hydrogen Production and Oxidation Electrodes with Hydrogenase Supported on Metallic Single-Wall Carbon Nanotube Networks. J Am Chem Soc. 2011;133(12):4299-306.
63. Krishnan S, Armstrong FA. Order-of-magnitude enhancement of an enzymatic hydrogen-air fuel cell based on pyrenyl carbon nanostructures. Chemical Science. 2012;3(4):1015-23.
64. Blakemore JD, Gupta A, Warren JJ, Brunschwig BS, Gray HB. Noncovalent Immobilization of Electrocatalysts on Carbon Electrodes for Fuel Production. J Am Chem Soc. 2013;135(49): 18288-91.
65. Adamska A, Silakov A, Lambertz C, Rüdiger O, Happe T, Reijerse E, et al. Identification and Characterization of the "Super-Reduced" State of the H-Cluster in [FeFe] Hydrogenase: A New Building Block for the Catalytic Cycle? Angew Chem Int Ed. 2012;51:11458 –62.
66. Moser CC, Keske JM, Warncke K, Farid RS, Dutton PL. Nature of biological electron transfer. Nature. 1992;355(6363):796-802.
67. Page CC, Moser CC, Chen X, Dutton PL. Natural engineering principles of electron tunnelling in biological oxidation-reduction. Nature. 1999;402(6757):47-52.
68. Reisner E, Powell DJ, Cavazza C, Fontecilla-Camps JC, Armstrong FA. Visible light-driven H_2 production by hydrogenases attached to dye-sensitized TiO_2 nanoparticles. J Am Chem Soc. 2009;131:18457-66.
69. Woolerton TW, Sheard S, Pierce E, Ragsdale SW, Armstrong FA. CO_2 photoreduction at enzyme-modified metal oxide nanoparticles. Energy & Environmental Science. 2011;4(7): 2393-9.
70. Chaudhary YS, Woolerton TW, Allen CS, Warner JH, Pierce E, Ragsdale SW, et al. Visible light-driven CO_2 reduction by enzyme coupled CdS nanocrystals. Chem Commun. 2012; 48:58-60.
71. Brown KA, Dayal S, Ai X, Rumbles G, King PW. Controlled Assembly of Hydrogenase-CdTe Nanocrystal Hybrids for Solar Hydrogen Production. J Am Chem Soc. 2010;132(28): 9672-80.
72. Chen Z, Dinh HN, Miller E. Photoelectrochemical Water Splitting: Standards, Experimental Methods, and Protocols: Springer; 2013.
73. Adams MW, Mortenson LE. The physical and catalytic properties of hydrogenase II of Clostridium pasteurianum. A comparison with hydrogenase I. J Biol Chem. 1984;259(11): 7045-55.
74. Lakadamyali F, Reynal A, Kato M, Durrant JR, Reisner E. Electron Transfer in Dye-Sensitised Semiconductors Modified with Molecular Cobalt Catalysts: Photoreduction of Aqueous Protons. Chem Eur J. 2012;18(48):15464-75.

Solar Photoelectrochemical Water Splitting with Bioconjugate and Bio-Hybrid Electrodes

Debajeet K. Bora, Artur Braun, and Krisztina Gajda-Schrantz

Abstract Biohybrid electrodes of different types have been described in the current chapter. The biohybrid photoanodes and photocathodes are new class of materials for solar energy application by utilizing the mother nature's photosynthetic units and its biomimetic counterparts. The physical characteristics and the performance of the biohybrid electrodes have been described in terms of charge transfer, hydrogen evolution, and stability in harsh electrochemical environment.

Keywords Hematite • Light harvesting protein • Surface functionalization • Bioconjugate

1 Introduction

This chapter deals with the artificial photosynthesis using photoelectrochemical cells (PEC), where the photoelectrodes are based on a semiconductor metal oxide backbone which is functionalized with a bioorganic component. This method is different from the conventional PEC which works with semiconductor photoelectrodes only, which is covered in chapter "Assessment of the Electronic Structure of Photoelectrodes with X-Ray and Electron Spectroscopy" of this book. It is also different from the particulate systems where semiconductor nanoparticles are functionalized with proteins, for example, as covered in chapter "Artificial Photosynthesis Producing Solar Fuels: Natural Tactics of Photosynthesis." The concept of attaching

D.K. Bora • A. Braun (✉)
Laboratory for High Performance Ceramics, Empa, Swiss Federal Laboratories for Materials Science and Technology, 8600 Dubendorf, Switzerland
e-mail: artur.braun@alumni.ethz.ch

K. Gajda-Schrantz
Laboratory for High Performance Ceramics, Empa, Swiss Federal Laboratories for Materials Science and Technology, 8600 Dubendorf, Switzerland

Environmental Chemistry Research Group, Institute of Chemistry, University of Szeged, 6723 Szeged, Hungary

Department of Inorganic and Analytical Chemistry, University of Szeged, 6720, Szeged, Hungary

E.A. Rozhkova, K. Ariga (eds.), *From Molecules to Materials*,
DOI 10.1007/978-3-319-13800-8_5

bioorganic components on electrodes dates back to the 1960s, when Rabinovich [1] attached chlorophyll on electrodes. Melvin Calvin and Tributsch in the 1970s [2] followed up on this with dye-sensitized solar cells.

2 The Inorganic Electrode Backbone Materials

The purpose and function of semiconducting electrodes for photoelectrochemical cells (PEC) are manifold. It must be a semiconductor with appropriate optical bandgap energy so as to serve as an efficient light absorber, based on which electron hole pairs are generated in the bulk of the electrode. In the photoanode, the holes are supposed to migrate to the electrolyte and oxidize water so that oxygen evolves as gas. The electrons migrate to the metal current collector. The photocathode is also a semiconductor with the same function, but here the electrons migrate to the electrolyte and reduce the protons to hydrogen. Electron holes migrate to the current collector. The photoelectrodes must be so constituted that they allow for the charge transfer within the electrolyte. This mandates that the electrode surface has a particular electronic structure. Also, the electrode surface must be resistant against any corrosion effects due to contact with the electrolyte. The geometrical area of the electrode determines how much of the incoming photon flux can be converted into electron hole pairs and thus into a photocurrent density. Limited diffusion lengths of holes and electrons in the electrode can impede the performance of photoelectrodes when the pairs recombine prematurely. Tailoring of the mesostructure is then a trick to give the holes in hematite, for example, the chance to diffuse to the surface at short length. Practically this amounts in making electrodes from nanoparticles films with a high surface area per volume. While hematite is a very well-studied and investigated system, it has not made it into the review [3] which lists a large number of well-performing compounds for PEC electrodes.

3 Development of Biohybrid Electrodes for Photoelectrochemical Water Splitting

3.1 Biohybrid Photoanodes

The integration of biomolecular components on PEC electrodes provides new ways of improving its performance. For instance, biological photosynthetic functionality can be integrated on the top of this kind of electrode, and the process of artificial photosynthesis can be realized which is nothing but the mimicking of photosynthetic machinery of nature [4]. The system of this kind is usually termed as biophotoelectrochemical cell as shown in Fig. 1. The organic–inorganic photo galvanic effect was first observed by Rabinowitch in 1940 in the thionine–iron system [5].

The development of integrated biomolecule—electrode assembly for artificial photosynthesis application—is not so much well progressed. The focus of this sec-

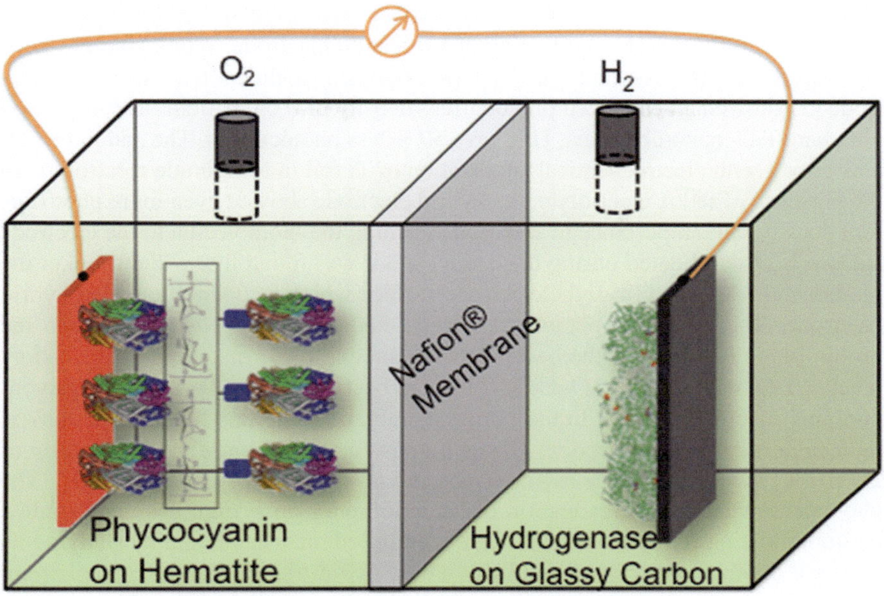

Fig. 1 A biophotoelectrochemical cell showing the phycocyanin-functionalized hematite as photoanode on the left and a hydrogenase-functionalized glassy carbon on the right separated by Nafion membrane in order to prevent the intermixing of evolved gases such as O_2 and H_2

tion is to generate an overview of the works progressing in this topic. The basic idea of developing this kind of electrodes is to mimic the synthetic analogs of light-harvesting complex, the active site of photosystem II as water oxidation catalysts and hydrogenases for hydrogen evolution reaction. Looking back into past decades, it is evident that the idea of making biomolecule-integrated electrodes was fairly conceived in 1971. During this time, photosynthesis pioneering chemists Calvin along with his colleague Tributsch developed hybrid semiconductor biological electrode for studying the photoelectrochemical properties of excited chlorophyll molecules on the single crystals of ZnO [2]. It was found that on optical excitation, chlorophyll molecules inject electrons into the conduction band of ZnO which gave rise to photoanodic current. Electron donors such as hydroquinone or phenylhydrazine helped chlorophyll moiety in transferring the electrons from its energy levels to the conduction band of semiconductor which acts as an electron acceptor. The electrode materials captured electrons in irreversible manner on applying an electrochemical potential gradient, and the chlorophyll molecule on its surface exhibits its usual physical properties such as unidirectional charge transfer and separation similar to those inside a photosynthetic reaction center. In a similar manner, the photogalvanic effect of chlorophyll was studied by attaching it to a metal electrode such as Pt inside a quinhydrone half-cell [6]. Here, an important observation was the formation of p-type semiconductor film by chlorophyll molecules on light excitation, and this gave rise to photocurrent on anodic sweeping. The particular feature was absent in case of chlorophyll-free electrode. From the action spectrum, the photocurrent obtained closely matched to visible wavelength of 700 nm where chlorophyll generally absorbs. Following this research, electrode of this kind has been developed recently. For example, bacteriorhodopsin-modified TiO_2 nanotube electrode [7] showed an enhanced photocurrent density of 0.65 mA/cm^2 in comparison with bare TiO_2 nanotube films. This gave 50 % less photocurrent. The nanotube film was grown with electrochemical anodization of Ti foil in formamide electrolyte. In the presence of redox electrolyte, the hybrid electrode showed even more photocurrent density. The importance of linker in attaching the biomolecule to the electrode surface is also validated during this study, and it was found that without linker the biomolecule detached from the electrode surface which leaded to degraded hybrid electrode. The enhanced photocurrent generated was believed to be due to the proton-pumping effect of the bacteriorhodopsin and close matching of the energy level of TiO_2 with the HOMO–LUMO level of bacteriorhodopsin. These are proton-pumping machinery found in the purple membrane bacteria where it functions as molecular membrane transporter. These proteins arrange themselves in the form of 2D nanocrystals, forming a uniform pattern known as purple membrane [8]. The photoelectrochemical investigation of bacteriorhodopsin system has been studied by immobilizing it on a gold substrate in controlled manner [9] and by studying its transient photocurrent response. In this study, two different types of photocurrent have been observed on exciting bacteriorhodopsin with light by making a light-chopped experiment. The first photocurrent was suggested to come from the pH response of the surface oxide layer caused by the proton release and uptake mechanism of bacteriorhodopsin. The second type of current arises due to the charging of the electrical double layer which has low magnitude in comparison with the first one. The mechanism for this transient behavior involved the shifting of the surface potential on the oxide-covered electrode caused by the proton release mechanism of the protein. This shift usually observed by looking at the change in flat band potential in anodic scan. The mechanism is illustrated in Fig. 2.

Effort has been given to build purple membrane-decorated lipid membrane by taking bacteriorhodopsin-containing purple membrane in order to mimic the natural plasma membrane of bacteria [10]. The membrane was deposited on porous alumina substrates with a pre-deposited gold layer in addition to chemisorption of 1,2-dipalmitoyl-sn-glycero-3-phosphothioethanol, 1,2-diphytanoyl-sn-glycero-3-phosphocholine, and octadecylamine dissolved in *n*-decane. The photocurrent measured under continuous light illumination is fairly stable in this system, and the device did not degrade in a long-term stability study. The conductivity of this lipid membrane was further increased by adding proton ionophore carbonyl cyanide-m-chlorophenylhydrazone, and the proton is believed to be diffused across the membrane in an improved fashion after following this step. Besides bacteriorhodopsin, also other types of photosynthetic machinery such as photosystem II, photosystem I, and light-harvesting complex were immobilized on electrode surface for photocurrent generation. A detailed account of each will be given in the later part of this chapter.

In continuation with the photoanode development, in our group we have utilized this strategy to make biofunctionalized hematite photoanode (Fig. 3) which shows two times increase in the photocurrent density [11].

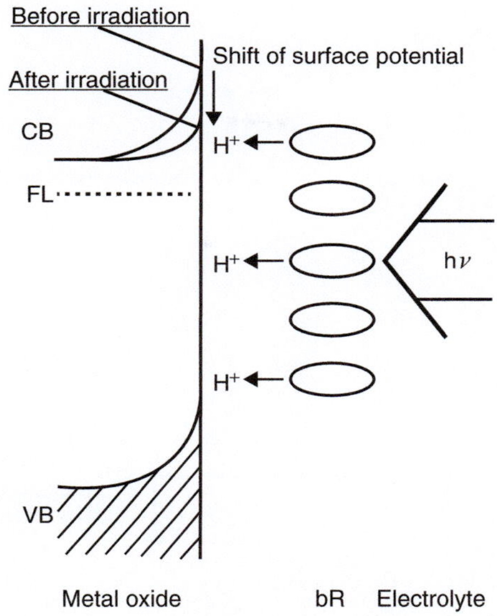

Fig. 2 Energy level diagram of a bacteriorhodopsin-coated electrode for the transient photocurrent response. Reproduce with permission from [9]. Copyright © 1999, American Chemical Society

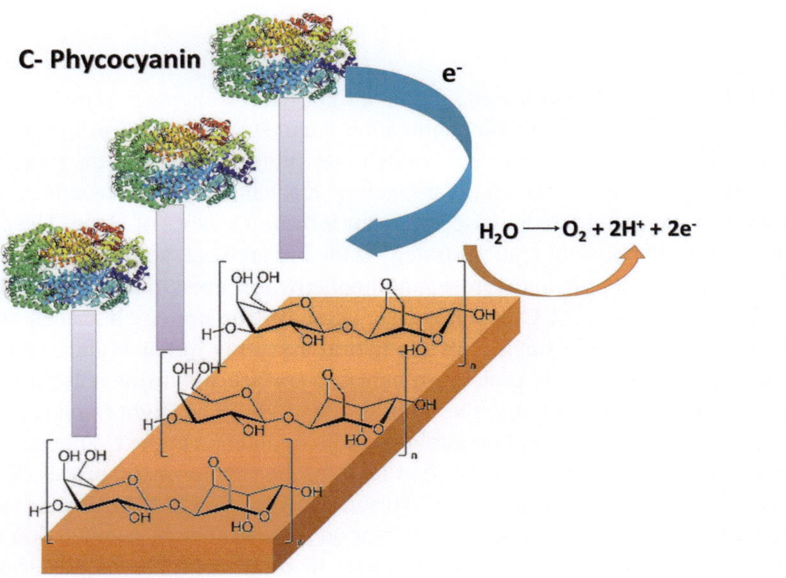

Fig. 3 Phycocyanin-functionalized hematite photoanode used for photoelectrochemical water splitting.

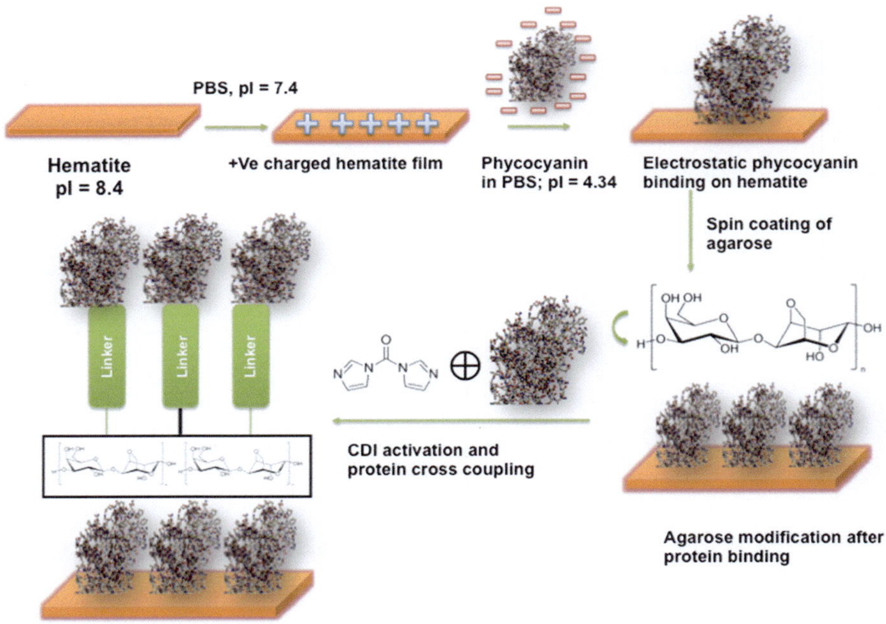

Fig. 4 Bioconjugation chemistry for making C-phycocyanin-functionalized hematite film

Here, we have incorporated section of light-harvesting membrane protein phycobilisome with the help of bioconjugation chemistry as shown in Fig. 4. We have taken the C-phycocyanin protein which constituted the part of the phycobilisome family and helps in the energy funneling mechanism to the reaction enter. The source of the protein in our case was from *spirulina* species of algae. The chemistry involved in the bioconjugation strategy is the hierarchical modification of the hematite surface with protein, agarose, and final covalent cross-linking of agarose layer with a second layer of C-phycocyanin. The covalent strategy has been adopted in order to preserve the stability of the hybrid assembly in harsh alkaline condition. The idea of putting this protein on hematite surface was to increase the panchromatic light harvesting by the hybrid electrode. The light-harvesting protein C-phycocyanin acts as a photon relay system or antenna in the cyanobacteria membrane, and in the present electrode, it helps in the panchromatic process by absorbing the visible spectrum at 620 nm. This method is believed to be an innovative way to increase the efficiency of water oxidation process. The bioconjugation of hematite has also been accomplished in order to understand the biophysicochemical interaction between cytochrome proteins with hematite surface [12].

3.2 Biohybrid Photocathodes

So far we have discussed about various types of biohybrid photoanodes. Now we turn our discussion to the development of biohybrid photocathodes which are responsible for the generation of hydrogen by photoelectrochemical water splitting. The choice of semiconductors applicable as photocathodes is limited [13] mainly due to their instability in aqueous solutions and costs ($GaInP_2$, GaAs, CdS, CdSe) and inappropriate conduction band-edge potential for proton reduction (Fe_2O_3, WO_3). Besides metal cathodes [14, 15], the researchers focus in the first place on Si [16], TiO_2 [17], and Cu_2O [18]. To improve the efficiency of these electrodes, most of the studies deal with their functionalization using photosystem I [19], hydrogenases [17, 20, 21], or bioinspired molecular cocatalysts [17, 22]. Immobilization of hydrogenase on electrodes has the advantage that the hydrogen evolution can be electrochemically controlled and spatially separated from the oxygen evolution, so that no oxyhydrogen gas is produced. An example on how hydrogenase is fixed on a silicon electrode is presented by [20]. The authors report an inexpensive and robust photoelectrochemical reduction system for the homogeneous catalytic reduction of protons to hydrogen, which consists of an illuminated and biased p-type Si photocathode, an [FeFe] electrocatalyst, a proton source ($HClO_4$), and a glassy carbon electrode. The [Fe_2 (m-bdt) $(CO)_6$] (bdt = benzene-1,2-dithiolate) complex has reversible reductive electrochemistry and maintains its structural integrity to a large extent even in the reduced and protonated state. Photoreduction of the [FeFe] complex was achieved at an illuminated p-type Si at a reduction potential of 500 mV less negative than that at a glassy carbon electrode. Homogeneous photocatalytic reduction of protons was achieved with 100 % faradic efficiency. Followed by this, a high current density of 38 mA cm^{-2} and 640 mV of open circuit voltage under 661 nm wavelength monochromatic illuminations with a light conversion efficiency of 8 % has been observed. It can be concluded from the above observations that the [FeFe] complex is fast enough to keep pace with incoming light intensity of up to 100 mW cm^{-2} (one sun).

Interest in the combination of inorganic semiconductors with hydrogenase has sparked again recently [21]. Their system was more complex than the previous one because in addition it contained a ruthenium sensitizer. In addition to it, hybrid electrode consisting of [NiFeSe]-hydrogenase with co-attached Ru-based photosensitizers on TiO_2 has been developed, and it shows good hydrogen gas production in aerobic environment which is detrimental to hydrogenase functionality [21]. Following the same synthetic route, another system of [NiFeSe]-hydrogenase from *Desulfomicrobium baculatum* (Db) attached to Ru dye-sensitized TiO_2 has been developed to study further the per site performance of the hydrogenase enzyme in making hydrogen [23]. On using triethanolamine as sacrificial electron donor, the hybrid system evolves hydrogen gas at a rate of 50 mol S^{-1} mol total hydrogenase^{-1} at pH = 7 and room temperature. The solar irradiation during this study was a typical northern European sky. The system is very robust for photocatalytic water splitting even in the sense of air exposure for the hydrogenase. For further examples and detailed hydrogenase immobilization techniques, the reader is referred to the reviews of Tran et al. [22] and Artero et al. [17].

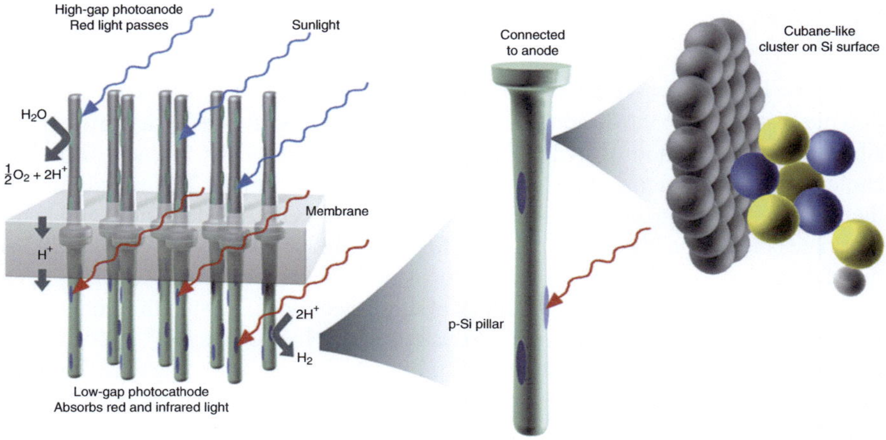

Fig. 5 Schematic of bioinspired molecular cubane-like cluster-functionalized silicon photocathode for hydrogen evolution. Adapted by permission from Macmillan Publishers Ltd: [Nature Materials] [25], copyright (2011)

This strategy is not only applicable to generate hydrogen. Some of the studies show the application toward photovoltaic activity [24]. In this study, effort has been made to understand the electronic coupling between proteins and metal electrodes by depositing a dried self-assembled monolayer of photosystem I on a gold surface. From the Kelvin probe force microscopy [KPFM] and surface photovoltage measurement of the hybrid electrode, an internal energy yield of 58 % and a photovoltage of 1 V with reasonable quantum efficiency have been obtained. The photosystem I is fairly stable even in the dried state in this hybrid system. For photoelectrochemical hydrogen generation, bioinspired molecular cocatalyst-functionalized silicon photocathodes have been developed recently [25]. In this study, hydrogen evolution rate close to that of platinum has been obtained with bioinspired molecular clusters of molybdenum and sulfur. These cubane-like clusters were immobilized on Si semiconductor in order to carry out the photoelectrochemical hydrogen evolution. A solar to hydrogen efficiency of 10 % has been obtained during this study. The hybrid photocathode is shown in Fig. 5.

Similar system, using Si electrode functionalized with ferrocenophane derivative, was reported by Muellerwesterhoff et al. [26]. Nann et al. [27] immobilized InP nanocrystals and 1,4-benzenedithiolate on a gold electrode with a “layer-by-layer” method; however, in neutral aqueous solution, the measured photocurrent was very low (250 nA/cm^2).

3.3 Hybrid Photocatalysts for Photoelectrochemical Water Splitting

In addition to the fabrication of bioconjugated semiconductor photoelectrodes, hybrid photocatalysts have also been developed in parallel for solar fuel production. A review of the functionalization of the photocatalyst surface with H_2 evolution catalysts, water oxidation catalysts, and carbon dioxide reduction catalysts has already been provided elsewhere [28]. We would like to discuss a few of them to give an overview of the function of hybrid photocatalysts. A very recent example is the development of Z-scheme-based hybrid photocatalyst system by the functionalization of TiO_2 nanoparticles with bacteriorhodopsin proton pump. The hybrid system evolves hydrogen significantly in the order of 5,275 μmol of H_2 (μmol protein)$^{-1}$ h^{-1} at pH 7 with methanol as sacrificial electron donor and at AM1.5 light [8]. Recently, hydrogen-evolving hybrid photocatalysts were fabricated by the non-covalent interaction of photosystem I with Pt nanoparticles [29]. In another case, a conjugate of photosystem I has been made with platinum catalyst. The hydrogen evolution rate obtained in this case was around 244 mmol H_2 per mg chlorophyll per hour in the presence of cytochrome C6 as soluble mediator and ascorbate as the sacrificial electron donor.

3.4 Hydrogenases as Molecular Hydrogen-Evolving Complexes for Making Biohybrid Photocathodes

The hydrogenases (Hases) catalyze the proton reduction toward hydrogen gas and are typically built from a pair of Fe and/or Ni atoms which are bridged by two sulfur atoms. [Fe–Ni]-hydrogenases are more active in catalyzing the reversible oxidation of molecular hydrogen according to the reaction $H_2 = 2H^+ + 2e^-$ [30]. A well-detailed review of the functional and structural assessment of the hydrogenases is presented in [31]. A bio-electrode assembly on a TiO_2 support with deposited FeFe hydrogenase has been presented recently by S. Morra et. al. [32]. There, immobilization of the [FeFe]-hydrogenase *Clostridium acetobutylicum* allowed for exchange of electrons with the TiO_2 electrode and yielded a catalysis of hydrogen production with 70 % efficiency. The same type of *Clostridium acetobutylicum* hydrogenase has been used years before in an electrocatalytic reformer which could convert a biosubstrate into hydrogen [33].

One of the earlier papers where bacterial hydrogenase was deposited on a metal oxide semiconductor was published by Nikandrov et al. [34]. They extracted hydrogenase from *Thiocapsa roseopersicina* and adsorbed it onto the Degussa P25 photocatalyst particulates (TiO_2 anatase) suspended in pH 7.2 buffer solutions, which contained also an organic electron donor. With the electrons generated upon illumination in the conduction band of TiO_2, hydrogen gas is evolved by this system. The gas production rate of this system (Fig. 6) is comparable to the rate when reduced methyl viologen is used as substrate. The quantum efficiency was around 10 %.

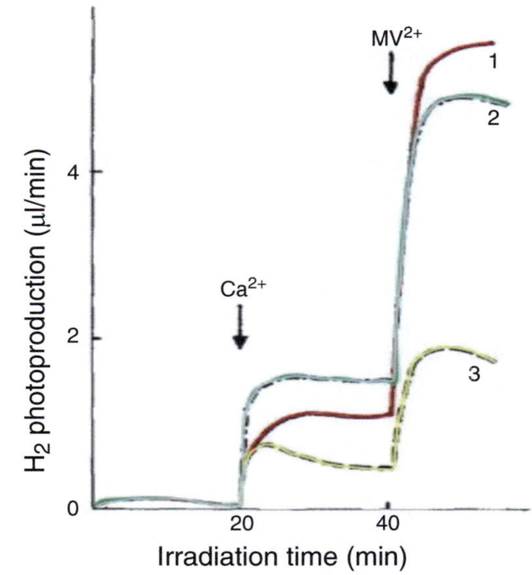

Fig. 6 Hydrogen photoproduction in the system Tris-TiO_2 hydrogenase. *Arrows* indicate additions of $CaC1_2$ (1, 0.4 mM; 2, 1.6 mM; 3, 12.5 mM) and $MV^{•+}$ (0.5 mM). Modified from [34]

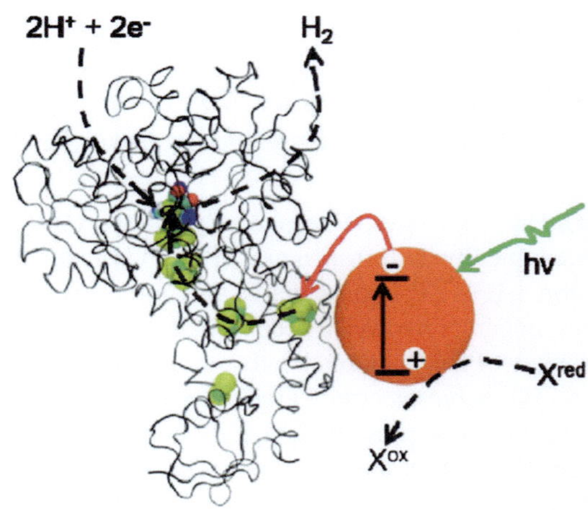

Fig. 7 Assembly of hydrogenase-conjugated CdSe nanocrystals. Reproduced with permission from [36]. Copyright © 1999, American Chemical Society

A route for the controlled self-assembly and covalent attachment of hydrogenase on carbon nanotubes was shown by McDonald et al. [35].

In another study, the self-assembly, charge transfer kinetics, and catalytic properties of [FeFe]-hydrogenase I isolated from *Clostridium acetobutylicum* conjugated with CdSe nanocrystal have been studied [36]. The interaction between the nanoparticle and hydrogenase is proposed as the electrostatic one and also it gives a stable complex. It has also been found that if the hydrogenase concentration is lower than that of nanocrystal concentration, maximum hydrogen production rate was obtained. This also depends on the photoluminescence quantum efficiency of CdSe nanocrystal which suggests the transfer of photogenerated electrons from the core of nanocrystals to the hydrogenase. Also in the presence of sacrificial electron donor such as ascorbic acid, the rate can be increased further to 9 % under monochromatic light. The assembly I is as shown in Fig. 7.

Till now, the development of biohybrid electrodes for photoelectrochemical water-splitting application has been described with reference to biohybrid photoanodes and photocathodes. In the following section, a more detailed discussion of electrodes with photosynthetic components is given in order to know about its physical properties and application.

4 Oxygen-Evolving Complexes from Photosynthesis and the Respective Biohybrid Electrodes for Photoelectrochemical Water Splitting

Photosystem II is responsible for the generation of molecular oxygen in a cyclic four-electron process with the help of Mn_4Ca cluster where water acts as a substrate [37]. The water oxidation process in PSII transit a cyclic steps (S0–S4) in which the Mn_4Ca cluster oxidized four times and two water molecules get deprotonated in parallel with this [38]. The O–O bond formation and release of O_2 molecule accomplished during the final state or S4 state.

Photosystem 2 (PSII) is also considered as a membrane-integrated redox enzyme complex mediating the electron transfer process in photosynthesis. By absorbing the visible light, it can form excited electronic states which are separating itself from the electron vacancy. The photooxidation property of PSII makes it as an effective candidate for making integrated biophotoelectrochemical device for converting visible light into chemical energy [39]. To achieve it, several challenges need to be face. For instance, the electron transfer pathway between the PSII and the electrode surface has to be designed in such a way that the electron acceptor site of photosystem II or Q_B-binding pocket is facing toward the electrode surface. Due to this, direct physical adsorption of the PSII on the electrode surface does not lead to effective photoelectrochemical performance. However, to accomplish it, several studies have been carried out. For example, a genetically engineered PSII with polyhistidine-tag was immobilized on the thiol-terminated gold surface bearing Ni(II)-nitrilotriacetic acid [40]. In this case, a redox mediator 2,6-dichloro-1,4-benzoquinone was used in order to improve the current density in a freely diffusive environment. In another study, the redox mediator poly(mercapto-*p*-benzoquinone) was co-immobilized along with PSII on the electrode surface to increase the current density, but the study was not so much successful [41]. To boost up the current density further, new methods have been developed by wiring photosystem II to gold electrode surface inside a redox hydrogel where PSII was affixed to the electrode surface via osmium-containing redox polymer such as poly(vinyl imidazole) [42]. In this case, the redox hydrogel acts as electron acceptor for the PSII, and upon illumination, a photocurrent density of 45 $\mu A/cm^2$ has been observed for the integrated assembly. The assembly showed an improved operational stability due to the fast electron transfer process between PSII and the redox polymer. The current density over time is fairly stable as can be observed in Fig. 8. The light-induced damage of the PSII is minimized by reducing the reactive oxygen species formed during the recombination process as mentioned elsewhere [43].

Recently, an integrated electrode of photosystem II immobilized on indium-doped tin oxide has been developed where it is proposed that two different competing pathways played a role in mediating the charge transfer process between PSII and electrode [44]. The photocurrent observed during this study was around 1.6 ± 0.3 mA cm^{-2}. The PSII was immobilized electrostatically on the electrode surface, whereas a covalent modification is better for making a stable assembly. So far we have discussed the PSII assembly on electrode surface. A biomimetic synthetic analogous of the active site (Mn_4Ca) of PSII-WOC has also been designed and hybrid assembly has been made [45]. In this study, the biomimetic cluster [Mn_4O_4 (MeOPh) $PO_2)_6$] was suspended on a Nafion membrane-coated electrode interface with dye-sensitized TiO_2 electrode. Applying visible light and a suitable potential of +1.2 V vs. SHE; oxygen gas can be generated in a sustainable manner followed by maintaining a significant amount of photocurrent density. The biomimetic catalysts are also renewed by the oxidation of the water molecules bind to a reactive intermediate formed by the application of applied bias and visible light illumination. The role of dye-sensitized TiO_2 is to inject the electron into the conduction band of TiO_2, flowing through an outer circuit. This lead to the generation of one-electron oxidant [RuIII $(bipy)_2(bipy\ (COO)_2)]^+$. This assembled multilayer system acts like a molecular PSII. The assembled system is shown in Fig. 9.

On matching the electrochemical potential of biomimetic cluster with that of redox dye and electron acceptor, it will be possible to mimic the photo-induced charge separation unit with water oxidation in parallel with the natural photosynthesis process.

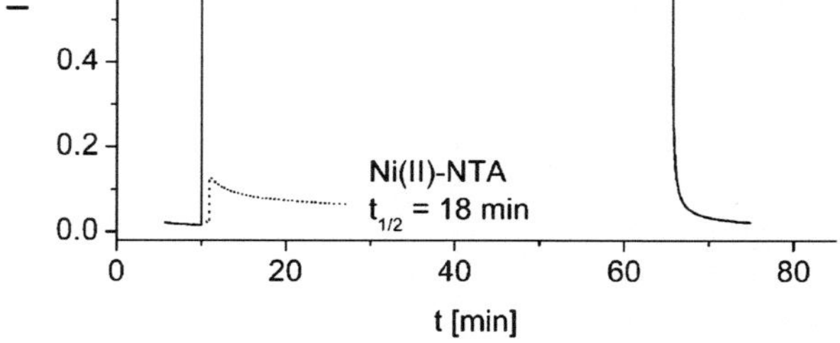

Fig. 8 Long-term operational stability of photosystem II/Au electrode embedded inside a redox hydrogel network. Reproduced with permission from [43]. Copyright © 2008 WILEY-VCH Verlag GmbH & Co. KGaA, Wienheim

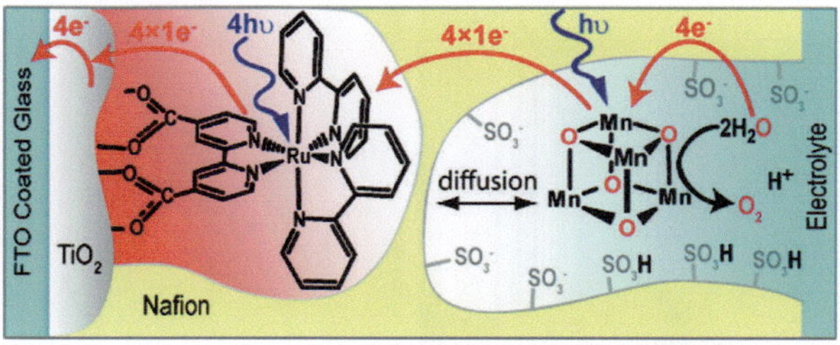

Fig. 9 A multilayer assembly of biomimetic PSII with dye-sensitized TiO_2 acts like a PSII for water oxidation reaction. Reproduced with permission from [45]. Copyright © 2010, American Chemical Society

5 Light Antenna Complex from Photosynthesis and Its Biohybrid Electrode

Phycobilisomes as mentioned in Section 3.1 are light-harvesting antenna complexes found in cyanobacteria, red algae, and cryptomonads. This molecular assembly constituted of biliproteins such as phycocyanin, phycoerythrin, phycoerythrocyanin, and allophycocyanin organized as staked disks. These proteins possess the light absorption capability relatively in a narrow wavelength of visible spectrum. In the stacked-like arrangement, proteins can effectively absorb sunlight and transfer to chlorophyll molecule in a Forster-like mechanism having an efficiency of 90 % [46, 47]. The function of these light-harvesting proteins is the absorption of sunlight in the regions of low chlorophyll and then participating in highly efficient exciton migration to the reaction center where it is converting into chemical energy. The energy absorbed by this protein also transduced to the photosystem I by a spillover process. These proteins are applied in several sectors besides energy generation process. For instance, a biomaterial containing phycoerythrin embedded inside sol gel matrix was developed in order to protect it from photodegradation and to apply it for biosensor application [48]. The protein is found to be preserving its optical property in embedded form meaning it is less prone to photodegradation. In another study, phycocyanin showed good optical properties even in dried condition which make it useful for photovoltaic application [49]. Similar to light-induced degradation, this protein or phycocyanobilins can be assembled to a cyclic configuration during denaturation [46]. It is found that the protein maintains an extended configuration in the native state and it remains stable in the denaturation step, by taking the cyclic configuration due to the relaxation of the apoprotein moiety on the billin counterpart. The diligence of these proteins for making these hybrid electrodes is relatively novel, though a few studies dealing with the biophysical interactions of proteins with semiconductor nanoparticles can be found in the literature. In this study, charge transfer from excited protein molecule to the conduction band of TiO_2 has been studied [50, 51]. Based on the light-harvesting action of C-phycocyanin, a new electrode has recently been developed by cross-coupling of phycocyanin with hematite with some chemical linker [11]. A brief overview of this hybrid electrode

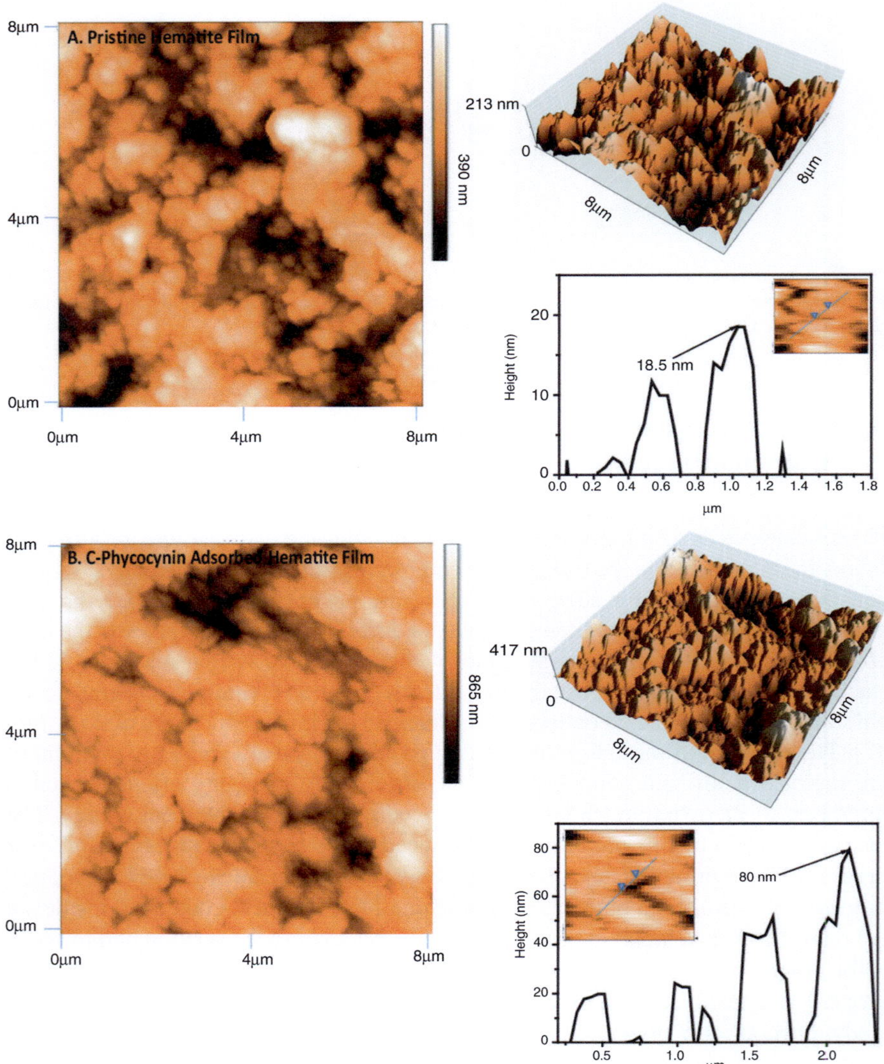

Fig. 10 AFM images of pristine hematite film (**a**) and corresponding protein-adsorbed film. (**b**) Inset showing the 3D view and cross-sectional height analysis. Reproduced with permission from [11]. Copyright © 2008 WILEY-VCH Verlag GmbH & Co. KGaA, Weinheim

has already been given in Sect. 3.1. Here, the morphology of this electrode (Fig. 10) and other properties are described.

The individual role of phycocyanin in enhancing the photocurrent density of hematite in this hybrid electrode has been studied carefully followed by its long-term operational stability. During this study, the photocurrent response of phycocyanin adsorbed on FTO and its bare analog have been determined by

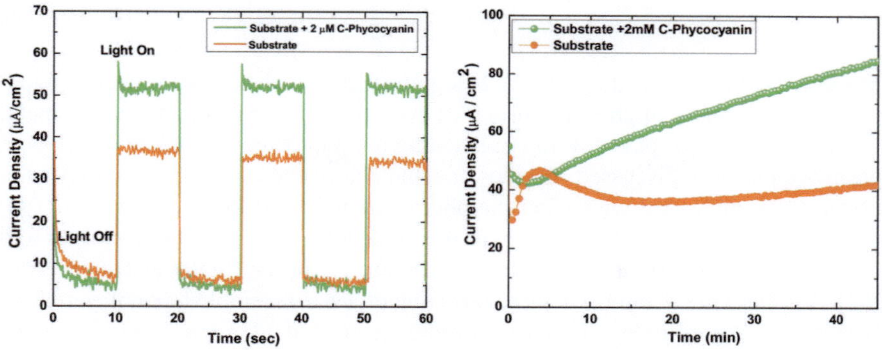

Fig. 11 *Left*—photocurrent response of phycocyanin-adsorbed film on conducting FTO substrate. *Right*—the long-term operational stability of phycocyanin-adsorbed FTO

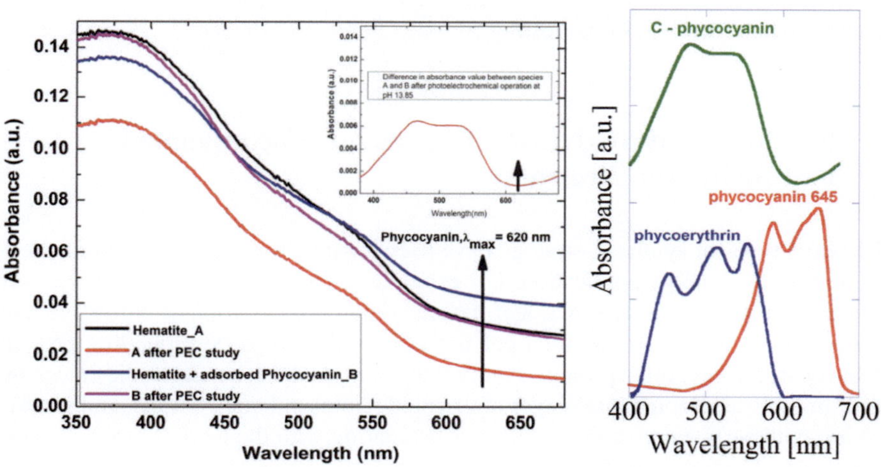

Fig. 12 The post-electrochemical stability of phycocyanin has been studied with the difference spectrum of absorbance data. Reproduced with permission from [11]. Copyright © 2008 WILEY-VCH Verlag GmbH & Co. KGaA, Wienheim

chronoamperometric cycling of the electrode in transient manner with a fixed bias of 100 mV as shown in Fig. 11 (left). The phycocyanin adsorbed on FTO gives higher photocurrent with respect to bare FTO and validates its contribution to enhanced photocurrent density of integrated hematite electrode. A similar type of result has been observed on LH1–RC complex integrated ITO electrode [52]. The presence of anodic spike in the phycocyanin referred to the recombination of excited state electrons in the bulk of FTO. The long-term stability of the phycocyanin on FTO has also been carried out and is shown in Fig. 11 (right). It is found that phycocyanin has good operational stability up to 1 h of chronoamperometric cycling.

To check further the stability of electrochemically processed phycocyanin-conjugated hematite electrode, the absorbance spectra have been recorded. The difference spectrum (Fig. 12) between the pre- and post-photoelectrochemically processed hybrid electrode gives rise to significance absorbance with spectral signature similar to those of phycoerythrin [53] even after the electrochemical treatment.

So far the role of phycocyanin in boosting the photocurrent density of hematite electrode has been discussed. Besides this, the chlorosomes are also found to be an effective candidate for enhancing the photocurrent density of dye-sensitized TiO_2 by the combined interaction of it with dye molecule. The device shows a 30-fold increase in photocurrent density in comparison to the one without chlorosomes [54]. Another kind of hybrid device consisting of a self-assembled monolayer of the light-harvesting complex (LHC) on amino-terminated ITO has been developed [52]. The photocurrent of this hybrid device showed maximum magnitude at the absorption band of light-harvesting complex combined with reaction center followed by illumination at 880 nm. Without the reaction center, the photocurrent generated at 770 nm only. The enhanced photocurrent observed at 880 nm is believed to be from the combined action of LHC and reaction center. This is due to the energy transfer from light-harvesting complex to reaction center.

6 Synthesis and Analytical Methods for Bioconjugated and Biohybrid Electrode

Different types of synthesis strategy have been adopted for making biohybrid electrode for solar energy application. For example, a co-immobilization method was utilized for making an artificial leaf device containing chlorophyll derivative and formate dehydrogenase on silica gel films for carbon dioxide reduction [55]. In other case, the integrated protein is immobilized on a solid state electronic device by forming self-assembled monolayer of oriented photosynthetic complex and stabilization with surfactant peptides followed by coating with the help of organic semiconductor. The function of organic semiconductor is to give a protective mechanism to the immobilized photosynthetic complex [56]. The integrated device shows an internal quantum efficiency of around 12 %. Due to the interfacial architectures of photosynthetic light-harvesting complex, it was found that it can sustain an open circuit voltage of +1.1 V. Chemical linker or anchor molecules are found effective for making stable biohybrid electrode. There are few studies dealing with the development of anchors (hydroxamate) for fixing dye molecules on the surface of TiO_2 [57]. This synthesis method leads to the formation of water-stable dye-sensitized TiO_2 complex. Normally in dye-sensitized cells, the dye molecules on the surface of semiconductor are not stable in aqueous environment when attached directly. But with the help of hydroxamate anchors, it is found that the dye molecules bind fairly well due to the firm linking of anchor to the surface of transition metal such as titanium. Following this kind of cross-coupling strategy, a photosystem I/Pt nanoparticle composite has been fabricated by the covalent functionalization of polymerizable thioaniline-functionalized Pt nanoclusters with photosystem I via bis-aniline bridging group [58]. On biasing with a potential of >0.1 V, the bis-aniline bridging group converted into quinoid state which acts as an electron acceptor and mediates

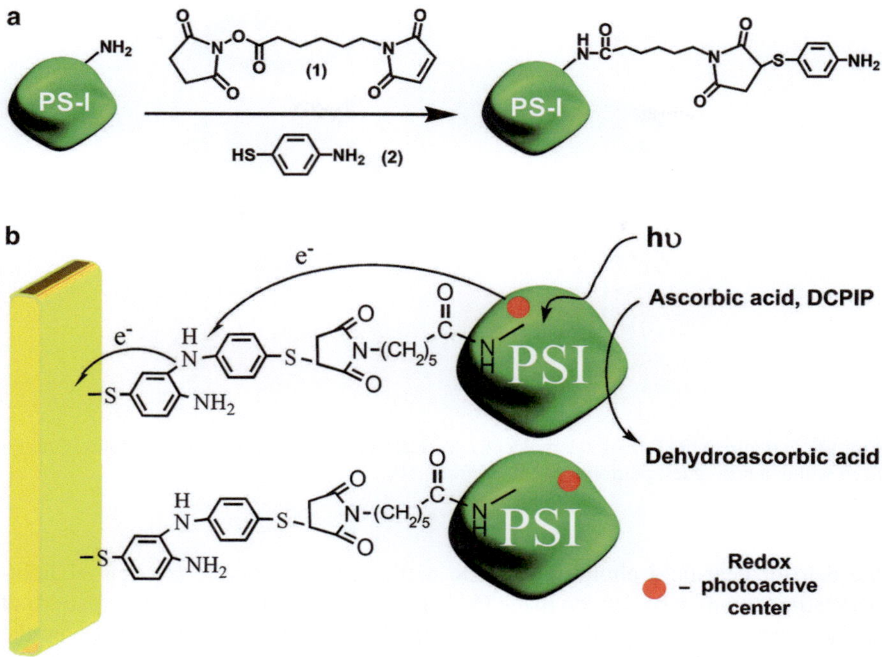

Fig. 13 Architecture of a hybrid electrode made of photosystem I with Pt nanoclusters via chemical cross-linking method. Reproduced with permission from [58]. Copyright © 2010, American Chemical Society

electron transfer to the electrode. Due to this, composite gives rise to tenfold higher photocurrent. The hybrid composite electrode is as shown in Fig. 13.

In another study, the chemical modification of cysteine residues in proteins helps in the functionalization process. Due to the nucleophilic nature of cysteine molecule, it is easy to make reaction with electrophilic reagent such as maleimide motif. Bromomaleimides are kind of linkers for functionalizing protein with three-point attachment [59]. For preserving the functionality of immobilized motifs on electrode surface, a molecular container has been designed. These molecular containers, for instance, cyclodextrin, have the ability to encapsulate the chromophores or redox-active compounds for immobilizing onto electrode surface [60]. The process is like a host–guest approach whereby host provides a platform for adsorbing the biological or organic motif on semiconductors and guest is shielded from the surrounding harsh environment. A typical host–guest electrode is shown in Fig. 14. These hybrid systems have higher performance in comparison to the molecule directly adsorbed on host surface or without any molecular container.

So far various types of synthesis methods described above may not be enough to get precise orientation of immobilized protein which is necessary for proper functioning of the hybrid electrode in some cases. The precision placement of the desired protein components on a suitable substrate is an essential requirement for this. With

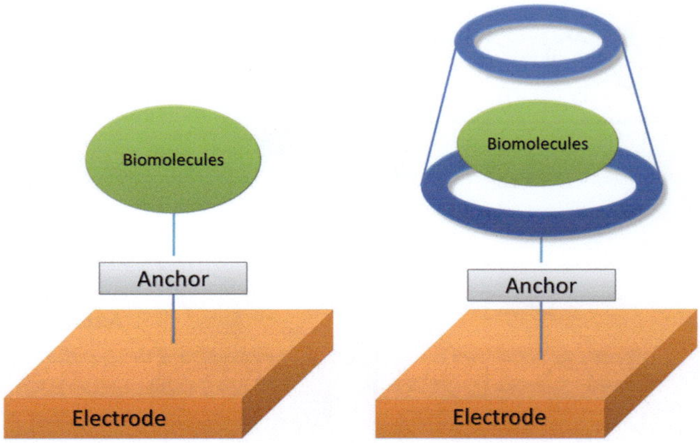

Fig. 14 *Left*, direct binding of dye molecules to electrode surface; *right*, dye molecules immobilized within a host–guest platform. Modified from [60]

the help of near-field photolithographic method, the nanoscale patterns of light-harvesting complex have been made [61] and immobilized on microscale pattern of self-assembled monolayer as shown in Fig. 15. A follow-up study involves the determination of the optical properties of light-harvesting complex in oriented form. It has been found that the assembly showed native-like optical spectra confirming its conformational stability in directed patterning form [62].

In order to study its physical and morphological properties, biohybrid electrodes described above need to undergo analytical characterization. For accomplishing it, various types of analytical methods have been developed in recent past and brief overviews of few of them are described here. For the photosynthesis research, near-field optical microscopy has been used to study the structure of photosynthetic membrane proteins such as chloroplasts and thylakoids [63]. This method has higher resolution in comparison with conventional optical microscopy and has added advantage of being able to take topography in simultaneous manner. Atomic force microscopy has also been extensively used to determine the topography of light-harvesting complex one and characterized further with near IR absorption and fluorescence spectra [64]. The atomic force micrograph shows a ringlike pattern of light-harvesting complex with different inner and outer diameters of 30 and 8 nm (Fig. 16 left). In line with this, single-particle electron microscopy has been utilized in order to determine the structure of light-harvesting phycobilisomes as mentioned above. A high-resolution image (Fig. 16 right) has been obtained by the two-dimensional projection and reveals new details about the attachment of funnel-like core to the membrane surface. The architecture is essential for the proper Forster-like energy funneling process [65].

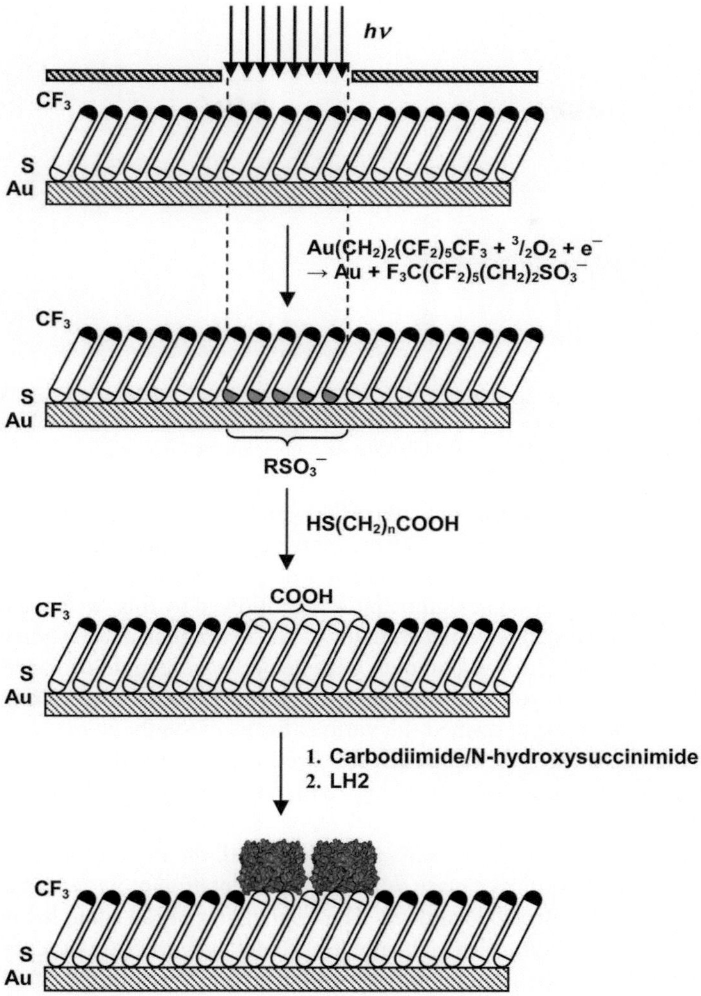

Fig. 15 Schematic of photolithographic patterning of light-harvesting complex by the subsequent activation of self-assembled monolayer. Reproduced with permission from [63]. Copyright © 2010, American Chemical Society

7 Conclusion and Outlook

Various types of biomolecule-integrated electrodes are described in the current chapter. The development of the electrodes needs to be further optimized for getting best performance. In the biohybrid photoanodes, we have discussed the conjugation of light-harvesting proton-pumping protein on semiconductor surfaces followed by

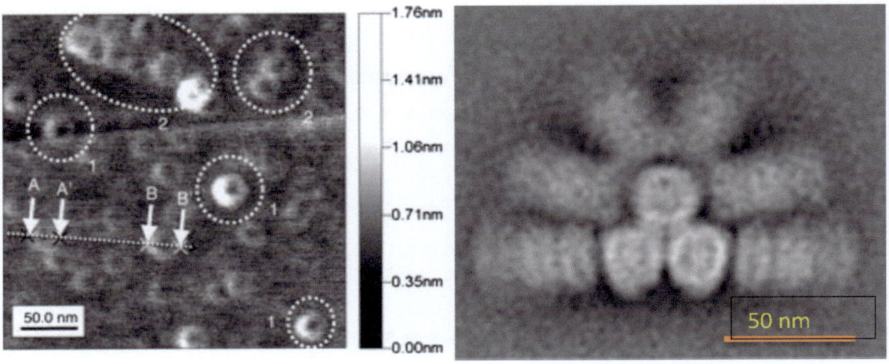

Fig. 16 *Left*: atomic force micrograph of a light-harvesting complex showing ringlike pattern. Reproduced with permission from [64]. Copyright 2010, American Chemical Society. *Right*: single-particle TEM imaging of a phycobilisome assembly, adapted from [65]

conjugation of light-harvesting antenna protein on hematite surface which gave rise to enhanced photocurrent density. For the biohybrid photocathodes, the integration of photosystem I on metal and semiconductor electrodes has been described which leads to hydrogen evolution in significant way. Followed by this, we have described the role of individual photosynthetic analogs such hydrogenases, oxygen evolution complex, and light-harvesting complex with respective example of functioning of this protein in an integrated photoelectrode during photoelectrochemical water splitting. Next, we have described the various forms of synthetic methods for getting biomolecule-integrated electrodes followed by analytical characterization to get information about the surface topology and morphology of each individual photosynthetic entity.

The future work in this area needs to be dealt with the deposition of protein on electrode surface in uniform pattern with correct orientation of its functional part in conjunction with the electrode surface. The application of the same for the photoelectrochemical water splitting can be made efficient by correcting the charge transfer pathway between the protein and electrode surface. These kinds of studies lay further foundation in the area of biohybrid electrode for solar fuel generation with photoelectrochemical water splitting.

Acknowledgment A.B. acknowledges the generous financial support from the Swiss National Science Foundation no° 206021-121306, the Swiss Federal Office of Energy contract no° 152316-101883 and 153613-102809 (for D.K.B.), SCIEX with Hungary #—10.013 (Sciex-NMSch—The Swiss contribution to EU enlargement) and the VELUX Foundation project no° 790 (both for K.G.-S.), Swiss National Science Foundation no° NANOTERA 20NA21-145936 (for D.K.B) and Swiss South African Joint Research project no° SNF IZLSZ2.149031, and EU COST Action TD1102 "PHOTOTECH."

References

1. Rabinovich E (1961) Photochemical utilization of light energy. Proc Natl Acad Sci USA 47: 1296-1303.
2. Tributsch H, Calvin M (1971) Electrochemistry of excited molecules: photo electrochemical reactions of chlorophylls. Photochem Photobiol 14: 95-112.
3. Osterloh F E (2008) Inorganic materials as catalysts for photochemical splitting of water. Chem Mater 20: 35–54.
4. Gust D, Moore T A, Moore A L (2012) Realizing artificial photosynthesis. Faraday Discuss 155: 9 -26.
5. Rabinowitch E (1940) The Photogalvanic Effect II. The photogalvanic properties of thionine-iron system. J Chem Phys 8: 560-566.
6. Fong F K, Winograd N (1976) In Vitro Solar Conversion after the Primary Light Reaction in Photosynthesis. Reversible Photogalvanic Effects of Chlorophyll - Quinhydrone Half-Cell Reaction. J Am Chem Soc 98: 2287-2289.
7. Allam N K, Yen C, Near R D et al (2011) Bacteriorhodopsin / TiO_2 nanotube arrays hybrid system for enhanced photoelectrochemical water splitting. Energ Environ Sci 4, 2909-2914.
8. Balasubramanian S, Wang P, Schaller R D et al (2013) High-Performance Bioassisted Nanophotocatalyst for Hydrogen Production. Nano Lett 13: 3365–3371.
9. Saga Y, Watanabe T, Koyama K E et al (1999) Mechanism of Photocurrent Generation from Bacteriorhodopsin on Gold Electrodes. J Phys Chem B 103, 234-238.
10. Horn C, Steinem C (2005) Photocurrents Generated by Bacteriorhodopsin Adsorbed on Nano-Black Lipid Membranes. Biophys J 89:1046–1054.
11. Bora D K, Rozhkova E A, Schrantz K et al (2012) Functionalization of Nanostructured Hematite Thin-Film Electrodes with the Light-Harvesting Membrane Protein C-Phycocyanin Yields an Enhanced Photocurrent. Adv Funct Mater 22: 490 – 502.
12. Eggleston C M, Khare N, Lovelace D M (2006) Cytochrome *c* interaction with hematite (α-Fe_2O_3) surface. J Electron Spectrosc Relat Phenom 150: 220 – 227.
13. Ibrahim N, Kamarudin S K, Minggu L J (2014) Biofuel from biomass via photo-electrochemical reactions: An overview., J Power Sources 259: 33-42.
14. Bae S, Shim E, Yoon J et al (2008) Photoanodic and cathodic role of anodized tubular Titania in light-sensitized enzymatic hydrogen production. J Power Sources 185: 439–444.
15. Joo H, Bae S, Kim C et al (2009) Hydrogen evolution in enzymatic photoelectrochemical cell using modified seawater electrolytes produced by membrane desalination process. Sol Energ Mat Sol C 93: 1555–1561.
16. Boettcher S W, Spurgeon J M, Putnam M C (2010) Energy-Conversion Properties of Vapor-Liquid-Solid–Grown SiliconWire-Array Photocathodes. Science 327: 185-187.
17. Artero V, Fontecave M (2011) Light-driven bioinspired water splitting: Recent developments in photoelectrode materials. C R Chimie 14: 799-810.
18. Paracchino A, Brauer J C, Moser J et al (2012) Synthesis and Characterization of High-Photoactivity Electrodeposited Cu_2O Solar Absorber by Photoelectrochemistry and Ultrafast Spectroscopy. J Phys Chem C 116: 7341-7350.
19. Janzen A F, Seibert M (1980) Photoelectrochemical conversion using reaction-center electrodes. Nature, 286: 584 - 585.
20. Kumar B, Beyler M, Kubiak C P et al (2012) Photoelectrochemical Hydrogen Generation by an [FeFe] Hydrogenase Active Site Mimic at a p-Type Silicon/Molecular Electrocatalyst Junction. Chem Eur J 18: 1295 – 1298.
21. Reisner E, Fontecilla - Camps J C, Armstrong F A (2009) Catalytic electrochemistry of a [NiFeSe]-hydrogenase on TiO_2 and demonstration of its suitability for visible-light driven H_2 production, Chem Commun 550 - 552.
22. Tran P D, Artero V, Fontecave M (2010) Water electrolysis and photoelectrolysis on electrodes engineered using biological and bio-inspired molecular systems, Energ Environ Sci 3: 727–747.

23. Reisner E, Powell D J, Cavazza C et al (2009) Visible Light-Driven H_2 Production by Hydrogenases Attached to Dye-Sensitized TiO_2 Nanoparticle. J Am Chem Soc 131: 18457-18466.
24. Carmeli I, Frolov L, Carmeli C et al (2007) Photovoltaic Activity of Photosystem I-Based Self-Assembled Monolayer. J Am Chem Soc 129: 12352-12353.
25. Hou Y, Abrams B L, Vesborg P C K et al (2011) Bioinspired molecular co-catalysts bonded to a silicon photocathode for solar hydrogen evolution, Nat Mater 10: 434 - 438.
26. Muellerwesterhoff U T, Nazzal A (1984) [l. l] Ferrocenophanes as Effective Catalysts in the Photoelectrochemical Hydrogen Evolution from Acidic Aqueous Media. J Am Chem Soc 106: 5381 - 5382.
27. Lakadamyali F, Reisner E (2011) Photocatalytic H_2 evolution from neutral water with a molecular cobalt catalyst on a dye-sensitised TiO_2 nanoparticle. Chem Commun 47: 1695.
28. Tran P D, Wong L H, Barber J et al (2012) Recent advances in hybrid photocatalysts for solar fuel production. Energy Environ Sci 5: 5902 – 5918.
29. Utschig L M, Dimitrijevic N M, Poluektov, O G et al (2011) Photocatalytic Hydrogen Production from Noncovalent Biohybrid Photosystem I/Pt Nanoparticle Complexes. J Phys Chem Lett 2 (3): 236-241.
30. Greening C, Berney M, Hards K et al (2014) A soil actinobacterium scavenges atmospheric H_2 using two membrane-associated, oxygen-dependent hydrogenases. Proc Acad Natl Sci USA 111(11): 4257-4261.
31. De Lacey A L, Fernandez V M, Rousset M (2007) Activation and Inactivation of Hydrogenase Function and the Catalytic Cycle: Spectroelectrochemical Studies. Chem Rev 107 (10): 4304–4330.
32. Morra S, Valetti F, Sadeghi S J et al (2011) Direct electrochemistry of an [FeFe]-hydrogenase on a TiO2 Electrode. Chem Commun 47: 10566.
33. Hambourger M, Gervaldo M, Svedruzic D et al (2008) [FeFe]-Hydrogenase-Catalyzed H2 Production in a Photoelectrochemical Biofuel Cell. J Am Chem Soc 130 (6): 2015–2022.
34. Nikandrov V V, Shlyk M A, Zorin N A et al (1988) Efficient photoinduced electron transfer from inorganic semiconductor TiO_2 to bacterial hydrogenase. Fed Eur Biochem Soc Lett 234 (1):111–114.
35. McDonald T J, Svedruzic D, Kim Y et al (2007) Wiring-Up Hydrogenase with Single-Walled Carbon Nanotubes. Nano Letters 7(11): 3528-3534.
36. Brown K A, Dayal S, Ai X et al (2010) Controlled Assembly of Hydrogenase-CdTe Nanocrystal Hybrids for Solar Hydrogen Production. J Am Chem Soc 132: 9672–9680.
37. Petrie S, Stranger R, Pace R J (2010) PSII, Mn-oxo complex: Location of Potential Substrate Water Binding Sites in the Water Oxidizing Complex of Photosystem II. Angew Chem Int Ed 49: 1 – 5.
38. Siegbahn P E M (2009) Structures and Energetics for O2 Formation in Photosystem II. Acc Chem Res 42(12) 1871-1880.
39. Lewis N S (2011) Light work with matter. Nature 414: 589
40. Maly J, Krejci J, Ilie M et al (2005) Monolayers of photosystem II on gold electrodes with enhanced sensor response—effect of porosity and protein layer arrangement. Anal Bioanal Chem 381:1558 - 1567.
41. Maly J, Masojidek J, Masci A (2005) Direct mediator less electron transport between the monolayer of photosystem II and poly(mercapto-*p*-benzoquinone) modified gold electrode—new design of biosensor for herbicide detection. Biosens Bioelectron 21: 923 - 932.
42. Badura A, Guschin D, Esper B et al (2008) Photo-Induced Electron Transfer Between Photosystem 2 via Cross-linked Redox Hydrogels. Electroanalysis 20(10): 1043 – 1047.
43. Hakala M, Tuominen I, Keranen M et al (2005) Evidence for the role of the oxygen-evolving manganese complex in photoinhibition of Photosystem II. Biochim Biophys Acta 1706: 68 - 80.
44. Kato M, Cardona T, Rutherford A W (2012) Photoelectrochemical Water Oxidation with Photosystem II Integrated in a Mesoporous Indium–Tin Oxide Electrode. J Am Chem Soc 134: 8332 - 8335.

45. Brimblecombe R, Koo A, Dismukes G C et al (2010) Solar Driven Water Oxidation by a Bioinspired Manganese Molecular Catalyst. J Am Chem Soc 132: 2892–2894.
46. Berns D S, MacColl R (1989) Phycocyanin in Physical-Chemical Studies. Chem Rev 89: 807-825.
47. Glazer A N (1985), Light harvesting by phycobilisomes, Annu Rev Biophys Biophys Chem 14: 47-77.
48. Chen Z, Samuelson L A, Akkara J et al (1995) Sol-Gel Encapsulated Light-Transducing Protein Phycoerythrin: A New Biomaterial. Chem Mater 7: 1779-1783.
49. Glazer A N, Stryer L (1984) Phycofluor probes, Trends Biochem Sci 9: 423-427.
50. Kathiravan A, Chandramohan M, Renganathan R et al (2009) Photoinduced electron transfer from phycoerythrin to colloidal metal semiconductor nanoparticles. Spectrochim Acta Part A 72: 496–501.
51. Kathiravan A, Renganathan R (2009) Photosensitization of colloidal TiO_2 nanoparticles with phycocyanin pigment. J Coll Interf Sci 335:1965.
52. Suemori Y, Nagata M, Nakamura Y et al (2006) Self-assembled monolayer of light-harvesting core complexes of photosynthetic bacteria on an amino-terminated ITO electrode. Photosynth Res 90: 17 – 21.
53. Berg J M, Tymoczko J L, Stryer L (2002) Biochemistry, W. H. Freeman USA.
54. Modesto-Lopez L B, Thimsen E J, Collins A M et al(2010) Electrospray-assisted characterization and deposition of chlorosomes to fabricate a biomimetic light-harvesting device. Energy Environ Sci 3:216 – 222.
55. Amao Y, Shuto N, Furuno K (2012) Artificial leaf device for solar fuel production. Faraday Discuss 155:289 – 296.
56. Das R, Kiley P J, Segal M et al (2004) Integration of Photosynthetic Protein Molecular Complexes in Solid-State Electronic Devices, Nano Lett, 4 (6) :1079-1083.
57. McNamara W R, Snoeberger III R C, Li G et al (2009) Hydroxamate anchors for water-stable attachment to TiO_2 nanoparticles. Energ Environ Sci 2:1173–1175.
58. Yehezkeli O, Wilner O I, Tel-Vered R et al (2010) Generation of Photocurrents by Bis-aniline-Cross-Linked Pt Nanoparticle / Photosystem I Composites on Electrodes. J Phys Chem B 114: 14383–14388.
59. Smith M E B, Schumacher F F, Ryan C P et al (2010) Protein Modification, Bioconjugation, and Disulfide Bridging Using Bromomaleimides. J Am Chem Soc 132: 1960–1965.
60. Freitag M, Galoppini E (2011) Molecular hostimides Bioconjugation, and Disulfide Bridging Nanoparticle Surfaces. Energ Environ Sci 4: 2482-2494.
61. Leggett G J, Hunter C N (2007) Directed Formation of Micro- and Nanoscale Patterns of Functional Light-Harvesting LH2 Complexes. J Am Chem Soc 129:14625-14631.
62. Escalante M, Maury P, Bruinink C M et al (2008) Directed assembly of functional light harvesting antenna complexes onto chemically patterned surfaces. Nanotechnology 19: 025101.
63. Shinkarev V P, Brunner R, Wraight C A (1999) Application of near-field scanning optical microscopy in photosynthesis Research. Photosynthes Res 61: 181–191.
64. Iida K, Inagaki J, Shinohara K et al (2005) Absorption and Fluorescence Spectra and AFM Observation of the Light-Harvesting 1 Complex on a Mica Substrate Refolded from the Subunit Light-Harvesting 1 Complexes of Photosynthetic Bacteria *Rhodospirillum rubrum.* Langmuir 21: 3069-3075.
65. Arteni A A, Ajlani G, Boekema E J (2009) Structural organisation of phycobilisomes from Synechocystis sp. strain PCC6803 and their interaction with the membrane. Biochimica et Biophysica Acta 1787: 272–279.

Hybrid (Enzymatic and Photocatalytic) Systems for CO_2-Water Coprocessing to Afford Energy-Rich Molecules

Michele Aresta, Angela Dibenedetto, and Wojciech Macyk

Abstract This chapter deals with selected aspects of man-made solar-driven photoreduction of CO_2. In particular, direct photosynthetic processes are discussed which include the use of man-made materials for solar energy capture and utilization in enzymatically driven CO_2 reduction to energy-rich C_1 molecules. The enzymatic reduction of CO_2 requires cofactors which are the energy vectors to enzymes which in turn reduce CO_2 to energy-rich C_1 molecules. The regeneration of the reduced form of such cofactors is a key step in the whole process. The level at which man-made systems may recycle the cofactor and support the reduction reaction is discussed, and barriers to a full exploitation of the option are highlighted. The use of visible light in the entire process is privileged.

The paper does not cover other uses of solar energy such as thermal devices (solar power concentrators) nor solar-driven pure electrochemical processes, such as the use of PV for water electrolysis and hydrogen generation employed for CO_2 chemical reduction.

Keywords Man-made photosynthesis • Hybrid systems • Enzymatic reduction of CO_2 • Cofactors for CO_2 reduction • Cofactors regeneration • Visible light • Use of water as electron donor • Organic electron donors

M. Aresta (✉)
Department of Chemical and Biomolecular Engineering, NUS, Singapore 119077, Singapore

Interuniversity Consortium on Chemical Reactivity and Catalysis (CIRCC),
via Celso Ulpiani, 27, 70126 Bari, Italy
e-mail: michele.aresta@uniba.it

A. Dibenedetto
CIRCC and Department of Chemistry, University of Bari,
Campus Universitario, 70126 Bari, Italy

W. Macyk
Faculty of Chemistry, Jagiellonian University, Ingardena 3, 30-060 Kraków, Poland

E.A. Rozhkova, K. Ariga (eds.), *From Molecules to Materials*,
DOI 10.1007/978-3-319-13800-8_6

1 Introduction

The quest for efficient technologies of solar to chemical energy conversion and reduction of carbon dioxide atmospheric level is boosting the development of several man-made systems mimicking photosynthetic pathways. In the natural process, a series of redox-active bio-systems facilitates the electron transfer from water to carbon dioxide. Numerous natural processes can be categorized in which CO_2 is converted either into a R–COO$^-$ moiety, by reaction with a carbanion R$^-$, or into a formal "HCOH" unit incorporated into sugars (C_5 to C_6 conversion) or else into other C_1 molecules richer in energy, such as HCOOH, CO, CH_2O, CH_3OH, and CH_4. In nature, the driving force of the overall reaction of CO_2 reduction is represented by solar energy, efficiently absorbed in photosystems I and II by chlorophylls. The latter must assure not only an efficient light absorption but also properly balanced redox properties, both in ground and excited states, optimal lifetimes of the excited states, and good thermal stability and photostability.

The components of photosynthetic systems, optimized in nature within last over three billion years, can be considered as reference blocks in the attempt to design a man-made photosynthetic system [1]. Selected redox-active bio-systems and enzymes can offer their unique properties, facilitating the reduction reactions or electron (energy) transfer processes. On the other hand, several components of photosystems I and II might be substituted with simpler molecules or materials, either organic or inorganic. For instance, wide bandgap semiconductors offer various spectral and redox properties that can be further tuned by either engineering their morphologies or by doping and modifying the surface.

This chapter summarizes photocatalytic and photoelectrocatalytic systems, which, coupled to enzymes and cofactors as redox-active components, can efficiently drive reactions for solar to chemical energy conversion based on carbon dioxide and water (both laying in a well of energy) transformations into energy-rich molecules. Water can be substituted in some specific cases with waste or very cheap organics.

2 Enzymatic Reduction of Carbon Dioxide

The enzymatic reduction reactions of CO_2 can be categorized into:

1. Carboxylation reactions in which a R–COO$^-$ or RNH–COO$^-$ is formed from R–H or RNH_2
2. Fixation into sugars as a "HCOH" unit
3. Conversion into other C_1 molecules richer in energy, such as HCOOH, CO, CH_2O, CH_3OH, and CH_4

Such processes may have different energy requirements, as one can deduce from Fig. 1 in which the Gibbs free energy of formation of various C_1 molecules is represented.

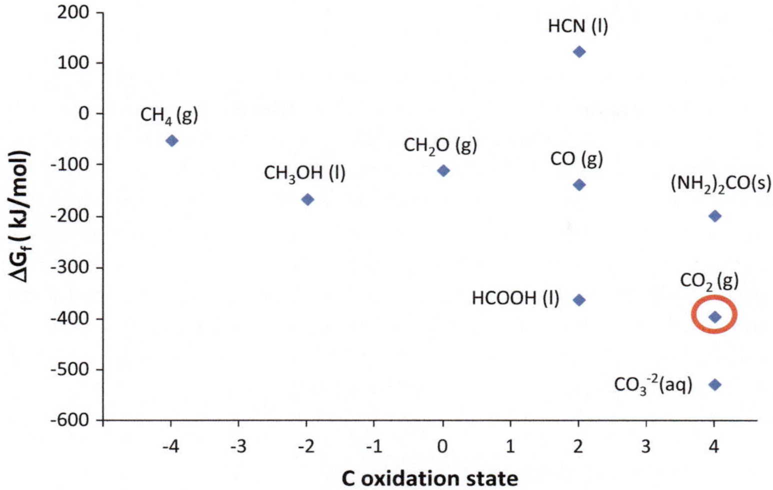

Fig. 1 The Gibbs free energy of formation of C_1 molecules in their standard state

It is evident that, with the exception of formation of inorganic carbonates (CO_3^{2-}) or organic carbonates (ROCOOR) (in the red circle in Fig. 1), any other transformation of CO_2 requires significant amounts of energy. Formic acid, which lies close to CO_2, indeed requires energy in the form of H_2 for its direct synthesis from CO_2. In our world, energy is mainly provided by fossil carbon (ca. 90 %) in its various forms (coal, oil, gas) that upon combustion is converted into CO_2. It is quite evident that if we wish to use CO_2 for producing energy-rich molecules, we should not use fossil carbon as source of energy, but more likely perennial energies, such as solar, wind, or geothermal energy or else hydropower. Renewable biomass can be better used for the production of chemicals or biofuels. Wind and geothermal energy and hydropower must first be converted into a usable form of energy for driving CO_2 reactions (most likely electric energy), while solar energy can be either directly (photochemical processes) or indirectly used (generation of PV or via photoelectrochemical processes). The use of solar energy to power reactions of CO_2 conversion catalyzed by enzymatic and model systems can be considered as a "man-made" photosynthetic process. The present paper makes an analysis of bottlenecks to the exploitation of such processes that may bring to the conversion of large volumes of CO_2 into useful chemicals and energy products.

2.1 *Enzymes Catalyzing the CO_2 Conversion*

The enzymes involved into the conversion of CO_2 are either "carboxylation enzymes" or "2e$^-$ redox" enzymes [1a].

2.1.1 Carboxylation Enzymes

Carboxylation enzymes are listed in Table 1 that is representative but not exhaustive of enzymes involved in carboxylation/decarboxylation reactions. Such list also indicates the direction of the reaction the enzyme catalyzes, whether carboxylation or decarboxylation. As Table 1 shows, in general enzymes require a *cofactor* that represents the *energy vector* and is converted during the reaction into a less energetic form. In the carboxylation reactions shown in Table 1, three different *cofactors* are used, namely, ATP, NADPH, or Ferredoxin$_{red}$.

In order to allow a continued process, the low-energy forms of cofactors must be back converted into the original ones for further use. Such conversion in nature is accomplished with the use of solar energy as primary energy source and the intermediacy of other enzymatic reactions or living organisms.

2.1.2 CO_2-Reducing Enzymes

In enzyme-based CO_2 reduction, the back conversion of the oxidized form of the cofactor into its reduced form is a key issue for the enzymatic reaction can be exploited. Cofactors also play a key role in CO_2 reduction processes. Table 2 lists some selected enzymatic reduction reactions of CO_2 into energy-rich C_1 molecules.

Table 1 Enzymes involved into "carboxylation"/"decarboxylation" enzymatic reactions

Enzyme name	Carbon species used	*Cofactor* required and its conversion	Thermodynamically preferred reaction
RuBisCO	CO_2	None	Carboxylation (not reversible)
PEP carboxylase	HCO_3^-	None	Carboxylation (not reversible)
Pyruvate carboxylase	HCO_3^-	ATP → ADP	Carboxylation
Acetyl-CoA carboxylase	HCO_3^-	ATP → ADP	Carboxylation
Propionyl-CoA carboxylase	HCO_3^-	ATP → ADP	Carboxylation
Methylcrotonyl-CoA carboxylase	HCO_3^-	ATP → ADP	Carboxylation
2-Oxoglutaratecarboxylase	HCO_3^-	ATP → ADP	Carboxylation
PEP carboxykinase	CO_2	ATP → ADP	Carboxylation/decarboxylation
Isocitrate dehydrogenase	CO_2	$NADPH + H^+ \rightarrow NADP^+$	Decarboxylation
Malic enzyme	CO_2	$NADPH + H^+ \rightarrow NADP^+$	Decarboxylation
Phosphogluconate dehydrogenase	CO_2	$NADPH + H^+ \rightarrow NADP^+$	Decarboxylation
Pyruvate oxidoreductase $NADP^+$	CO_2	$NADPH + H^+ \rightarrow NADP^+$	Decarboxylation
Pyruvate synthase[a]	CO_2	$2Fd_{red} \rightarrow 2Fd_{ox}$	Decarboxylation
2-Oxobutyrate synthase[a]	CO_2	$2Fd_{red} \rightarrow 2Fd_{ox}$	Decarboxylation
2-Ketoglutarate synthase[a]	CO_2	$2Fd_{red} \rightarrow 2Fd_{ox}$	Decarboxylation
2-Ketoisovalerate oxidoreductase[a]	CO_2	$2Fd_{red} \rightarrow 2Fd_{ox}$	Decarboxylation

[a]These enzymes are oxygen sensitive

Table 2 Enzymes involved in the reduction of CO_2 to C_1 energy-rich molecules

Name of the enzyme	Acronym	Starting compound	Product of reduction	Metal center involved
Formate dehydrogenase	$F_{ate}DH$	CO_2	$HCOO^-$	W
Formaldehyde dehydrogenase	$F_{ald}DH$	HCOOH	H_2CO	Zn
Alcohol dehydrogenase	ADH	H_2CO	CH_3OH	Zn
Carbon monoxide dehydrogenase	CODH	CO_2	CO	Ni
Tetrahydrofolate	THF	CO_2	$-CH_3$	None directly

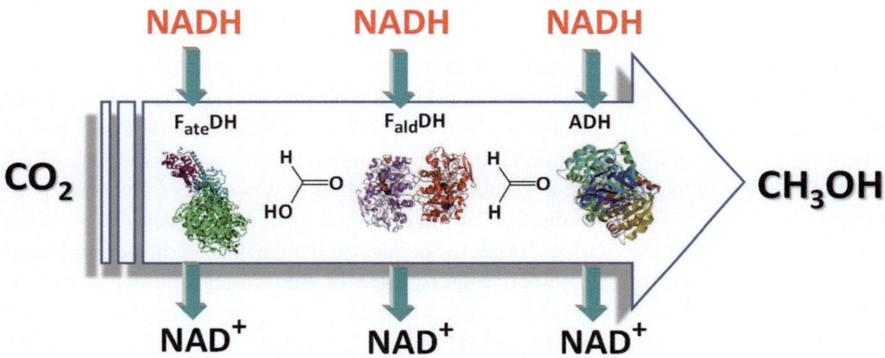

Fig. 2 Conversion of CO_2 into methanol mediated by three enzymes: each step is a $2e^-$ step and consumes one molecule of NADH

The stepwise reduction of CO_2 into $HCOOH > CO > CH_2O > CH_3OH > CH_4$ is an interesting case as if it is driven by enzymes, the reactions are very selective and fast. One can think to exploit such reaction for the production of methanol that is a large-scale chemical used either as a chemical commodity or as a fuel (direct combustion or in fuel cells) [2]. The conversion of CO_2 into methane is one of the biological routes of synthesis of the latter, and it occurs in biogas production processes from organics. Conversely, the conversion into methanol requires more sophisticated technologies that imply the use of enzymes and cofactors under specific conditions, as discussed below.

The sequence that brings from CO_2 to CH_3OH uses three enzymes, namely, *formate dehydrogenase* ($F_{ate}DH$), *formaldehyde dehydrogenase* ($F_{ald}DH$), and *alcohol dehydrogenase* (ADH) is a quite interesting one. It has been shown to work since long time [3], and, very interestingly, the three enzymes are commercial products. As it is shown in Fig. 2, the process requires 3 mol of NADH per each mol of CO_2 converted into CH_3OH. This is an unfortunate event that makes the process not economically viable: the back conversion of NAD^+ into NADH is highly advisable for an exploitation of such system.

As a matter of fact, the couple NAD^+/NADH has a great interest from the point of view of potential application in synthetic chemistry (it could find a use in any reduction reaction, also in the synthesis of optically active molecules). Therefore, among many cofactors that can be found in nature (NAD^+/NADH, $NADP^+$/NADPH, FAD/$FADH_2$, FMN/$FMNH_2$, among others), the NAD^+/NADH couple is one of most investigated for it may find application in several synthetic processes.

2.2 *Man-Made Systems for the Back Conversion of Oxidized Forms of Cofactors*

As mentioned above, several cofactors exist in nature, each being characterized by a different redox potential. For instance, for NAD^+/NADH and $APAD^+$/APADH, the redox potentials are −0.32 [4] and −0.218 V vs. SHE [5], respectively. The value of the potential of a couple determines its application in a redox process. At the same time, each couple may require a suitable semiconductor as source of electrons generated by solar light for the reduction of the oxidized form may occur. Often, if the reduction reaction of the oxidized cofactor occurs on the surface of the semiconductor, the reaction may not be selective as isomers of the reduced cofactor may be formed as well as dimers. However, in a man-made system, for a selective reduction of the oxidized cofactor may be carried out, a "mediator" is required which can fasten and make selective the e^- transfer from the semiconductor to the oxidized substrate. $[Cp^*Rh(bpy)(H_2O)]^{2+}$ has been shown to fulfill quite effectively such role [6]. The hydrido form of such complex, $[Cp^*Rh(bpy)H]^+$, is formed at −0.53 V vs. SHE [6]. This potential assures a facile electron transfer to the oxidized form of the cofactor with subsequent regeneration of the reduced form.

However, a man-made system enabling carbon dioxide reduction into energy-rich molecules, based on NAD^+/NADH redox couple as final e^- transfer agent, results in a multicomponent system made of the parts categorized below:

1. A semiconductor able to harvest photons in the visible light range and convert them into *excitons*
2. A mediator for the electron transfer to the oxidized form of the cofactor
3. A cofactor that may transfer energy to the enzymatic systems
4. An enzymatic system that may carry out a fast and selective CO_2 reduction
5. A reducing agent (water or any other inexpensive chemical whose oxidized form would have a large use) that may provide electrons necessary for CO_2 reduction

In the sequence (1–5), the weak points are represented by step (2), the regeneration of the reduced form of the cofactor, and step (5) the electron donor. Water should be used as electron source if a large-scale economic implementation of the technology is wished. Alternatively, organic waste can be used that are converted into useful chemicals, with a large-scale market. In the following paragraphs, the state of the art of the regeneration of cofactor NAD^+ as a most suited e^-/H^+ transfer agent will be discussed. The regeneration of NADH from NAD^+ can be carried out in a chemical, electrochemical, and photochemical way. The latter is quite interesting, as it may not affect enzymes, and is relevant to the concept of this paper.

3 Photocatalytic Systems Cooperating with Enzymes

Two alternative approaches involving semiconductors, enzymes, and redox-active cofactors are possible. In the first case, two separated steps will operate. First, the cofactor NADH is generated in a photocatalytically driven NAD^+ reduction. Subsequently and separately, the catalytic reduction of carbon dioxide is driven by the chemical energy stored in NADH coupled to the enzyme(s). Such *two-pot* system and its energy diagram are presented in Fig. 3. The left part of such scheme is the key issue and most essential: it will be discussed in the next paragraphs.

In a second case, a *one-pot* system combines all components together (Fig. 4). The electron flow from the electron donor to CO_2 is realized through a series of processes initiated by a photon absorption. The excited photosensitizer is reduced by an electron donor while providing electrons to enzymatic centers for CO_2 reduction. Such electron transfer can be mediated by a mediator (or by multiple mediators) and must always involve reversible redox processes. The role of mediators can conveniently be played by the rhodium complex, $[Cp^*Rh(bpy)(H_2O)]^{2+}$, that is known to be able to facilitate the electron transfer from the photosensitizer to the cofactor, e.g., NAD^+. The reduced cofactor supplies electrons to an enzymatic subsystem which performs the CO_2 reduction. The energy diagram is presented under the system scheme in Fig. 4.

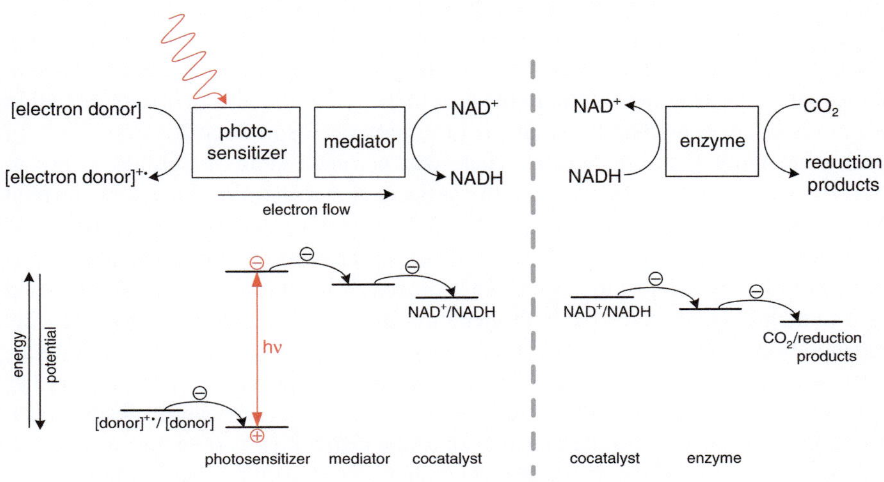

Fig. 3 The general scheme and energy diagram of hybrid photocatalytic systems of NAD^+ reduction. The photocatalytically generated NADH is then used in a separate compartment for CO_2 reduction carried out by enzymes: NADH is oxidized to NAD^+

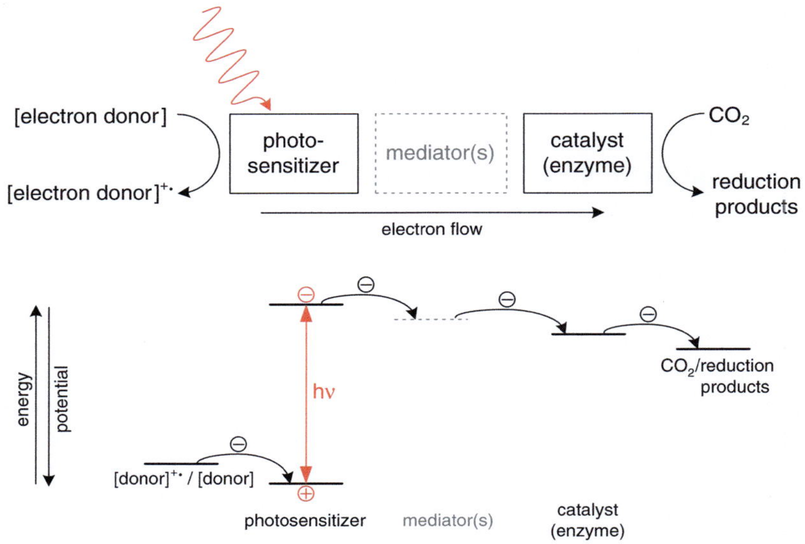

Fig. 4 The general scheme and energy diagram of hybrid photocatalytic/enzymatic systems of carbon dioxide reduction (*one-pot system*)

3.1 Photocatalytic Regeneration of the Cofactor NAD$^+$

The regeneration of a cofactor (e.g., NADH from NAD$^+$) is essential for the conversion of CO_2 to any energy-rich molecule may be exploited. We focus on the NADH/NAD$^+$ couple because such system is active in several reactive systems and can play a key role also in synthetic organic chemistry in the synthesis of optically active molecules. It is also the cofactor in the enzymatic reduction of CO_2 into methanol in water under ambient conditions. Such conversion occurs in three 2e$^-$ steps (Fig. 2), each consuming 1 mol of NADH. Therefore, per each mole of CH_3OH, 3 mol of NADH are consumed: this makes the process noneconomic, unless the formed NAD$^+$ could be transformed back into NADH, mimicking nature. As said above, the back conversion of NAD$^+$ into NADH, if implemented on a large scale and at low cost, would be of interest to several other processes in which NADH could be used as a reducing agent (selective hydrogenation of organic substrates). However, the regeneration of NADH from NAD$^+$ is an issue of large industrial interest and has, thus, been explored for long time by several research groups. In principle, either chemical or electrochemical or photochemical processes are candidates for such reduction. The chemical route may present some difficulties as reducing agents may affect the enzymes. The electrochemical reduction of NAD$^+$ has been attempted for long time, but the results are not exciting. The photochemical recycling seems to be quite appealing as it would require simple devices. The target is to use visible light. In the next paragraphs, the attempts made so far are described.

3.1.1 UV Light-Induced Processes

Graetzel and Cuendet reported the NADH regeneration in the presence of TiO_2 and $[Rh(bpy)_3]^{3+}$ complex [7]. The reaction rate was enhanced by methanol used as an electron donor. The anatase form of titanium dioxide, excited with ultraviolet light, can reduce NAD^+, $NADP^+$, and NMN^+ to their corresponding reduced forms even in the absence of the rhodium complex [8]. The process has been realized in slightly alkaline solutions, in the presence of amino acids or EDTA as electron donors. The formation of biologically inactive dimers as by-products, e.g., $(NAD)_2$, was observed. Noteworthy, such dimers are not effective in the electron transfer and therefore in the reduction of CO_2.

When compared to titanium dioxide, sulfides of transition metals are characterized by lower reduction potentials. For example, zinc sulfide has a potential of the conduction band edge as low as −1.8 to −2.0 V vs. NHE (to be compared to ca. −0.6 V vs. NHE for TiO_2). Moreover, holes photogenerated in ZnS particles are weaker oxidants than holes formed within TiO_2 particles. This lowers the possibility of uncontrolled photocatalytic oxidation of organic components of the system by hydroxyl radicals that are efficiently produced upon excitation of titanium dioxide with UV light. It has been recently demonstrated that NADH regeneration in the presence of zinc sulfide can be achieved upon illumination with UV-A radiations, around 390 nm [3c]. In such reaction, various electron donors were employed. Among others, bio-glycerol appeared to be particularly efficient. These results open new perspectives for substituting commonly used, expensive electron donors, like TEOA and EDTA, by cheap by-products of the chemical industry, like bio-glycerol. The next step in the development of this approach may encompass the coupling of oxygen evolving cocatalysts to enable use of water as electron donor.

3.1.2 Visible Light-Induced Processes

As mentioned above, the exploitation of NADH as reducing agent will be possible if visible light will be employed for its regeneration from NAD^+. Recently, Park and coworkers [9] have reviewed the visible light-induced processes of NADH cofactor regeneration. The paper covers photocatalysts belonging to various classes of materials: organic dyes, porphyrins, semiconductors, and metal nanoparticles. All systems utilize $[Cp^*Rh(bpy)(H_2O)]^{2+}$ complex as electron mediator for electron transfer from the photocatalyst to the cofactor. Table 3 presents selected examples of photosensitizers as reported in the literature.

Visible light-absorbing organic dyes can operate as photosensitizers in the processes of reductive regeneration of cofactors. It is the case for eosin Y [15, 10, 11], porphyrins [12], and other photosensitizers. Moreover, the same process can be applied to the reduction of a whole family of oxidized cofactors, e.g., NAD^+, $APAD^+$, $PAAD^+$, $TNAD^+$, and $NAAD^+$ [16].

The efficient quenching of eosin Y fluorescence by $[Cp^*Rh(bpy)(H_2O)]^{2+}$ complex results from the electron transfer from the excited dye to the rhodium species.

Table 3 Some examples of molecular and colloidal photosensitizers used in NADH photoregeneration

System	Photosensitizer	Redox potential E_{red} (NHE)	TOF/h NADH regeneration	References
Organic				
Xanthene	Eosin Y	–0.92 V	1,690	[10, 11]
Porphyrin	ZnTPPS	–0.96 V	0.46	[12]
Others	Proflavine	na	127.8	[13]
Inorganic				
Semiconductors	CdTe (nanoparticles)	–0.6 V	0.54	[14]
Polyoxometalate	$W_2Fe_4Ta_2O_{17}$	na	0.37	[15]

na not available

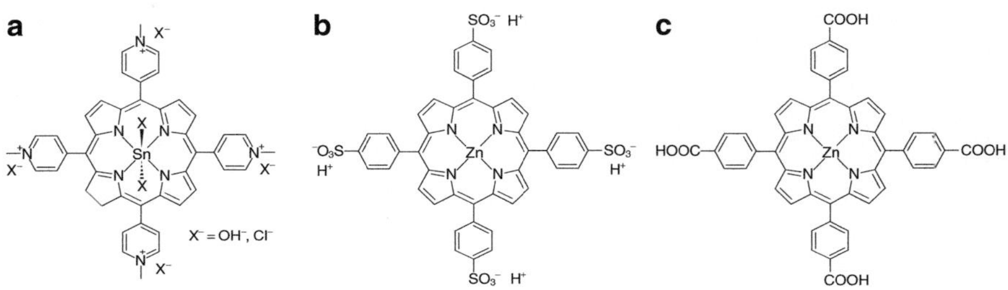

Fig. 5 The structures of tin(IV)-*meso*-tetrakis(*N*-methylpyridinium)-chlorin (**a**); ZnTPPS (**b**) and ZnTPPC (**c**)

Then the hydride transfer to the oxidized cofactor takes place yielding the corresponding reduced form. In the case of eosin Y-sensitized reduction of NAD^+, the reaction yield was only 48 %, while in the case of $APAD^+$ and $PAAD^+$, the yields were as high as 77 and 56 %, respectively, after 30 min of irradiation with visible light ($\lambda > 420$ nm, $[\text{cofactor}]_0 = 1$ mM). On the other hand, reduction of $TNAD^+$ and $NAAD^+$ was much less effective, as the conversion to the corresponding reduced products amounted only to 8.6 and 2.1 %, respectively.

An example of an analogous chlorin-based system has been presented recently by Knoer et al. [17] who have explored a new class of photosensitizers, such as tin(IV)-*meso*-tetrakis(*N*-methylpyridinium)-chlorin (Fig. 5a). In his studies, Knoer used EDTA or TEOA as electron donors and the rhodium complex as the electron mediator and observed reduction of various cofactors such as NAD^+ and $BNAD^+$. The measured quantum yield of NADH production amounted at 0.014 (irradiation at 592 nm). Such value has to be compared with quantum yields in the range of 0.01–0.06 achieved in natural photosynthetic systems. Interestingly, the above tin complex appeared relatively stable upon prolonged irradiation.

A similar approach has been investigated by Park et al. [12] who have used the Zn-coordinated porphyrins shown in Fig. 5b, c. In the presence of $[Cp^*Rh(bpy)(H_2O)]^{2+}$ and TEOA, NADH production was observed. Irradiation of the systems yielded 20 % conversion of NAD^+ to NADH in the case of ZnTPPS, while a slightly lower conversion efficiency has been reported for ZnTPPC (1 h irradiation, λ>400 nm, $[NAD^+]_0$=1 mM). For comparison, the corresponding H_2TPPS and H_2TPPC porphyrins appeared less effective (7 and 12 % conversion under similar conditions, respectively), while the manganese complexes (MnTPPS, MnTPPC) were almost photocatalytically inactive. This difference has been ascribed to a favorable back electron transfer from the reduced rhodium mediator to Mn^{III} center (the redox potential of Mn^{III}/Mn^{II} couple is –0.21 V vs. Ag/AgCl).

NADH has been photoregenerated in the photocatalytic system containing carbon-doped titanium dioxide (a photosensitized TiO_2 material showing photoactivity upon visible light irradiation), $[Cp^*Rh(bpy)(H_2O)]^{2+}$ complex, and various electron donors in a water solution [18]. In the absence of any electron donor (using water as electron donor), the photoregeneration of NADH reached ca. 26 %. The NAD^+ conversion reached 74 and 94 % when thioethanol or H_2 were added, respectively.

A very elegant photocatalytic system made of CdS/TiO_2 photoelectrode has been proposed by Park and coworkers [19]. Tubular or particulate titanium dioxide film was covered with cadmium sulfide, deposited by SILAR (successive ionic layer adsorption and reaction) method. Visible light absorbed by cadmium sulfide results in the exciton (electronhole pair) generation. The lifetime of the latter is prolonged by an efficient electron transfer to the conduction band of titanium dioxide, where it is consumed in NAD^+ reduction mediated by the $[Cp^*Rh(bpy)(H_2O)]^{2+}$ complex. On the other hand, holes localized at CdS nanoparticles oxidize TEOA. The authors demonstrated four times higher efficiency of NADH regeneration at tubular nanofilms than at the particulate films. This effect has been attributed to a better diffusion of reactants along tubes and to a better collection and transport of electrons by nanotubular titanium dioxide (Fig. 6). NADH regeneration reached ca. 80 % after

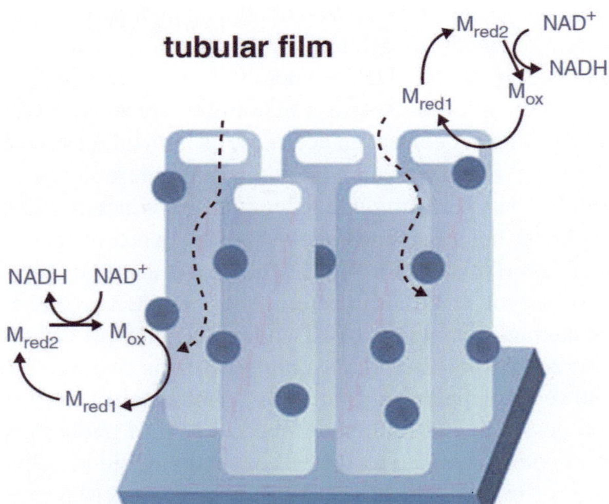

Fig. 6 The photocatalytic system composed of TiO_2 nanotubes and CdS nanoparticles for an efficient NADH photoregeneration under visible light illumination (λ>420 nm) [15, 19]

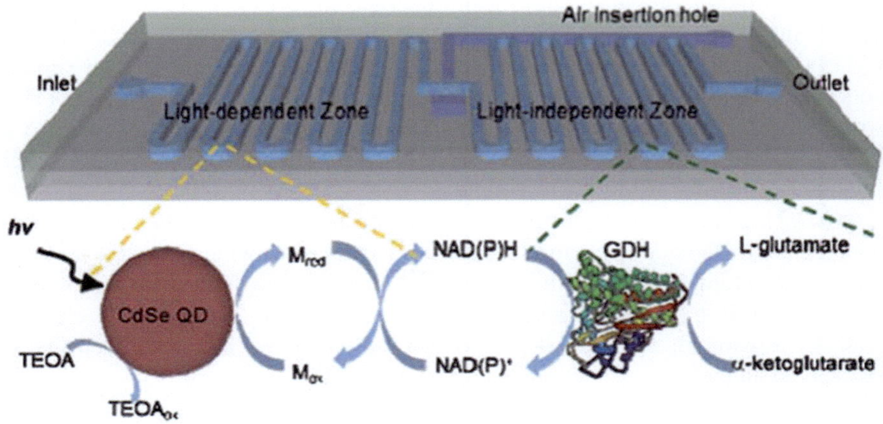

Fig. 7 The microfluidic artificial photosynthetic platform [21, 22]

2 h irradiation with visible light ($\lambda > 420$ nm; $[NAD^+]_0 = 1$ mM). A similar process of NADH regeneration has been reported to proceed also at CdS nanoparticles deposited on a silica support by SILAR method [20]. In this case, the regeneration of NADH reached 80 % after 1 h irradiation.

Cadmium selenide in the form of quantum dots performs a similar reaction. The use of such photosensitizer was proposed by Park et al. [21] in a microfluidic photosynthetic platform (Fig. 7) that utilizes the TEOA/CdSe(QD)/$[Cp^*Rh(bpy)(H_2O)]^{2+}$/NAD^+ system to drive enzymatic conversion of α-ketoglutarate to L-GLUTAMATE. The microdevice is divided in two parts: in the first one, the photocatalytic regeneration of the coenzyme takes place, while in the second one, in dark, the enzymatic production of L-GLUTAMATE proceeds. Depending on the size of applied quantum dots, i.e., on their bandgap energy, the NADH regeneration yield reached approximately 27 and 17 % for CdSe quantum dots with a maximum emission at 655 and 565 nm, respectively. No significant decrease in activity was noticed for repeated photoreactions up to five times. The photocatalytic part of the microdevice can possibly be coupled with another part performing, e.g., enzymatic reduction of CO_2 to formate.

The development of heterogeneous photocatalysts in recent years has shown that the photoinduced interfacial electron transfer processes, responsible for a huge majority of photocatalytic reactions in heterogeneous systems, can be realized using organic semiconductors. The works of Kisch and Beranek proved the applicability of the poly(tri-*s*-triazine) polymer, $(C_3N_4)_n$, in sensitizing titanium dioxide [23, 24]. This material, however, appeared also an active photocatalyst itself in water photodetoxification process. Liu and Antonietti reported NADH photoregeneration in the presence of such material with up to 100 % conversion yield (irradiation with LED 420 nm, $[NAD^+]_0 = 1$ mM, $[Cp^*Rh(bpy)(H_2O)]^{2+}$ as the electron mediator) [25]. In particular, the authors compared diatom frustule-structured graphitic C_3N_4 (DE-g-C3N4; 5 $m^2\ g^{-1}$), mpg-C_3N_4 (286 $m^2\ g^{-1}$), and g-C_3N_4 (5 $m^2\ g^{-1}$). The highest conversion rates were achieved in the case of mpg-C_3N_4 (ca. 80 % conversion in 45 min), but the maximum conversion yield was observed for DE-g-C_3N_4 (100 % conversion in 120 min, but only 50 % in 45 min). Only 35 % conversion rate was observed after 2 h irradiation in the presence of g-C_3N_4. It should be pointed that the biologically active 1,4-NADH isomer was exclusively formed upon irradiation.

Also metal nanoparticles (NPs) were successfully employed as photosensitizers capable of visible light absorption. Platinum NPs in the presence of TEOA (triethanolamine) as the electron donor afforded maximum 86 % regeneration of NADH from 0.2 mM solution of NAD^+ after 12 h of irradiation with visible light [26]. The optimal particle size was 5 nm (5, 10, and 30 nm diameter particles were tested). The estimated quantum yield amounted at 0.043 for 420 nm. Similar effects were not observed when palladium or gold nanoparticles were used instead of platinum.

3.2 *Reduction of Carbon Dioxide in Multicomponent Photocatalytic/Enzymatic Systems*

Photoenzymatic regeneration of NADPH was also achieved when graphene-based photosensitizers were applied. Choudhury and coworkers have described a photoenzymatic reduction of CO_2 to formate (Fig. 8) [27] and acetophenone derivatives to chiral 1-phenylethanols [28]. In the described photocatalytic cycle, the covalently bonded chromophore, a multianthraquinone substituted porphyrin with the chemically converted graphene (CCGCMAQSP), plays the role of the photosensitizer and assures an effective charge separation. Also in this system, the rhodium complex, $[Cp^*Rh(bpy)(H_2O)]^{2+}$, acts as the electron mediator. Reported amounts of regenerated NADH reached ca. 45 % after 2 h of irradiation with visible light ($\lambda > 420$ nm, $[NAD^+]_0 = 1.24$ μmol/3.1 mL). In addition, the production of formate under identical conditions amounted at 110 μmol. The authors tested also other photosensitizers

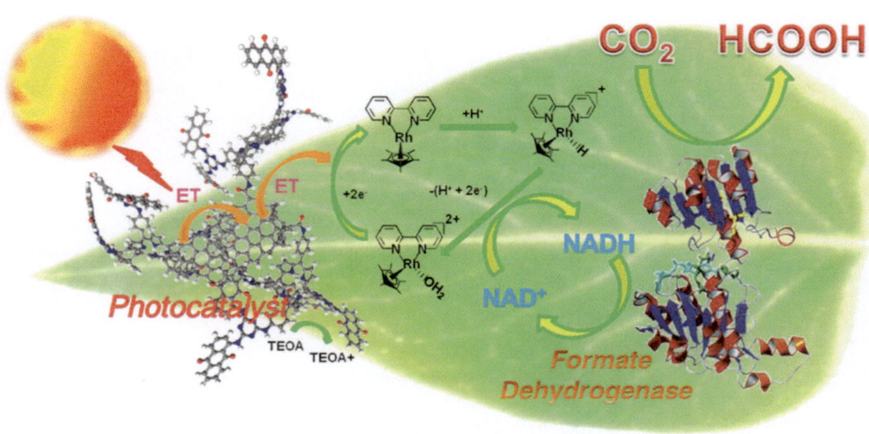

Fig. 8 CCGCMAQSP-driven photocatalytic carbon dioxide reduction [28]

instead of CCGCMAQSP. However, utilization of MAQSP (multianthraquinone substituted porphyrin) or $W_2Fe_4Ta_2O_{17}$ [15] resulted in ca. 2.4 and 7.7 times lower production of formate, respectively.

An interesting approach (Fig. 9) has been undertaken by Fujita et al. ([29] and papers cited therein). The authors synthesized a ruthenium complex, $[Ru(bpy)_2(pbn)]^{2+}$, acting as a photosensitizer upon MLCT excitation, with NADH-inspired ligand (pbn). The photogenerated $[Ru(bpy)_2(pbnHH)]^{2+}$ (Fig. 9; pbnHH = 1,5-dihydro-2-(2-pyridyl)-benzo[b]-1,5-naphthyridine) is a hydride donor, but not sufficiently strong to reduce directly carbon dioxide. However, the product of a further reduction step, $[Ru(bpy)_2(pbnHH)]^{+}$, should facilitate this process, as predicted by theoretical calculations [30].

The concept proposed by Kurayama et al. [31] presents one of the most complex systems, thought as a *one-pot* reactor. Titanium dioxide, upon absorption of UV light, should oxidize water to molecular oxygen, while the electrons are transferred through a multistep cascade (Fig. 10). Instead of the rhodium complex, methyl viologen salt is used. This component should transfer the electrons through a

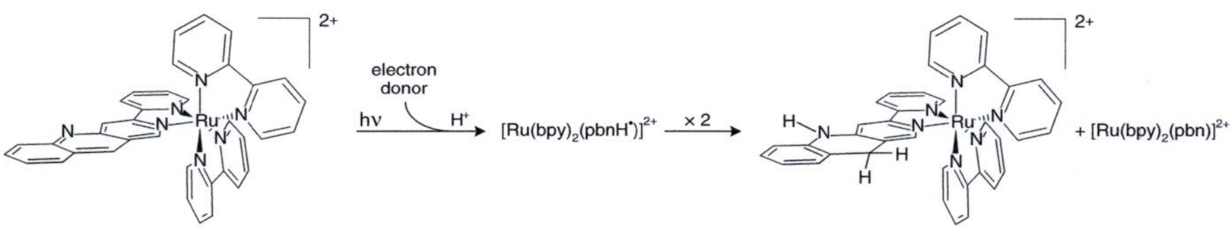

Fig. 9 The $[Ru(bpy)_2(pbn)]^{2+}$ complex and its photoreduction to $[Ru(bpy)_2(pbnHH)]^{2+}$

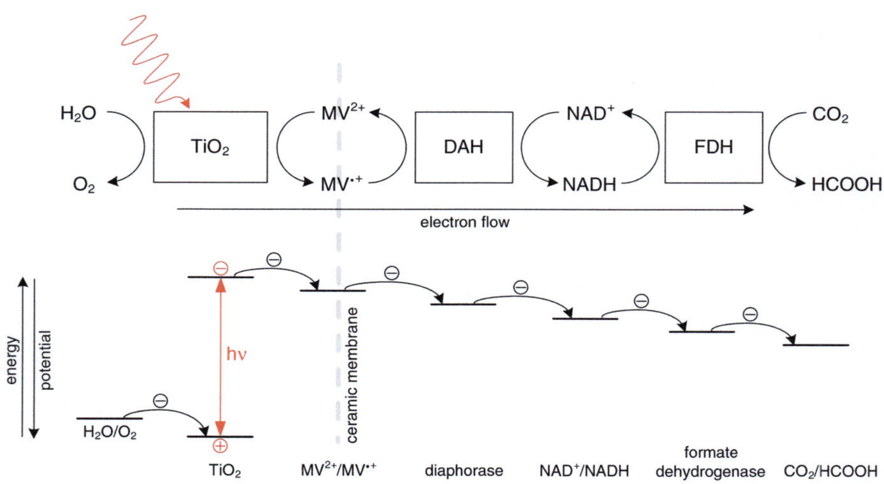

Fig. 10 The hybrid photocatalytic/enzymatic system of CO_2 to formate reduction proposed by Kurayama et al. [31]

ceramic membrane, which divides the reactor into two parts: illuminated and dark. The role of the ceramic membrane is to avoid reoxidation of formate to CO_2 by holes generated in the valence band of titania. Moreover, the dark part of the reactor must be protected from light to suppress photoinduced degradation of enzymes. The dark part contains two enzymes (diaphorase, DAH, and formate dehydrogenase, FDH) and the NADH cofactor. Since methyl viologen does not operate directly on FDH, diaphorase from pig heart together with NADH is introduced in the cycle, driving the enzymatic reduction of carbon dioxide to formate. The authors demonstrated the photocatalytic formate production by using the two-compartment reactor. Although measurable amounts of formate were formed in the absence of any additional electron donor, significantly higher yields of the process were recorded when ethanol was added to the photocatalytic part of the system. The formate production reached the level of ca. 0.8 mmol dm^{-3} after 45 min.

The same concept of a two-compartment reactor [31] has been used by Aresta and coworkers who have used TiO_2 modified with $[CrF_5(H_2O)]^{2-}$ as visible light absorption system [32c] for NAD^+ reduction to NADH mediated by $[Cp^*Rh(bpy)(H_2O)]^{2+}$ (or its Ir analog) that acts as electron transfer to the oxidized form of the cofactor. In the illuminated compartment, the photoreduction of NAD^+ to 1,4-NADH occurs. The latter is used in the dark compartment of the reactor for the enzymatic reduction of CO_2 to methanol (Fig. 2). The NADH occurs in water. Although the latter was shown to be a potential source of electrons, the yield in NADH greatly increased when bio-glycerol was added to the water solution and used as e^- donor. Interesting yields of NAD^+ reduction were observed using the device described above. The $TiO_2/[CrF_5(H_2O)]^{2-}/[Cp^*Rh(bpy)(H_2O)]^{2+}$/glycerol-water system was used for repetitive cycles, and a CH_3OH/NADH molar ratio above 1,000 was reached making this system quite interesting for further developments. The critical issue in the hybrid photocatalytic/enzymatic system is the different rate of NADH consumption in the enzymatic reduction of CO_2 and NADH regeneration in the photocatalytic reaction. The former reaction is some >10 times faster than the latter. Such reaction rates must be equalized for a practical application that requires more efficient photocatalytic systems for NADH regeneration.

4 Photoelectrocatalytic Systems for NAD^+ Reduction to NADH

Figure 11 shows the primary photochemical system active in photosynthesis. The electron transfer from H_2O to a cofactor (e.g., $NADP^+$) is enabled by photoexcitation of chlorophylls in photosystems PS-I and PS-II. Several redox couples are active in such process.

The use of chlorophylls in man-made systems has always been a challenge for mimicking nature. Hybrid systems that coupled chlorophylls and semiconductors have been attempted since long time. Pioneering studies to use hybrid photoelectrochemical systems have been done in the early 1970s [33, 34] using either a

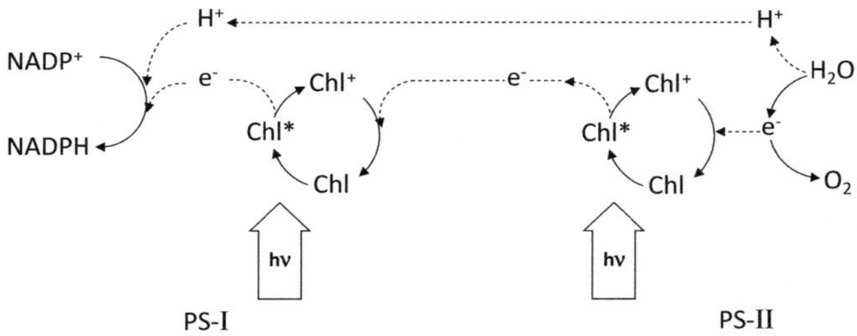

Fig. 11 Electron transfer scheme of primary photochemical process in photosynthesis

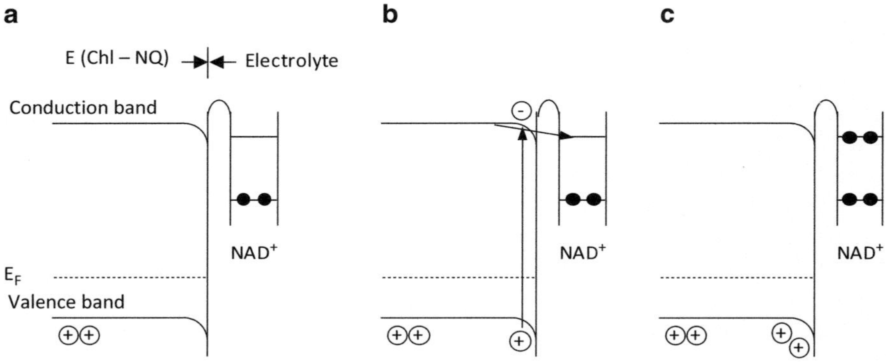

Fig. 12 Energy potential schemes of the photoelectrochemical reactions at E(Chl-NQ)

semiconductor electrode on which chlorophyll was deposited [34] or metal-chlorophyll-metal sandwich devices [35, 36]. Low photocurrents were observed, often of the order of 10^{-7} A cm^{-2} or lower. Moreover, a fast decay was observed. Later on, Takahashi and Kikuchi [37] developed new concept chlorophyll electrodes able to achieve stable photocurrents of the order of 4×10^{-6} A cm^{-2}, with the photoexcitation of chlorophyll reversibly cycled. The photoexcitable electrodes were prepared by coating a Pt electrode with a thin layer of chlorophyll-quinone/hydroquinone composite. The cathode was made of chlorophyll/naphthoquinone (Chl-NQ) and acting as PS-I, while the anode was made of chlorophyll-anthrahydroquinone (Chl-AHQ) and acting as PS-II under illumination. The two electrodes made a photoelectrolytic cell: E(Chl-NQ)/NAD^+//$[Fe(CN)_6]^{4-}$/E(Chl-AHQ).

Later on, the same authors [38] have published a detailed study on the use of the naphthoquinone composites (Chl-NQ) as a cathode and anthraquinone (Chl-AQ) as anode for NAD^+ reduction. NAD^+ was reduced without any external voltage application. The E(Chl-NQ) electrode, where the reduction of the cofactor occurred, was in the range +70 to −50 mV vs. SCE that avoided any one-electron transfer to NAD^+. The behavior of the E(Chl-NQ) electrode can be compared to that of a *p*-type semiconductor electrode, and the valence and conduction bands can be described as in Fig. 12a.

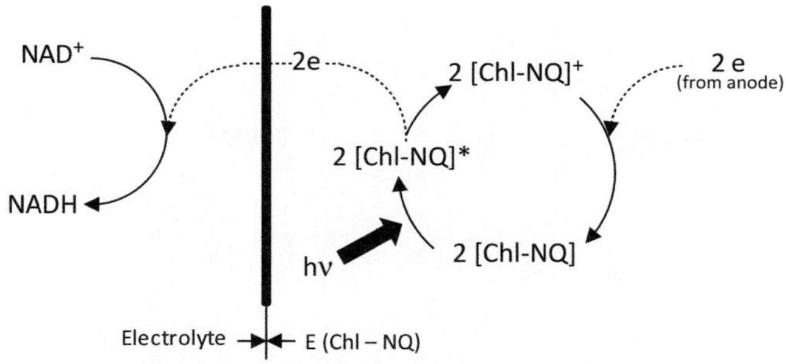

Fig. 13 Electron transfer at the E(Chl-NQ) in the photoelectrochemical reduction of NAD^+ to NADH

The energy bandgap between the valence and conduction band was not measured, but estimated to be around 1.9 eV on the basis of the fact that chlorophyll has a maximum of absorbance at 663 nm both in acetonitrile solution and in the composite Chl-NQ. When photoexcited, the electron is transferred to the lowest vacant energy level of NAD^+ as presented in Fig. 12b. Most probably a tunneling effect plays a key role. Figure 12c shows the effect of two electrons flowing to NAD^+ with proton transfer from the electrolyte. The hole that remains in the valence band can be neutralized, and the initial ground state regenerated, with an electron from the counter electrode. The whole process is depicted in Fig. 13.

The behavior of NQ is explained as an effect of composite formation (Chl-NQ) that allows the system to behave as PS-I system.

5 Other Systems

Another approach to man-made photosynthesis utilizes inorganic solid materials (e.g., silica) serving as a porous framework for whole photosynthetic living cells [39–43] or cell components [44, 45]. The use of natural photosynthetic systems, developed by algae and photosynthetic plant cells, enables profiting from relatively high quantum yields of carbon dioxide conversion. Appropriate design of the inorganic matrix, assuring an optimal transport of reagents and nutrients together with a high biocompatibility, may offer efficient, relatively stable, and controllable photosynthetic systems for solar CO_2-to-fuel and solar CO_2-to-chemical conversion (Fig. 14) [46]. The results of research in this field are very promising, since encapsulation restrains degradation of cells, while their photoactivity can be maintained for months.

Immobilization does not have to be limited to whole cells. Su et al. demonstrated a photocatalytic activity of thylakoids isolated from spinach, immobilized in porous silica gels [44–46]. The entrapped thylakoids showed a significantly enhanced stability when compared to isolated, unprotected ones. The hybrid systems were active in photoinduced oxygen production.

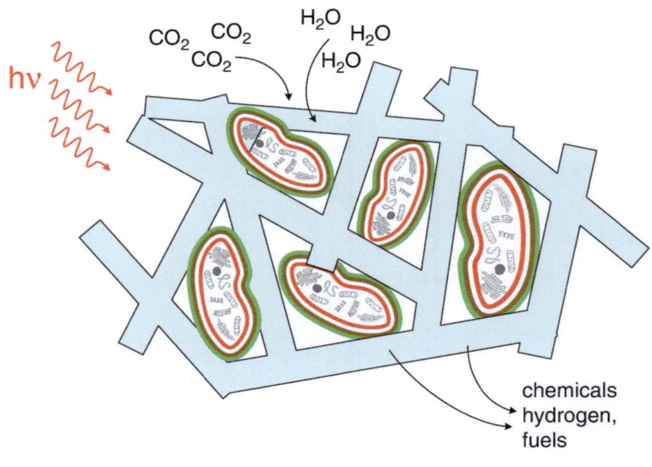

Fig. 14 Algae or plant cells encapsulated into inorganic matrix

6 Conclusions and Outlook

The utilization of CO_2 is today considered with much attention as it could contribute to cycling carbon, mimicking nature that makes thousands of compounds from atmospheric CO_2, or to reduce the climate change impact (CCI) by using CO_2 in place of chemicals that may have a much more serious impact. An example is the replacement of chlorofluorocarbons (CFCs) that may have a CCI also 10,000 times higher than CO_2. The conversion of CO_2 into added value products can be categorized as (1) production of chemicals, (2) production of fuels, (3) and enhanced biological utilization for aquatic biomass production. The utilization of CO_2 has recently acquired a novel perspective justified by the new attitude to exploit the conversion of perennial energy such as solar-wind-geothermal (SWG) into other forms of energy, e.g., electrical or chemical. Such new opportunity is intensified by the perspective lower cost and longer life of devices such as PV, opening, thus, a new perspective for the conversion of *spent carbon* into *working carbon*. Large volumes of CO_2 can be converted into fuels making possible a step toward *cycling carbon by mimicking nature*. This would represent an epochal change as it would move the C-economy from a linear to a cyclic trend. However, the utilization of CO_2 would allow to reduce its emission into the atmosphere and to reduce the extraction of fossil-C, part of which would be substituted by recycled-C (Fig. 15).

As a matter of fact, the natural C-cycle that runs some >720 Gt_{CO2}/y is not able to buffer the anthropogenic emissions (30–32 Gt_{CO2}/y) that causes the accumulation of CO_2 into the atmosphere with an ever growing reinforcement of the natural greenhouse effect. The parallel growing trend of the curves representing the (1) world population, (2) energy consumption, (3) CO_2 emissions, and (4) CO_2 level

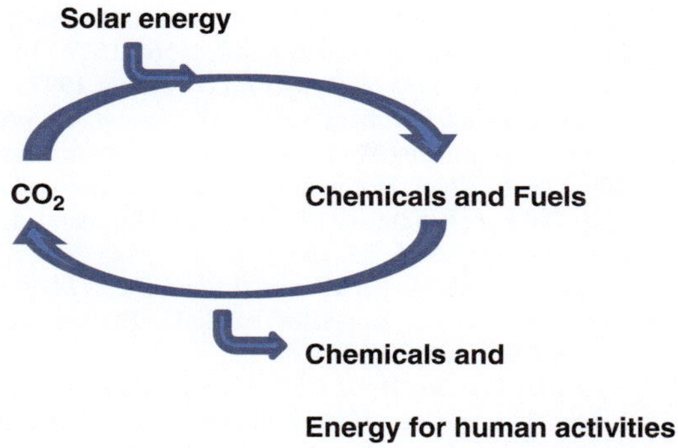

Fig. 15 C-cycling: expanding the man-made cycle will reduce the use of natural C-resources

into the atmosphere seems to be an evidence of an existing relationship between CO_2 atmospheric level and CC. Therefore, the limitation of the emission of CO_2 into the atmosphere is a must if *no-return points* have to be avoided. The utilization of CO_2 may become a key technology for our future, supposed that large volumes of CO_2 are converted. The actual utilization of CO_2 is close to 200 Mt/y that represents a very minor fraction (0.62 %) of the emission. To become significant such fraction must be increased at least by one order of magnitude or more, rising to at least 10 %: this means that 3,300 Mt/y must be avoided; the conversion of *spent-C* into *working-C* by means of man-made or hybrid synthetic procedures will play a key role. This big jump needs the development of innovative conversion strategies. The use of solar energy and coupling photochemistry and bio-systems has a great potential for the implementation of technologies that may convert large volumes of CO_2. The man-made photosynthesis may be the technology of the future for working side by side to nature for a better management of carbon.

Acknowledgments This work is a part of the "Activation of small molecules in photocatalytic systems" project, realized within the TEAM program awarded by the Foundation for Polish Science, cofinanced by the European Union, Regional Development Fund (TEAM/2012-9/4), and of the IC^2R srl program on "Carbon Recycling."

References

1. a) Michele Aresta and John V. Schloss Eds, "*Enzymatic and Model Carboxylation and Reduction Reactions for CO_2 Utilization*" NATO Series, Kluwer Acad. Publ., 1990. b) Michael Hambourger, Gary F. Moore, David M. Kramer, Devens Gust, Ana L. Moore, Thomas A. Moore *Chem. Soc. Rev.* **2009**, *38*, 25-35.
2. Michele Aresta, Angela Dibenedetto, Liang-N. He, *Analysis of demand for captured CO_2 and products from CO_2 conversion.* Report exclusively for Members of the Carbon Dioxide Capture and Conversion (CO_2CC) Program of The Catalyst Group, **2012**.
3. a) Robin Obert, Bakull C. Dave *J. Am. Chem. Soc.* **1999**, *121(51)*, 12192-12193; b) Angela Dibenedetto, Paolo Stufano, Wojciech Macyk, Tomasz Baran, Carlo Fragale, Mirco Costa, Michele Aresta *ChemSusChem* **2012**, *5*, 373-378.

4. a) Eberhard Steckhan, Sabine Herrmann, Romain Ruppert, Eva Dietz, Markus Frede, Elke Spika *Organometallics* **1991**, *10*, 1568-1577; b) Gottfried Unden, Johannes Bongaerts *Biochimica et Biophysica Acta-Bioenergetics* **1997**, *1320(3)*, 217-234.
5. Corinna Gallus, Bernhard Schink *Archives of Microbiology* **1998**, *169*, 333-338.
6. a) Frank Hollmann, Andreas Schmid, Eberhard Steckhan *Angew. Chem. Int. Ed.* **2001**, *40*, 169-171. b) Frank Hollmann, Po-Chi Lin, Bernard Witholt, Andreas Schmid *J. Am. Chem. Soc.* **2003**, *125*, 8209-8217. c) Franck Hollmann, K. Hofstetter, Andreas Schmid *Trends in Biotechnology* **2006**, *24*, 163-171.
7. Pierre Cuendet, Michael Graetzel *Photochem. Photobiol.* **1984**, *39*, 609-612.
8. Elzbieta Bojarska, Krzysztof Pawlicki, Barbara Czochralska *J. Photochem. Photobiol. A: Chem.* **1997**, *108*, 207-213.
9. Sahng Ha Lee, Jae Hong Kim, Chan Beum Park *Chem. Eur. J.* **2013**, *19*, 4392-4406.
10. Sahng Ha Lee, Dong Heon Nam, Chan Beum Park *Adv. Synth. Catal.* **2009**, *351*, 2589-2594.
11. Sahng Ha Lee, Dong Heon Nam, Jae Hong Kim, Jin-Ook Baeg, Chan Beum Park *ChemBioChem* **2009**, *10*, 1621-1624.
12. Jae Hong Kim, Sahng Ha Lee, Joon Seok Lee, Minah Lee and Chan Beum Park *Chem. Commun.* **2011**, *47*, 10227-10229.
13. Dong Heon Nam, Chan Beum Park *ChemBioChem* **2012**, *13*, 1278-1282.
14. Dong Heon Nam, Sahng Ha Lee, Chan Beum Park *Small* **2010**, *6*, 922-926.
15. Chan Beum Park, Sahng Ha Lee, Esakkiappan Subramanian, Bharat. B. Kale, Sang Mi Lee and Jin-Ook Baeg *Chem. Commun.* **2008**, 5423-5425.
16. Sahng Ha Lee, Hye Jung Lee, Keehoon Won, and Chan Beum Park *Chem. Eur. J.* **2012**, *18*, 5490-5495.
17. Kerstin T. Oppelt, Eva Wöß, Martin Stiftinger, Wolfgang Schöfberger, Wolfgang Buchberger, and Günther Knör *Inorg. Chem.* **2013**, *52*, 11910-11922.
18. Zhongyi Jiang, Chenqiu Lu, Hong Wu *Ind. Eng. Chem. Res.* **2005**, *44*, 4165-4170.
19. Jungki Ryu, Sahng Ha Lee, Dong Heon Nam, Chan Beum Park *Adv. Mater.* **2011**, *23*, 1883-1888.
20. Sahng Ha Lee, Jungki Ryu, Dong Heon Nam, Chan Beum Park *Chem. Commun.* **2011**, *47*, 4643-4645.
21. Jae Hong Kim, Dong Heon Nam, Chan Beum Park *Current Opinion in Biotechnology* **2014**, *28*, 1-9.
22. Joon Seok Lee, Sahng Ha Lee, Jae Hong Kim, Chan Beum Park *Lab Chip*, **2011**, *11*, 2309-2311.
23. Dariusz Mitoraj, Radim Beranek, Horst Kisch *Photochem. Photobiol. Sci.* **2010**, *9*, 31-38.
24. Michal Bledowski, Lidong Wang, Ayyappan Ramakrishnan, Radim Beranek *J. Mater. Res.* **2013**, *28*, 411-417.
25. Jian Liu, Markus Antonietti *Energy Environ. Sci.* **2013**, *6*, 1486-1493.
26. Shrikant S. Bhoware, Ka Young Kim, Jin Ah Kim, Qiong Wu, Jinheung Kim, *J. Phys. Chem., C* **2011**, *115*, 2553-2557.
27. Sumit Choudhury, Jin-Ook Baeg, No-Joong Park, Rajesh K. Yadav *Angew. Chem. Int. Ed.* **2012**, *51*, 11624-11628.
28. Rajesh K. Yadav, Jin-Ook Baeg, Gyu Hwan Oh, No-Joong Park, Ki-jeong Kong, Jinheung Kim, Dong Won Hwang, Soumya K. Biswas *J. Am. Chem. Soc.* **2012**, *134*, 11455-11461.
29. Mark D. Doherty, David C. Grills, James T. Muckerman, Dmitry E. Polyansky, Etsuko Fujita *Coord. Chem. Rev.* **2010**, *254*, 2472-2482.
30. James T. Muckerman, Patrick Achord, Carol Creutz, Dmitry E. Polyansky, Etsuko Fujita *PNAS* **2012**, *109*, 15657-15662.
31. Fumio Kurayama, Tatsushi Matsuyama, Hideo Yamamoto *Advanced Powder Technol.* **2004**, *15*, 51-61.
32. a) Piero Giannoccaro, Ms Thesis, Dept. of Chemistry, University of Bari, 2008; b) Michele Aresta, Angela Dibenedetto, Tomasz Baran, Antonella Angelini, Przemyslaw Labuz, Wojciech Macyk, invited paper, BJOC, 2014; c) Michele Aresta, Angela Dibenedetto, Wojciech Macyk, Tomasz Baran, IT Patent MI20131135.

33. Nguyen T. Van, H. Ti. Tien *J. Phys. Chem.* **1970**, *74*, 3559-3568.
34. Helmut Tributsch, Melvin Calvin *Photochem. Photobiol.* **1971**, *14*, 95-112.
35. Ching W. Tang, Andreas C. Albrecht *J. Phys. Chem.* **1975**, *62*, 2139-2149.
36. Ching W. Tang, Andreas C. Albrecht *J. Phys. Chem.* **1975**, *63*, 953-961.
37. Fujio Takahashi, Ryoichi Kikuchi *Biochim. Biophys. Acta-Bioenergetics* **1976**, *430*, 490-500.
38. Fujio Takahashi, Masuo Aizawa, Ryoichi Kikuchi, Shuichi Suzuki *Electrochim. Acta* **1977**, *22*, 289-293.
39. Alexandre Leonard, Philippe Dandoy, Emeric Danloy, Gregory Leroux, Christophe F. Meunier, Joanna C. Rooke, Bao-Lian Su *Chem. Soc. Rev.* **2011**, *40*, 860-885.
40. Alexandre Leonard, Joanna C. Rooke, Christophe F. Meunier, Hugo Sarmento, Jean-Pierre Descy, Bao-Lian Su *Energy Environ. Sci.* **2010**, *3*, 370-377.
41. Christophe F. Meunier, Philippe Dandoy, Bao-Lian Su *Journal of Colloid and Interface Science* **2010**, *342*, 211-224.
42. Christophe F. Meunier, Joanna C. Rooke, Alexandre Leonard, Pierre Van Cutsem, Bao-Lian Su *J. Mater. Chem.* **2010**, *20*, 929-936.
43. Joanna Claire Rooke, Alexandre Leonard, Hugo Sarmento, Christophe F. Meunier, Jean-Pierre Descy, Bao-Lian Su *J. Mater. Chem.* **2011**, *21*, 951-959.
44. Christophe F. Meunier, Joanna C. Rooke, Alexandre Leonard, Hao Xie, Bao-Lian Su *Chem. Commun.* **2010**, *46*, 3843-3859.
45. Christophe F. Meunier, Pierre Van Cutsem, Young-Uk Kwon, Bao-Lian Su *J. Mater. Chem.* **2009**, *19*, 1535-1542.
46. Joanna Claire Rooke, Alexandre Léonard, Christophe F. Meunier, Hugo Sarmento, Jean-Pierre Descy, Bao-Lian Su *Journal of Colloid and Interface Science* **2010**, *344*, 348-352.

Current Challenges of CO_2 Photocatalytic Reduction Over Semiconductors Using Sunlight

Víctor A. de la Peña O'Shea, David P. Serrano, and Juan M. Coronado

Abstract The possibility of producing fuels and chemicals from CO_2 using semiconductors and sunlight holds the promise of storing abundant, despite diluted, solar power in the form of energy-rich chemicals. In principle, this process can operate under very soft conditions, and this may facilitate a distributed production of solar fuels. However, photocatalytic reduction of CO_2 over semiconductors has proved to be a very challenging process, and the yields currently attained are rather limited. Accordingly, a number of drawbacks should be overcome if this process is expected to become a feasible alternative for generation of sustainable fuels and chemicals in the midterm. First, the conversion rate of CO_2 should be boosted to reach values in the order of millimoles per hour and gram of semiconductor. In addition, several relevant engineering challenges should be conveniently solved before this technology becomes viable. In this chapter, we evaluate thoroughly the most recent advances on the understanding of this photocatalytic process, and at the same time, we try to define the most relevant aspects that scientific research should address in the near future.

Keywords Carbone dioxide activation • Photocatalytic conversion • Semiconductor • Artificial photosynthesis • Solar Photreactors

V.A. de la Peña O'Shea • J.M. Coronado (✉)
Thermochemical Processes Unit, IMDEA Energy Institute,
Avenida Ramón de la Sagra, 3, 28935 Móstoles, Madrid, Spain
e-mail: juanmanuel.coronado@imdea.org

D.P. Serrano
Thermochemical Processes Unit, IMDEA Energy Institute,
Avenida Ramón de la Sagra, 3, 28935 Móstoles, Madrid, Spain

Chemical and Environmental Engineering Group, ESCET, Rey Juan Carlos University,
C/Tulipán s/n, 28933 Móstoles, Madrid, Spain

E.A. Rozhkova, K. Ariga (eds.), *From Molecules to Materials*,
DOI 10.1007/978-3-319-13800-8_7

1 Routes for the Activation of the CO_2 Molecule

As expected for a molecule that constitutes the end product of multitude of chemical reactions, from fermentation to combustion, CO_2 is highly stable. This is reflected in the high negative value of its free energy of formation, $\Delta G_f° = -396$ kJ/mol, which is remarkably larger than that of water. Therefore, activation of this molecule requires a considerable supply of energy. In addition, being a gas, if the process implies the incorporation of the CO_2 into a solid, there is a significant entropy penalty to be considered [1].

Accordingly, the uses of CO_2 in chemical industry are currently rather limited, and many of the possible applications are related to use in the food and beverage industries, or exploit its capacity as supercritical solvent. Regarding large-scale chemical transformations of CO_2, the synthesis of urea, H_2NCONH_2, with a global production of about 115 Mt/y, and salicylic acid, with a world demand of >20 kt/y, are among the most relevant commercial processes [1, 2].

The development of transformation routes of CO_2 into fuels using other primary energy sources is highly appealing because it has the potential of establishing a closed, despite imperfect, cycle of carbon use and recovery, as depicted schematically in Fig. 1. Obviously, the efficiency in cutting down the greenhouse emissions will depend critically on the source of the energy input and the efficiency of the process. In this respect, there is a clear opportunity to couple the overproduction of electricity by renewable technologies (basically wind and PV) with the synthesis of chemicals that can be utilized as energy vectors. A suitable process could be the reduction of CO_2 with H_2 to methanol over Cu/ZnO catalysts operating at 260 °C and 80 bars of pressure. Under these conditions, a CO_2 conversion per pass of 35–40 % is obtained using a H_2/CO_2 ratio of 3:1 at 10,500 h^{-1} GHSV, and the selectivity to

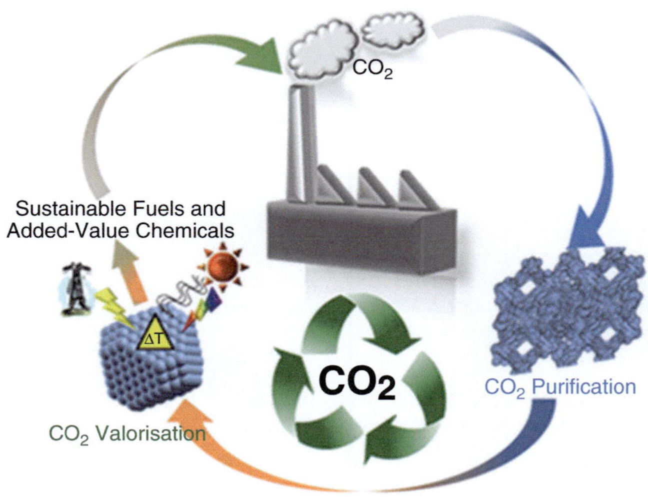

Fig. 1 Pictorial representation of the possible routes for the thermal, photocatalytic, or electrocatalytic transformations of CO_2 into fuels and chemicals

methanol reaches about 64 % [3]. Although the methanol yield obtained with a syngas feed is higher, these results show the relevant potential of CO_2 hydrogenation for fuel production. On this regard, a simple estimation indicates that using PV panels to generate electricity, the global energy efficiency of the process could be higher than 10 % and even better if direct co-electrolysis of CO_2 and H_2O is considered [1]. Nevertheless, there are significant uncertainties about the integration of the different processes, and exploitation costs are still far from being competitive [1].

In this context, the possible use of sunlight for CO_2 activation constitutes a remarkably attractive, despite very challenging, way for storing solar energy, because the use of this renewable source may enhance the reduction of greenhouse gas emissions. Very recently, a thermosolar route based on the use of CeO_2 as a redox mediator has been reported [4]. In this process, the initial partial reduction of the solid oxide at high temperature and inert atmosphere is followed in a second step by the reoxidation of CeO_{2-x} by CO_2 to yield CO. Alternatively, in this last stage, a mixture of CO_2 and H_2O can be fed to a solar reactor to produce syngas, which can be further transformed via Fischer–Tropsch into liquid fuels. The reported yield for the solar energy storage is 1.73 %, which sets a notable mark for the processing of solar fuels. In fact, this value is comparable to that of natural photosynthesis although the energy density of syngas is lower than, for example, cellulose. However, due to the land requirements of concentrating solar plants and the complexity of working at high temperatures, the development of alternative and softer routes for the production of solar fuels is still attracting a great deal of interest.

In this respect, the possibility of taking advantage of renewable electricity, especially during the periods of low consumption and high production, makes highly attractive the electrochemical reduction of CO_2 as a convenient way to regulate the demand. In fact, good faradaic efficiencies have been reported for this process, and it is known that product distribution can be tuned to a certain extent by selecting the metal used in the electrodes (e.g., Cu facilitates the formation of hydrocarbons). Nevertheless, the high overpotentials required, the low solubility of CO_2 in water, and the complex separation of the products are still important barriers for the implementation of this electrochemical process [5].

In this context, the photocatalytic reduction over semiconductors stands out as a very promising route for CO_2 transformation using solar energy in a very direct way. As it will be described in detail in the following sections, simplicity of the process and the very mild operation conditions required constitute the most relevant incentives for further investigating this technology, which is not free of important limitations [5–7].

2 Fundamentals of Semiconductor Photocatalysis

The increasing interest in photocatalysis for energy and environmental applications, which is clearly reflected by the growing number of articles published every year (more than 4,100 works in 2013 according to SCOPUS), relies on the possibility of employing sunlight, the paradigmatic renewable energy, to promote reactions. In contrast with most industrial catalytic processes, which usually entail high

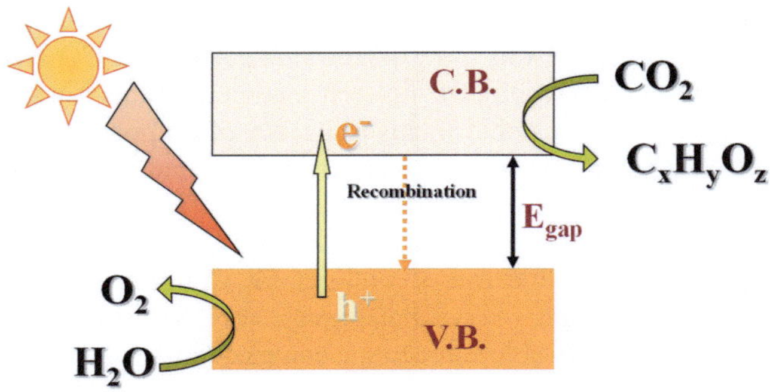

Fig. 2 Scheme of the photocatalytic reduction of CO_2 process on an irradiated semiconductor

temperature and/or elevated pressure, photocatalysis operates under ambient conditions provided that an adequate radiation source is available. Moderate pressure and temperature may have a potentially high impact in reducing the cost of different processes, but if these expectations are to be fulfilled, a significant boost of the efficiency of the photocatalytic processes is crucial. The reason for this is that, as it will be further discussed, the energy efficiency of many photoactivated processes is usually lower than 5 % but is even lower in the case of CO_2 activation. However, in order to completely understand the interest and limitations of this technology, the fundamentals of the photocatalytic production of solar fuels using semiconductors will be briefly reviewed in this section.

A scheme of the basic processes that take place on photocatalytic semiconductors in the specific case of CO_2 photoreduction is displayed in Fig. 2. First, if the wavelength of the photon energy exceeds that of the bandgap, light absorption induces the electron promotion from the valence to the conduction band. Formally, this electronic leap creates an unoccupied state or electron hole in the valence band, which can be conveniently described as a particle on its own [8, 9]. Most of these electron–hole pairs recombine releasing the excess energy as light or, more frequently, as heat. In fact, one of the advantages of semiconductors as photocatalysts, as compared to photoactive coordination compounds, is that charge carriers formed upon illumination of these solids present a relatively high stability and mobility [9]. Subsequently, the transference of the photogenerated carriers to the reactant molecules should take place in order to initiate the photocatalyst process. Conduction band electrons can reduce electron acceptors, which in the most common case of aerobic reaction are oxygen molecules. On the other hand, valence band holes act as oxidants, and they attack donor species, which in the case of fuel production should be ideally water molecules. Migration of electrons and holes through the interface requires semiconductors with certain electronic characteristics, in particular, a good alignment of the energy states of the adsorbed species and those of the photocatalyst. Consequently, chemical potential of both electron and holes depends on the relative positions of the conduction and the valence band, which are determined by the physicochemical nature of the semiconductor, and these parameters establish what reactions are thermodynamically achievable using a specific semiconductor.

Thus, for example, water splitting requires an energy of at least 1.23 eV (without considering overpotentials), but this reaction can only take place if electronic structure of the semiconductors is such that it allows simultaneously the reduction of protons ($E_{NHE}(H^+/H_2) = 0.0$ eV) and the oxidation of water ($E_{NHE}(O_2/H_2O) = 1.2$ eV). These conditions impose a significant constraint to the electronic properties of effective photocatalysts.

The efficiency of a photocatalytic reaction is determined by the quantum yield, which is defined as the ratio between the rate of chemical transformation and the flux of absorbed photons. The values of the quantum yield vary with the process considered, but they are usually lower than 1 % for reactions in solution, although they can be significantly higher for some specific processes in the gas phase [10]. However, the estimation of absorbed light is frequently difficult in real systems due to extensive photon scattering and reflection. Consequently, for the sake of simplicity, many studies assume that all the radiation is absorbed, and accordingly, an alternative efficiency index, known as photonic efficiency, is defined as the quotient between the rate of the photocatalytic events and the total photon flux.

Over 200 different semiconductors based on metal oxides (*ZnO*, WO_3, $NaTaO_3$, $BiWO_3$, etc.), oxynitrides ($(Ga_{1-x}Zn_x)(N_{1-x}O_x)$), metal sulfides (CdS, ZnS, etc.), and nonmetallic phases ($g\text{-}C_3N_4$), among other chemical families, have been assayed as potential photocatalysts, mainly for water splitting. However, despite the continuous search for novel photoactive materials, TiO_2 still remains as the most widely used photocatalyst in many applications due to its relatively good activity, high stability, low cost, and limited toxicity. Nevertheless, the reduced overlapping with the solar spectrum represents an important drawback of TiO_2, and this fact has stimulated the research in visible-light photoactivation of semiconductors.

3 Artificial Photosynthesis for CO_2 Reduction

The expression *artificial photosynthesis* frequently denotes any photocatalytic process powered by sunlight to produce fuels by analogy with the natural process. In this way, the production of hydrogen via water splitting is often considered a type of photosynthesis [11]. However, in this chapter, we will refer exclusively to the photocatalytic valorization of CO_2, using preferentially water as electron donor. In fact, photosynthesis in green plants refers to the sunlight-driven CO_2 reduction to yield carbohydrates via the multistep Calvin cycle, which is finely controlled by series of enzymes with very specific functions [12]. Similarly, the photochemical transformation of CO_2 into fuels and chemicals using sunlight constitutes an interesting route for storing chemical energy in the form of simple molecules (e.g., methane or methanol), and accordingly in the future, this technology may have an impact in reducing the use of fossil resources if fully deployed. Currently, this area of research constitutes one of the hottest topics among the field of energy and catalysis at worldwide level, with specific research programs in the USA and Japan [13].

However, in comparison with the significant advances in the development of new photocatalysts for the removal of pollutants and, especially, for the production of

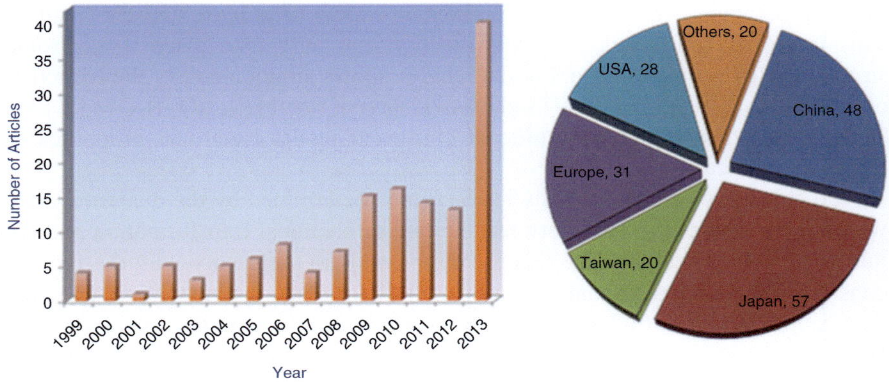

Fig. 3 Yearly evolution of the number of publication related to CO_2 photocatalytic reduction and country distribution of the researchers involved. Data source: SCOPUS

hydrogen by water splitting, CO_2 valorization has received less attention. The still-moderate number of studies is probably related to the inherent difficulty of the photocatalytic activation of CO_2, which has greatly limited the progress in this field. However, the interest in this process seems to be growing as judged by the number of publications, as it can be appreciated in Fig. 3. In this section, the specific barriers for the development of this technology will be discussed.

In artificial photosynthesis, the photocatalyst is expected to play several key roles such as absorbing light, generating electron–hole pairs, and allowing the separation and migration of the generated charges to redox sites of the surface, where CO_2 conversion can take place. Multielectrónic reduction of CO_2, as described by the chemical equations (1)–(7) (see below), potentially can lead to the formation of C_1 molecules with different degrees of reduction, ranging from formic acid to methane:

$$CO_2 + \mathrm{e}^- \rightarrow \mathrm{CO_2}^{\bullet -} \quad E = -0.49\mathrm{V} \tag{1}$$

$$\mathrm{CO_2 + H^+ + 2e^- \rightarrow HCO_2^-} \quad E = -0.49\mathrm{V} \tag{2}$$

$$\mathrm{CO_2 + 2H^+ + 2e^- \rightarrow CO + H_2O}\, E = -0.53\mathrm{V} \tag{3}$$

$$\mathrm{CO_2 + 2H^+ + 2e^- \rightarrow HCO_2H}\, E = -0.61\mathrm{V} \tag{4}$$

$$\mathrm{CO_2 + 4H^+ + 4e^- \rightarrow HCHO + H_2O} \quad E = -0.48\mathrm{V} \tag{5}$$

$$\mathrm{CO_2 + 6H^+ + 6e^- \rightarrow CH_3OH + H_2O}\, E = -0.38\mathrm{V} \tag{6}$$

$$\mathrm{CO_2 + 8H^+ + 8e^- \rightarrow CH_4 + 2H_2O}\, E = -0.24\mathrm{V} \tag{7}$$

In fact, most of these products of CO_2 reduction have been detected experimentally in variable concentration depending on the reaction conditions [7]. Besides, in order to keep the overall electroneutrality of the process, other sacrificial compounds (acting as an electron donor) should be oxidized. Ideally, water molecule could be used as a source of protons in artificial photosynthesis, as it occurs in the natural process in green plants. This process takes place according to

$$H_2O \rightarrow 1/2O_2 + 2H^+ + 2e^- \; E = +0.82V \quad (8)$$

and, accordingly, it should generate molecular oxygen as a by-product of the reaction. On the other hand, it is also worth noting that the photoreduction of CO_2 competes with that of protons to yield H_2 or H_2O_2 due to the following processes:

$$2H^+ + 2e^- \rightarrow H_2 \quad E = -0.41V \quad (9)$$

$$O_2 + 2H^+ + 2e^- \rightarrow H_2O_2 \quad E = +0.30V \quad (10)$$

Anyhow, the selection of water molecule as reductant has important environmental and economic benefits, because water is widespread, inexpensive, and does not generate residues. Nevertheless, the use of other electron donors such as methanol or amines has been also considered in some works [14], particularly in those related to the use of photoactive metal complexes.

As it has been mentioned above, one of the main advantages of the photocatalytic reductions of CO_2 is the possibility of powering the process using sunlight as sustainable energy source. The energetic position of the valence and conduction bands, along with the local structure at the surface of the photocatalyst, determines the feasibility of these processes [15]. In addition, any process inhibiting electron–hole recombination would increase the rates of CO_2 photoreduction, since it has to be taken into account that electron–hole recombination is several orders of magnitude faster than any other electron transfer process [16]. Additional factors such as surface area of the semiconductor, or operation parameters such as the CO_2:H_2O ratio, and light intensity have also a notable influence on the efficiency of the process.

Initially, the development of efficient CO_2 reduction routes was mainly focused on the use of homogeneous catalytic reactions based on different metallic compounds in solution. However, although these molecular complexes are still investigated, current advances in artificial photosynthesis are mainly based on semiconductor photocatalysts. In particular, most commonly used materials for this application are transition metal oxides, coupled with noble metal as cocatalyst and, in some cases, with photosensitizer such as metal complexes. In 1979 Inoue, Fujishima, and coworkers reported for the first time the photocatalytic reduction of CO_2 in aqueous solution to produce formaldehyde, formic acid, methanol, and traces of methane. These researchers used different semiconductors, such as WO_3, TiO_2, ZnO, CdS, GaP, and SiC, illuminated with both xenon and mercury lamps [17]. Since the publication of this pioneering work, many research groups have tried to boost the efficiency of the photocatalytic reduction of CO_2 using a variety of modified semiconductors.

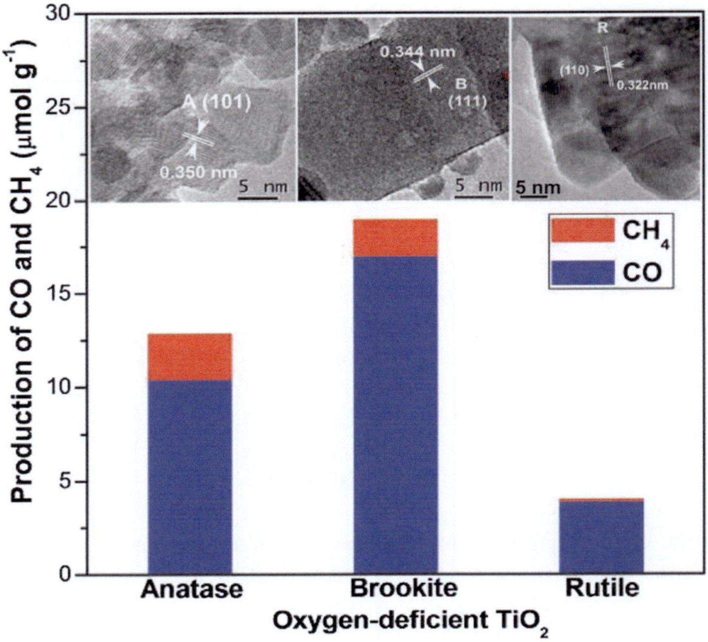

Fig. 4 Photocatalytic activity rate for CO_2 reduction to CO and CH_4 over different TiO_2 crystalline phases. Reproduced with permission from 20

The most frequently used semiconductor for this process is still TiO_2, which constitutes the benchmark photocatalysts widely applied to photooxidation reactions of organic compounds for water and air purification. However, the main drawback of TiO_2-based systems is that they have a bandgap of 3.2 eV, and consequently, this semiconductor can be activated exclusively under UV irradiation, which represents only a small fraction (about 5 %) of the solar spectrum. Moreover, the relatively high recombination rate of photoinduced electron–hole pairs and the low quantum yield for the production of O_2 also limits the efficiency of TiO_2 in photocatalytic reactions for fuel production [18, 19].

The reports on the use of titanium dioxide photocatalysts for CO_2 valorization have employed any of the main crystalline phases of this material: anatase, rutile, or brookite [20] or, very often, a mixture of two of them. This is the case of the frequently used Degussa (Evonik) P25 commercial sample, which is constituted by anatase and rutile in an approximately 8:2 ratio [47]. An example of the photoactivity of the different titania crystallographic phases in the gas phase can be observed in Fig. 4, which shows the rate of the two main products of the reaction, CO and CH_4 as a function of the crystalline phase. This study also shows that formation of Ti^{3+} induced by a soft thermal treatment in He atmosphere results in enhancement of the reaction rate. Following this activation process, photoactivity measurements reveal the limited efficiency of rutile, while brookite phase presents the higher activity and anatase the larger methane yield [20].

In addition, several morphological modifications of TiO_2 in the nanoscale, such as nanotubes [21] or nanorods [22], have been also proposed as efficient photocatalysts. In this respect, recently, Ohno et al. have reported that shape-controlled brookite nanorods with larger proportion of exposed {210} crystal planes have higher activity for methanol production from aqueous saturated solutions of CO_2 [23]. These authors have suggested that this enhancement of the photoactivity is most likely due to the physical separation of oxidation and reduction sites in different facets of brookite. Similarly, different reactivity of the anatase facets for the photocatalytic reduction of CO_2 has been also observed. Thus, samples with higher proportion of exposed {010} facets are more efficient for the generation of methane, and this has been attributed to preferential adsorption of CO_2 on those surfaces. Remarkably, this tendency is reversed after Pt deposition, and in the presence of this cocatalyst, the production of methane is more efficient on the {001} facets [24]. This fact has been attributed to the enhanced electronic transference from the semiconductor to the metal particles located on the {001} planes. On the other hand, a TiO_2 film constituted by elongated single crystals with preferential (112) orientation deposited on indium tin oxide (ITO) substrate has shown a remarkable boost in the methane production rate after Pt incorporation [25]. This work reported a record production of methane of 1,361 μmol/g-cat/h with a quantum yield of 2.41 %.

Some titanates with layered perovskite structures and general formula $ALa_4Ti_4O_{15}$ (A=Ca, Sr, and Ba) have been also assayed for the photocatalytic reduction of CO_2 after Ag incorporation. With this family of semiconductors, the most active was $BaLa_4Ti_4O_{15}$, which allows the formation of CO and lower amounts of formic acid from a CO_2-saturated aqueous solution [26].

Although strictly they are not semiconductors but single-site photocatalysts, interesting results have been obtained with Ti dispersed on zeolites and mesostructured materials [9]. In this way, titanium oxide species supported on Y or β zeolites show a significant photoactivity for the production of methanol and methane from CO_2 and H_2O. Interestingly, yields to these products decrease with the hydrophobicity of the aluminosilicate that can be modified by adsorption of fluoride ions [27, 28]. Similarly, good performance for the photocatalytic reduction of CO_2 can be obtained with Ti-SBA-15. Remarkably, using that mesostructured photocatalysts, water has proved to be a more efficient electron donor than hydrogen [29]. On the other hand, the combination of TiO_2 with carbon nanostructures ranging from graphene to nanotubes can improve appreciably the rate of CO_2 photocatalytic reduction. Such improvements can be attributed to the contribution of different factors. Thus, modification of the optical properties allows the widening of the spectrum of photons harvesting, whereas the improved adsorption capacity of these composite photocatalyst may also promote CO_2 photoreduction [30]. In a similar way, composites of the clay montmorillonite (MMT) and TiO_2 have been reported to present a remarkable photoactivity for CH_4 and CO production [31].

However, apart from some occasional record claims, most of the reported CO_2 conversion rates with TiO_2-based photocatalysts vary from 0.5 to 50 μmol/gh depending on the specific reaction media and the power of light source, among other operation conditions. This limited photoactivity is in sharp contrast with the results obtained with TiO_2 for water splitting, which even in the absence of sacrificial agent shows a rate of hydrogen production of 373 μmol/g h [32]. Furthermore, under sunlight irradiation, the CO_2 conversion to CO is not higher than 3 μmol/g h [7].

Accordingly, there is a significant awareness in the scientific community about the need of developing more efficient alternative photocatalyst for CO_2 valorization. Thus, in order to further improve the yield of semiconductors, two main general approaches should be considered: (1) maximizing charge separation and the transport of charge carriers and (2) enhancing the capacity of harvesting solar radiation at all wavelengths. This last aspect is paramount for the development of solar applications and, as it has been mentioned above, is one of the most important drawbacks of using Ti-based photocatalysts.

In this context, one of the most successful strategies to improve photocatalytic behavior is the modification of the semiconductor by introducing small concentrations of nonmetal elements, in particular N, but also halogens like I or F [21, 33, 34]. The incorporation of these dopants in the semiconductor framework can lead to a variation in the electronic structure of the systems that enhances the visible-light absorption. Following this approach, one of the higher rates of CO_2 reduction was obtained by Varghese et al. using a thin film of Cu/N-TiO_2 nanotubes. Under solar radiation and using only water vapor as electron donor, this material achieved a rate of 4.4 μmol/g h [21]. The main products of the reaction were H_2 and CO, but a significant production of methane and minor concentrations of other hydrocarbons with longer chain were also detected. Nitrogen doping has been also applied to other semiconductors such as $InTaO_4$ as route for shifting the bandgap to lower energy and to enhance the CO_2 photoreduction activity under visible-light illumination [34]. Besides, bandgap engineering by N-doping can be combined with the incorporation of sensitizer to the semiconductor surface. This approach was adopted for sensitization of nitrogen-doped Ta_2O_5 with Ru(II) bipyridine complexes, and it allows to achieve a quantum yield of 1.9 % with a selectivity for HCOOH larger than 75 % using triethanolamine as sacrificial agent for the photocatalytic reduction of CO_2 [35]. Alternatively, TiO_2 can be modified with both Ru complexes [36] and the CO_2-reducing enzyme CODH I, extracted from the anaerobic microorganism *Carboxydothermus hydrogenoformans*, to achieve a very selective photocatalytic conversion of CO_2 to carbon monoxide under visible-light illumination [37] (Fig. 5).

Other semiconductor oxides, such as ZnO, are also active for the photocatalytic reduction of CO_2 in the gas phase producing mainly CO and CH_4, although the efficiency is poor compared to TiO_2 [38]. More recently, mesoporous Fe-doped CeO_2 materials have proved to be very promising alternative photocatalysts for this application. In fact, the production of methane (ca. 4 μmol/g h) and CO (ca. 17 μmol/g h) can be comparable to those obtained with TiO_2 [39]. Another potentially interesting photocatalysts is Zn_2GeO_4, especially in the form of nanoribbons and using RuO and Pt as cocatalysts. Nevertheless, the large bandgap of this material (3.8 eV) constitutes a serious limitation for practical applications [40]. A large band is also a significant drawback of the photocatalysts based on Mg and In double-layered hydroxides, which, nevertheless, show a significant activity for the decomposition of CO_2 into CO and O_2 [41].

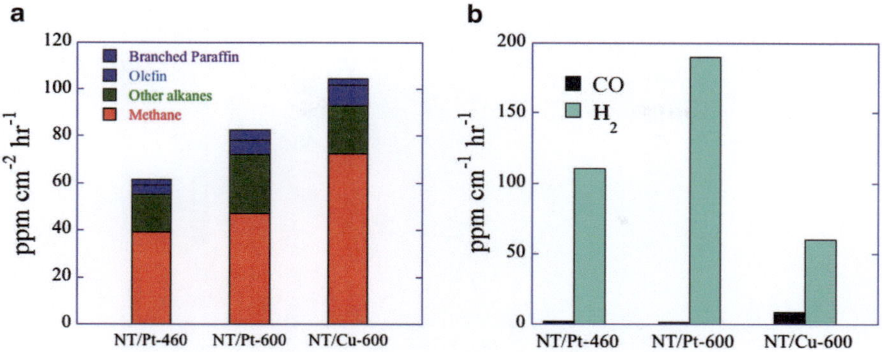

Fig. 5 Photocatalytic CO_2 reduction activity over nitrogen-doped nanotube array films with Pt (NT/Pt) and Cu (NT/Pt) cocatalysts, C annealed 460 or 600° as indicated, under sunlight illumination: (**a**) hydrocarbon generation rate and (**b**) H_2 and CO generation. Reproduced with permission from [21]

More promising are the results obtained with monoclinic $BiVO_4$ because it presents good photoactivity under visible-light illumination ($\lambda > 400$ nm) with a remarkable selectivity toward the ethanol production in aqueous solution kept at 0 °C to increase CO_2 solubility [42]. A smaller bandgap semiconductor based on metallic sulfides and selenides (e.g., ZnS, ZnSe, and CdSe) has been also considered as alternative photoreduction catalysts under visible-light irradiation. However, these chalcogenide photocatalysts are not very stable, and they require specific sacrificial agents (i.e., S^{2-}/SO_3^{-}) to suppress the irreversible oxidation of these semiconductors. Up to now, only a few studies have been performed using these materials although interesting results have been obtained using heterostructures formed by CdSe quantum dots on TiO_2. In these photocatalysts, the sulfide acts as sensitizer, and accordingly, a relatively high production of CH_4 and CH_3OH is obtained under visible-light illumination [43]. Another promising non-oxidic semiconductor is g-$CN_{3,}$ which has been recently assayed in CO_2 photocatalytic reaction to produce CO using Co*Ox* as cocatalysts and triethanolamine as sacrificial agent. Nevertheless, using these materials, the CO production rate under illumination at 400 nm is still quite low [44].

In addition to chemical and morphological modification of the semiconductors, as it has been already implicitly mentioned, most of these studies consider also the use of either noble metals (Pt, Pd, Rh, Au) or transition metal oxides (CuO*x*, NiO*x*, CoO*x*, etc.) to promote photocatalytic activity. Monometallic and bimetallic transition metals used as cocatalysts (either in metallic or oxide form) deposited on the semiconductor act, in general, as electron scavengers, and therefore, they contribute to inhibit electron–hole recombination. They may also facilitate the electronic transference from the adsorbed reactants or the coupling of the intermediates with reduced protons [9]. However, in order to perform efficiently those functions, cocatalysts must be well dispersed on the surface of the photocatalysts to simultaneously maximize the contact with the semiconductor and the exposure to the reagents.

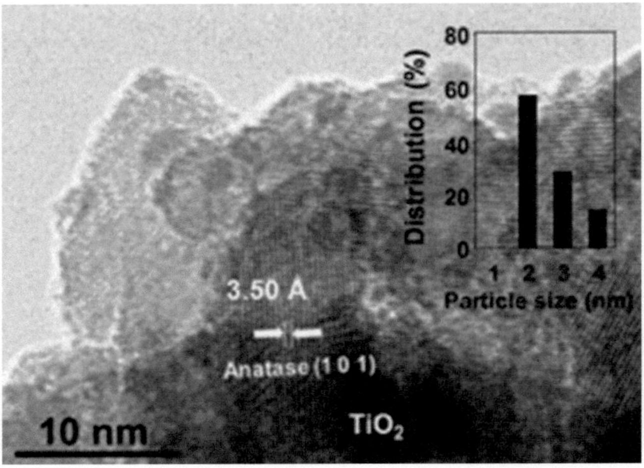

Fig. 6 TEM images of a Ag/TiO_2 sample showing the anatase and Ag nanoparticles as well as their size distribution. Reproduced from [38]

Frequently, this is conveniently achieved by photocatalytic deposition of the metal from a precursor solution because this route ensures an efficient transference of photogenerated carriers from the semiconductor.

In general terms, the use of these cocatalyst leads to a decrease of CO production and to the subsequent enhancement of products with higher electron demand such as CH_4 and CH_3OH (see (5)–(7)). [25, 38, 45]. In addition, the selectivity is also modulated by the type of metal and its oxidation state. Tseng and coworkers have shown that using a Cu/TiO_2 catalyst, the methanol formation depends on the oxidation state of Cu species [46]. Thus, a higher yield of this alcohol is obtaining in the case of well-dispersed Cu^+ species, while Cu^0 is more selective to hydrocarbon production. In the same way, the studies carried out by Slamet et al. with Cu/TiO_2 catalysts reveal also the existence of a relationship between Cu oxidation state and methanol production, but in this case, an increase in the CH_3OH selectivity is observed in the following order: Cu^{2+}, Cu^+, and Cu^0 [49]. Moreover, in situ IR characterization studies have also corroborated that methanol is produced on these Cu active sites via formate and methoxy intermediates [20, 45]. Furthermore, Cu has been defined as a promoter of the CO_2 multielectronic photoreduction process, due to its capabilities to facilitate the trapping of photoexcited electrons and to decrease the recombination rate [20] (Fig. 6).

In addition, the incorporation of metal nanoparticles of some specific metals can be a convenient way to further modify the electronic structure of the semiconductor. On this regard, particles of Au and Ag and to a lesser extent Cu, which can be prepared in a surface plasmon resonance (SPR) state, have drawn remarkable attention in the last years due to their unique electronic properties. Accordingly, they have a wide range of potential applications in solar cells, visible-light-responsive photocatalysis, and light-emitting diodes [50]. The nanoparticles in an SPR state are excited when photons of a specific wavelength, which varies with the metal and the particle dimensions and the morphology, impinge on the material, and this process causes the plasma of conduction electrons to oscillate collectively. One consequence of photoactivating the localized SPR is the generation of locally enhanced electromagnetic (EM) fields at the nanomaterial's surface. These boosted electromagnetic fields are believed to enhance visible-light absorption and significantly improve the efficiency of photochemical reactions on the surfaces of these materials. Although SPR particles can be advantageously used for the harvest and conversion of solar energy, it is necessary to keep in mind that only nanoparticles of three metals, Au, Ag, and Cu, within a narrow-size interval can be efficiently excited by visible light.

In summary, a large number of semiconductors, modified chemically and morphologically and often promoted with specifically designed cocatalysts, have been tested for the photocatalytic reduction of CO_2 using water as preferred electron donor. However, despite the remarkable research effort undertaken in the last years, the obtained rates of CO_2 transformation are still extremely low. These current results call for an additional research effort, which should be at the same time imaginative and systematic as to design of innovative photocatalytic materials that can efficiently realize the CO_2 reduction under near ambient conditions, using affordable reducing and sustainable agents like water.

4 Additional Challenges of CO_2 Photoreduction Over Semiconductors

As it has been outlined above, the photocatalytic reduction of CO_2 still requires a notable research effort. Currently, the main drawback of this process is the very low quantum yield that keeps the technology in the lab bench. Therefore, most likely, research effort will be focused on the short term in the design of more efficient photoactive materials. Nevertheless, there are some additional barriers to overcome, ranging from the need of gaining a more complete understanding of the process at molecular scale to the technical difficulties of monitoring the progress of the reaction and powering the process with sunlight. Not all these aspects are equally important, nor are in the same state of developments, but they all need to be adequately addressed in order to fully develop artificial photosynthesis as a factual alternative for fuel productions. The most relevant of these challenges for the photocatalytic reduction of CO_2 are briefly described in the following subsections.

4.1 Analysis of the Products

Although at first glance the analysis of the products of CO_2 photocatalytic reduction may not be perceived as a major limitation, the fact is that simultaneous detection of low concentration of polar (e.g., CH_3OH) and nonpolar (e.g., CH_4) chemicals may be rather challenging. Yet, accurate determination of products is essential for the evaluation of photocatalyst performance. Recently, the optimization of the different methods of analysis has been thoroughly approached by a combination of GC and HPLC measurements [51]. For determining the composition of the products in the gas phase, a high concentration of CO_2 is a serious drawback, so it has to be separated using a first column (HayeSep Q) and avent valve to avoid damaging the molecular sieve used for the separation of other gases. Overall, a two-channel–three-detector (FID, TCD, and a methanizer provided with Ni catalysts) gas chromatograph is necessary to monitor O_2, CH_4, ethane, CO, and CO_2 simultaneously. An additional column and an FID detector are necessary to determine small alcohols such as methanol or ethanol. On the other hand, short-chain aldehydes and ketones (formaldehyde, acetaldehyde, propionaldehyde, and acetone) can be determined by HPLC after derivatization with 2,4-dinitrophenylhydrazine (DNPH). In contrast, liquid chromatography can be applied directly for the analysis of formic and acetic acid. Using these methods, detection limits for the photoreduction products range from 0.08 μmol/L for CO to 3.9 μmol/L for HCOOH.

Despite these advances, simplification of the methodology and decreasing the time of analysis, which in some cases can last more than 15 min, will be necessary to facilitate a continuous monitoring of the processes. In particular, if a significant improvement of the efficiency is achieved, quicker response of the analytical method will be increasingly relevant. Finally, it is worth mentioning that production of oxygen is also expected as a by-product, but many reports do no pay attention to the possible formation of this gas.

4.2 Dealing with Impurities

Closely related to the previous point is the fact that, considering the very lowconcentration of products obtained, the possible effect of carbon-containing impurities is not at all negligible in the study on CO_2 photocatalytic reduction. Dealing with this aspect requires a particularly careful manipulation of the photocatalysts to avoid the exposure to any source of contamination. In addition, performing systematic cleaning pretreatment before activity assays is highly recommended. Furthermore, it is crucial to perform all possible blank assays to discard spurious contributions of uncontrollable carbon residues, which can be highly misleading. In fact, many literature claims of high rates for the photocatalytic reduction of CO_2 over semiconductors are suspicious of being affected by undetected impurities.

These problems were thoroughly approached by C. C. Yang et al., who performed a systematic FTIR study of the photoreduction using isotopically labeled CO_2 [52]. The scheme shown in Fig. 7 indicates how using $^{13}CO_2$ is possible to distinguish real photoreduction process from the steam reforming of carbon residues present on the photocatalyst surface. In fact, this investigation has revealed that generation of CO may also be originated from coke residues on the surface of the

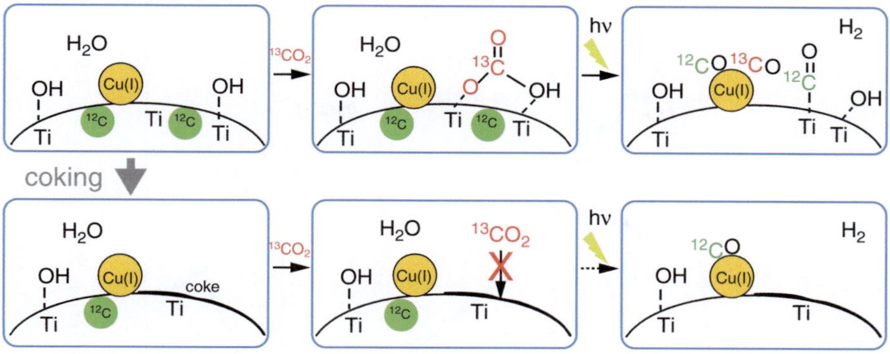

Fig. 7 Reaction scheme of the expected products of the photocatalytic reduction of isotopically labeled $^{13}CO_2$ in the presence or absence of coke residues on the TiO_2 surface. Reproduced with permission from [51]

Cu/TiO_2 photocatalysts. These results highlight the importance of undertaking isotopically labeled probe molecules whenever doubts about the photocatalytic conversion of CO_2 arise.

4.3 Stability of the Photocatalysts

The lifetime of the catalysts is obviously a crucial parameter which affects the engineering aspects but also the economy of the processes. In the field of the photocatalytic production of hydrogen, the relevance of maintaining the stability of the used semiconductor has been recognized in a number of reports. In this respect, a long-term study carried out with GaN:ZnO has revealed that the photoactivity does not change significantly for the first 3 months, but is almost halved after 6 months of operation due to the loss of cocatalyst and the degradation of the semiconductor [53]. In contrast, the durability and recycling of the photocatalysts used for CO_2 transformation have been scarcely approached, despite its significance for the future implementation of this technology. However, most reports showing a temporal evolution of the photocatalytic reduction of CO_2 reveal a progressive decline in the rate. An example of this behavior can be appreciated in Fig. 8, which shows how the production of CO and CH_4 goes through a maximum after about 5 h of irradiation and afterward progressively declines. This has attributed to saturation of the absorption sites. Interestingly, a substantial recovery of the activity is attained by simply exposing the photocatalyst to the air, without further treatment [54].

These observations call upon the need to establish the reasons at molecular scale for this decay in photoactivity in order to delay or reverse the deactivation process. This is a paramount aspect for the practical development of this technology because any progress in enhancing the efficiency may result unsuccessful if the photocatalysts remain active for only a few hours.

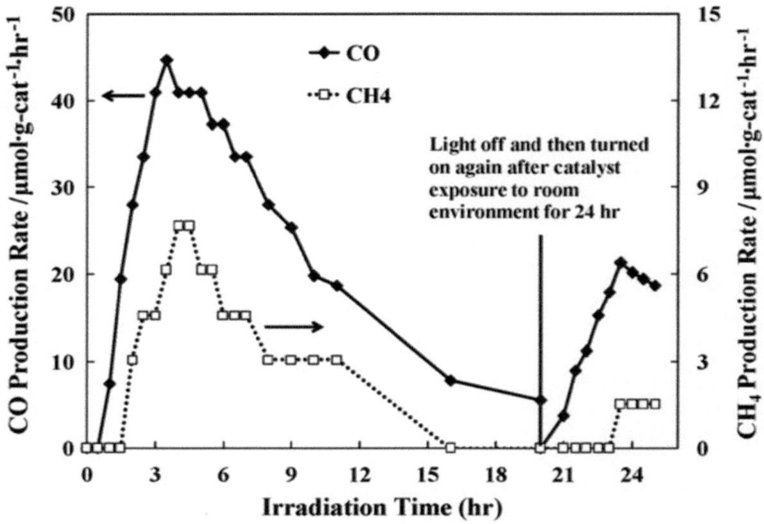

Fig. 8 Dependence on the production rates of CO and CH_4 over 0.5 % Cu/TiO_2–SiO_2 with irradiation time in a continuous flow reactor. Reproduced with permission from [54]

4.4 Designing Efficient Solar Photoreactors

Although the irradiation on the Earth's surface varies widely depending on latitude, season (month, day, and hour), and atmospheric effects (reflection, absorption, and scattering), the magnitude of the solar constant, 1,353 ± 21 W m^{-2}, provides a clear indication of the magnitude of Sun power which is readily available [9]. In this context, the dream of an efficient transformation of CO_2 into fuels and chemicals using sunlight requires not only highly active, durable, and affordable photocatalysts but also an efficient solar reactor to harness sunlight. Although a number of innovative designs for photoreactors have been assayed for photocatalytic reduction of CO_2, real [59] solar tests have been performed using only simple batch-type reactors with a window for illumination, which is shown in Fig 9a [21]. Nevertheless, these simple tests have probed that unconcentrated solar light may suffice to induce these challenging reactions.

In fact, examples of the photocatalytic production of any kind of fuels powered by sunlight are still uncommon. Recently, photocatalytic hydrogen generation at pilot-plant scale was accomplished in a system based on compound parabolic collector (CPC) under direct solar irradiation and using different sacrificial agents such as wastewater [55]. These experiments were performed at the facilities of the Plataforma Solar de Almeria, by taking advantage of the ample experience acquired in the solar photocatalytic treatment of pollutants on this type of facilities.

Although simple flat photoreactors may be adequate for small-scale or modular applications, photoreactor based on CPC technology is currently most suited for

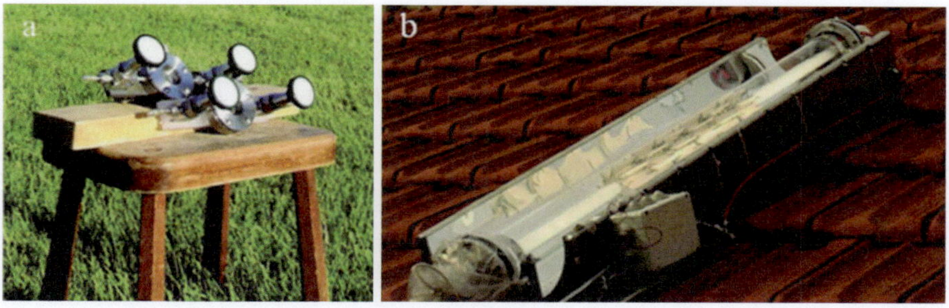

Fig. 9 (**a**) Batch-type photoreactor used for CO_2 reduction under solar illumination. Reproduced with permission from [21] and (**b**) hybrid continuous flow photoreactor provided with both solar CPC and lamp illumination for gas treatments. Reproduced with permission from [57]

operation at pilot scale. The great advantage of this light harvesting technology is that the reflector geometry allows indirect light to be reflected onto the central tube, which acts as a chemical reactor and encloses the photocatalysts (see Fig 9b). This characteristic is particularly important in cloudy days because diffuse light can be efficiently captured and driven to the central reactor. In addition, no solar tracking is required and this results in simpler and cheaper systems. However, in order to maximize light harvesting, CPC photoreactors must be oriented to the south and tilted an angle equivalent to the latitude [10]. Not less relevant to applicability of this kind of solar reactors is the fact that either aqueous solution or gases can be treated in these systems [56, 57]. In general, the dimensions of solar collectors are selected in such a way that the acceptance half angle is 90°, and this results in CPC with a concentration factor of one (one Sun). This configuration allows circumventing the problems associated with overheating of the sample. Furthermore, this geometry leads, in general terms, to a higher efficiency of the process because at high photon flux, it may increase electron–hole recombination [58]. However, a certain degree of concentration (<5 suns) may be more adequate for sluggish processes as those of CO_2 photoreduction, which may benefit for higher irradiance and temperature. However, all these aspects will need extensive investigations in order to adapt this solar technology to the CO_2 processing.

In addition to the design of the solar collector, further engineering aspects should be addressed for real-world application. First, the need of excluding oxygen from the inlet flow requires a stringent system for CO_2 purification, possibly using amine-based absorbents [1]. This aspect makes a fundamental difference with photosynthesis in green plants, which can capture selectively CO_2 from air. Besides, downstream of the photoreactor, separation of a generally complex mixture of chemicals will be also necessary. In addition, considering the limited rates of photocatalytic CO_2 reduction, recirculation of the unconverted products will be most likely required and accordingly should be considered in the photoreactor design. All these aspects have important implications for both the economy and efficiency of the overall process, because only the separation of the products from the CO_2 stream could account for 6–17 % of the total energy required for the process [5].

5 Concluding Remarks

Although the interest in developing efficient photocatalyst for CO_2 valorization is clearly growing due to the great concern about the consequences of global warming, a significant research effort is still necessary to increase the efficiency of the process. Although the current results are encouraging because they confirm the feasibility of powering this energetically demanding process with light, real-world applications will require a significant enhancement of the performance of both materials and photoreactors. In this respect, it has been recently proposed that an efficiency improvement of at least three orders of magnitude is indeed required to turn artificial photosynthesis into a viable alternative for the production of chemicals and fuels [5]. In addition, some important engineering problems will require further investigations. Therefore, the development of efficient photocatalytic system, capable to overcome the current limitations, will constitute a significant challenge in the near future, which will require a multidisciplinary research approach to be overcome.

References

1. Aresta M, Dibenedetto A, Angelini A (2014) Catalysis for the Valorization of Exhaust Carbon: from CO_2 to Chemicals, Materials, and Fuels. Technological Use of CO_2. Chem Rev 114(3):1709–1742
2. Song C (2006) Global challenges and strategies for control, conversion and utilization of CO_2 for sustainable development involving energy, catalysis, adsorption and chemical processing. Catal Today 115:2–32
3. Pontzen F, Liebner W, Gronemann V et al (2011). CO_2-based methanol and DME – Efficient technologies for industrial scale production. Catal. Today 171:242–250
4. Furler P, Scheffe JR, Gorbar M, et al (2012) Syngas production by simultaneous splitting of H_2O and CO_2 via ceria redox reactions in a high-temperature solar reactor. Energy Fuels 26:7051–7059
5. Kondratenko EV, Mul G, Baltrusaitis J et al (2013) Status and perspectives of CO_2 conversion into fuels and chemicals by catalytic, photocatalytic and electrocatalytic processes. Energy Environ Sci 6:3112–3135
6. Usubharatana P, McMartin D, Veawab A et al (2006) Photocatalytic Process for CO_2 Emission Reduction from Industrial Flue Gas Streams. Ind Eng Chem Res 45:2558–2568
7. Izumi Y (2013) Recent advances in the photocatalytic conversion of carbon dioxide to fuels with water and/or hydrogen using solar energy and beyond. Coord Chem Rev 257:171–186
8. Kubacka A, Fernández-García M, Colón G (2012) Advanced Nanoarchitectures for Solar Photocatalytic Applications. Chem Rev 112:1555–1614
9. Coronado JM, Fresno F; Hernández-Alonso MD et al (eds) (2013) Design of Advanced Photocatalytic Materials for Energy and Environmental Applications. Springer, London.
10. M. Romero, J. Blanco, B. Sánchez, A. Vidal, S. Malato, A. Cardona and E. García, (1999) Solar Photocatalytic Degradation of Water and Air Pollutants: Challenges and Perspectives Sol. Energy, 66(2), 169.
11. Tachibana Y, Vayssieres LD, Durrant JR (2012) Artificial photosynthesis for solar water-splitting. Nature Photonics 6(8):511–518

12. Gust D, Moore TA, Moore AL (2009). Solar Fuels via Artificial Photosynthesis. Accounts Chem Res 42:1890–1898
13. Gray HB (2009) Powering the planet with solar fuel. Nature Chemistry 1, 7
14. Richardson RD, Holland EJ, Carpenter BJ (2011). A renewable amine for photochemical reduction of CO_2. Nature Chem 3:301–303
15. Indrakanti VP, Kubicki JD, Schobert HH (2009) Photoinduced activation of CO_2 on Ti-based heterogeneous catalysts: Current state, chemical physics-based insights and outlook. Energ Environ Sci 2:745–758.
16. Zhou H, Fan T, Zhang D (2011) An Insight into Artificial Leaves for Sustainable Energy Inspired by Natural Photosynthesis. ChemCatChem 3:513–528.
17. Inoue T, Fujishima A, Konichi S et al (1979) Photoelectrocatalytic reduction of carbon dioxide in aqueous suspensions of semiconductor powders. Nature 277: 637-638
18. Tang J, Durrant JR, Klug DR (2008) Mechanism of photocatalytic water splitting in TiO_2. Reaction of water with photoholes, importance of charge carrier dynamics, and evidence for four-hole chemistry. J Am Chem Soc 130:13885–13891
19. Cowan AJ, Tang J, Leng W et al (2010) Water Splitting by Nanocrystalline TiO_2 in a Complete Photoelectrochemical Cell Exhibits Efficiencies Limited by Charge Recombination. J Phys Chem C 114:4208–4214
20. Liu, L.; Zhao, H.; Andino, J.M.; Li, Y. (2012) Photocatalytic CO_2 Reduction with H_2O on TiO_2 Nanocrystals: Comparison of Anatase, Rutile, and Brookite Polymorphs and Exploration of Surface Chemistry. ACS Catal., 2, 1817–1828
21. Varghese OK, Paulose M, LaTempa TJ, et al (2009) High-Rate Solar Photocatalytic Conversion of CO_2 and Water Vapor to Hydrocarbon Fuels. Nano Lett 9(2):731–737
22. Tan JZY, Fernández Y, Liu D et al (2012) Photoreduction of CO_2 using copper-decorated TiO_2 nanorod films with localized surface plasmon behaviour. Chem Phys Lett 531:149–154
23. Ohno T, Higo T, Murakamia N et al (2014) Photocatalytic reduction of CO_2 over exposed-crystal-face-controlledTiO_2 nanorod having a brookite phase with co-catalyst loading. Appl Catal B 152–153:309–316
24. Mao J, Ye L, Li K, Zhang X, Liu J, Peng T, Zan L (2014) Pt-loading reverses the photocatalytic activity order of anatase TiO_2{0 0 1} and {0 1 0} facets for photoreduction of CO_2 to CH_4. Appl. Catal. B 144:855–862
25. Wang W N, An W J, Ramalingam B et al (2012) Size and Structure Matter: Enhanced CO_2 Photoreduction Efficiency by Size-Resolved Ultrafine Pt Nanoparticles on TiO_2 Single Crystals. J Am Chem Soc 134:11276–11281
26. Iizuka K, Wato T, Miseki Y et al (2011) Photocatalytic Reduction of Carbon Dioxide over Ag Cocatalyst-Loaded $ALa_4Ti_4O_{15}$ (A=Ca, Sr, and Ba) Using Water as a Reducing Reagent. J Am Chem Soc 133(51):20863–20868.
27. Anpo M, Yamashita H, Ichihashi Y et al (1997) Photocatalytic Reduction of CO_2 with H_2O on Titanium Oxides Anchored within Micropores of Zeolites: Effects of the Structure of the Active Sites and the Addition of Pt. J Phys Chem B 101(14):2632–2636
28. Ikeue K, Yamashita H, Anpo M et al (2001) Photocatalytic Reduction of CO_2 with H_2O on $Ti-\beta$ Zeolite Photocatalysts: Effect of the Hydrophobic and Hydrophilic Properties. J Phys Chem B 105(35):8350–8355
29. Yang CC, Vernimmen J, Meynen V et al (2011). Mechanistic study of hydrocarbon formation in photocatalytic CO_2 reduction over Ti-SBA-15. J Catal 284(1):1–8
30. Sun H, Wang S (2014) Research Advances in the Synthesis of Nanocarbon-Based Photocatalysts and Their Applications for Photocatalytic Conversion of Carbon Dioxide to Hydrocarbon Fuels. Energy Fuels 28(1):22–36
31. Tahir M, Amin NS (2013) Photocatalytic reduction of carbon dioxide with water vapors over montmorillonite modified TiO_2 nanocomposites. Appl Catal B. 142–43:512–522
32. Tabata S, Nishida H, Masaki Y et al (1995) Stoichiometric photocatalytic decomposition of pure water in Pt/TiO2 aqueous suspension system. Catal Lett 34(1–2):245–249
33. Zhang, Q, Li Y, Ackerman EA et al (2011) Visible light responsive iodine-doped TiO_2 for photocatalytic reduction of CO_2 to fuel. Appl Catal A 400:195–202

34. Tsai C, Chen H, Liu R, et al (2011) Ni@NiO Core–Shell Structure-Modified Nitrogen-Doped $InTaO_4$ for Solar-Driven Highly Efficient CO_2 Reduction to Methanol. J Phys Chem C 115:10180–10186.
35. Sato S, Morikawa T, Saeki S et al (2010) Visible-Light-Induced Selective CO_2 Reduction Utilizing a Ruthenium Complex Electrocatalyst Linked to a p-Type Nitrogen-Doped Ta_2O_5 Semiconductor. Angew Chem 49(30):5101–5105
36. Nguyen, T.-V.; Wu, J. C. S.; Chiou, C.-H. (2008) Photoreduction of CO_2 over Ruthenium dye-sensitized TiO_2-based catalysts under concentrated natural sunlight. Catal. Comm., 9, 2073–2076
37. Woolerton TW, Sheard S, Reisner E et al (2010) Efficient and Clean Photoreduction of CO_2 to CO by Enzyme-Modified TiO_2 Nanoparticles Using Visible Light. J Am Chem Soc 132:2132–2133
38. Collado L, Jana P, Sierra B et al (2013) Enhancement of hydrocarbon production via artificial photosynthesis due to synergetic effect of Ag supported on TiO_2 and ZnO semiconductors. Chem Eng J 224:128–135.
39. Wang Y, Wang F, Chen F, Zhang D, Li B, Kang S, Li X, Cui L (2014), Enhanced photocatalytic performance of ordered mesoporous Fe-doped CeO_2 catalysts for the reduction of CO_2 with H_2O under simulated solar irradiation Appl. Catal. B: 147 602–609
40. Liu Q, Zhou Y, Kou J et al(2010) High-Yield Synthesis of Ultralong and Ultrathin Zn_2GeO_4 Nanoribbons toward Improved Photocatalytic Reduction of CO_2 into Renewable Hydrocarbon Fuel. J Am Chem Soc 132:14385–14387
41. Teramura K, Iguchi S, Mizuno Y et al (2012) Photocatalytic Conversion of CO_2 in Water over Layered Double Hydroxides. Angew. Chem. Int. Ed. 51:8008–8011.
42. Liu Y, Huang B, Dai Y et al (2009) Selective ethanol formation from photocatalytic reduction of carbon dioxide in water with $BiVO_4$ photocatalyst. Catal Comm 1: 210–213
43. Wang C, Thompson RL, Baltrus J et al (2010) Visible Light Photoreduction of CO_2 Using CdSe/Pt/TiO_2 Heterostructured Catalysts. J Phys Chem Lett 1:48–53
44. Lin J, Pan Z, Wang X (2014) Photochemical Reduction of CO_2 by Graphitic Carbon Nitride Polymers ACS Sustainable Chem Eng 2(3):353–358
45. Liu L, Gao F, Zhao H, et al (2013) Tailoring Cu valence and oxygen vacancy in Cu/TiO_2 catalysts for enhanced CO_2 photoreduction efficiency. Appl Catal B 134–135:349–358
46. Tseng IH, Wu JCS, Chou HY (2004) Effects of sol–gel procedures on the photocatalysis of Cu/TiO_2 in CO_2 photoreduction. J Catal 221:432–440
47. Tseng, IH, Chang WC, Wu JCS (2002) Photoreduction of CO_2 using sol–gel derived titania and titania-supported copper catalysts. Appl Catal B 37:37–48
48. Liu D, Fernandez Y, Ola O et al (2012) On the impact of Cu dispersion on CO_2 photoreduction over Cu/TiO_2. Catal Commun 25:78-82
49. Slamet, H. W. Nasution, E. Purnama, S. Kosela, J. Gunlazuardi. (2005) Photocatalytic reduction of CO_2 on copper-doped Titania catalysts prepared by improved-impregnation method. Catal Comm 6:313–319
50. Christopher P, Xin H, Marimuthu A et al (2012) Singular characteristics and unique chemical bond activation mechanisms of photocatalytic reactions on plasmonic nanostructures Nature Mater 11:1044–1050
51. Hong J, Zhang W, Ren J et al (2013) Photocatalytic reduction of CO_2: a brief review on product analysis and systematic methods. Anal Methods 5:1086–1096
52. Yang CC, Yu YH, van der Linden B et al (2010).Artificial Photosynthesis over Crystalline TiO_2-Based Catalysts: Fact or Fiction?. J Am Chem Soc 132(24):8398–8406
53. Ohno T, Bai L, Hisatomi T et al (2012) Photocatalytic water splitting using modified GaN: ZnO solid solution under visible light: long-time operation and regeneration of activity. J Am Chem Soc 134:8254–8259
54. Li Y, Wang WN, Zhan Z et al (2010) Photocatalytic reduction of CO_2 with H_2O on mesoporous silica supported Cu/TiO_2 catalysts. Appl Catal B 100:386–392

55. Villa K, Domenech X, Malato S et al (2013) Heterogeneous photocatalytic hydrogen generation in a solar pilot plant. Inter J Hydrogen Energy 38(29):12718–12724
56. Coronado JM, Fresno F, Suárez S, et al (2008) Influence of Catalyst Properties and Reactor Configuration on the Photocatalytic Degradation of Trichloroethylene Under Sunlight Irradiation. J Sol Energy Eng 130(4):041012–041015
57. Portela R, Tessinari RF; Suárez S et al (2012) Photocatalysis for Continuous Air Purification in Wastewater Treatment Plants: From Lab to Reality. Environ Sci Technol 46:5040–5048.
58. Braham RJ, Harris AT (2009) Review of Major Design and Scale-up Considerations for Solar Photocatalytic Reactors. Ind Eng Chem Res 48:8890–8905
59. Nguyen T-V, Wu JCS (2008) Photoreduction of CO_2 in an optical-fiber photoreactor: Effects of metals addition and catalyst carrier. Appl Catal A Gen 335:112–120

Functionalized Nanocarbons for Artificial Photosynthesis: From Fullerene to SWCNT, Carbon Nanohorn, and Graphene

Osamu Ito and Francis D'Souza

Abstract This chapter describes the recent advances in photosensitized electron transfer processes on nanocarbon surfaces functionalized with electron-donating or electron-accepting molecules. As nanocarbons, well-characterized fullerenes, size (diameter)-sorted single-walled carbon nanotubes (SWCNTs), carbon nanohorns (CNHs), graphene, and graphene oxide are mainly employed. These nanocarbons act as excellent electron acceptors, when they are linked to the light-absorbing electron donors such as porphyrins or phthalocyanines. Employing diameter-sorted SWCNT makes it possible to reveal the relationship between nanotube size and electronic structure, which strongly affect the photoelectron transfer properties. In these studies, for the confirmation of electron transfer processes, time-resolved spectroscopic methods have been widely used. The kinetic data obtained in solution are found to be quite useful to understand the electron transfer mechanisms. Photovoltaic cells constructed by employing these photosensitizing electron transfer systems are indicative of their potential uses to construct artificial photosynthetic systems for the light-to-electricity and light-to-fuel productions.

Keywords Light-induced electron transfer • Fullerenes • SWCNT • Carbon nanohorn • Graphene • Photosynthesizer • Supramolecular Hybrids • Solar cell

O. Ito
Tohoku University (Research Adviser for JST (PRESTO) and MEXT for Artificial Photosynthesis Projects), Sendai 981-3215, Japan
e-mail: ito@tagen.tohoku.ac.jp

F. D'Souza (✉)
Department of Chemistry, University of North Texas,
1155 Union Circle, #305070, Denton, TX 76203, USA
e-mail: Francis.DSouza@UNT.edu

E.A. Rozhkova, K. Ariga (eds.), *From Molecules to Materials*,
DOI 10.1007/978-3-319-13800-8_8

1 Introduction

In natural photosynthesis, efficient light-induced energy transfer (EnT) and electron transfer (ET) events occur among the well-organized pigments [1–3]. Thus, it is essential to design artificial photosynthetic systems for efficient light-energy conversion into electric and chemical energies by mimicking the natural photosynthesis systems. For example, in the bacterial photosynthetic reaction centers (Fig. 1), it has been revealed that the electron donors, acceptors, and mediators are arranged via non-covalent incorporation into a well-defined protein matrix [2]. Conversion of solar energy in photosynthesis is based on the formation of long-lived charge-separated (CS) states. Strategy for the lifetime extension of the CS states in natural photosynthetic reaction centers is via distant isolation of radical cation (hole) and radical anion (electron) of the CS states, which involves sequential electron hopping using electron mediators that are positioned appropriately. This strategy thus eliminates the energy-wasting back electron transfer [3].

For the design of efficient artificial photoenergy conversion devices, development of relatively simple composite molecular systems with electron donor (D) and electron acceptor (A) (Fig. 2) has been one of the important goals of chemistry during the past two decades [4]. For simple electron donor and acceptor (D–A) molecular systems, the purpose is to establish fast photoinduced charge separation (CS) process to generate the hole–electron pair ($D^{\bullet+}$–$A^{\bullet-}$) and subsequent slow charge recombination (CR) process to retain the electron and hole separately for

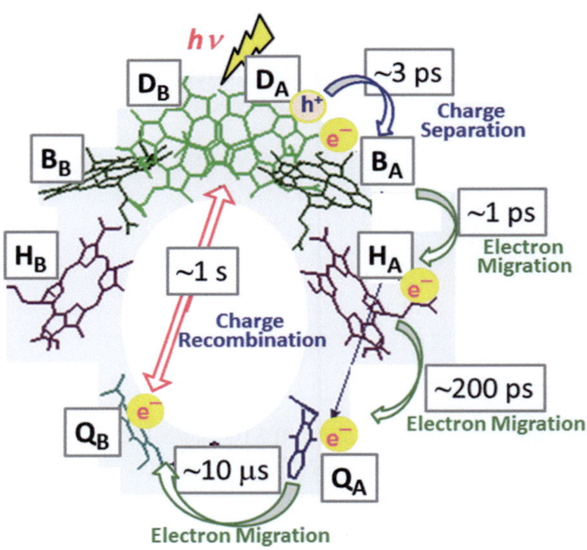

Fig. 1 Electron transfer events around the reaction center in a typical bacterial photosynthesis system; D_A, D_B, special dimer pair (chlorophyll dimer); B_A, B_B, bridged chlorophyll monomer; Φ_A, Φ_B, pheophytins; Q_A, Q_B, phylloquinones. *Numbers in box* are the time constants (reciprocal of the rate constants). Reconstructed using figures from reference [2]

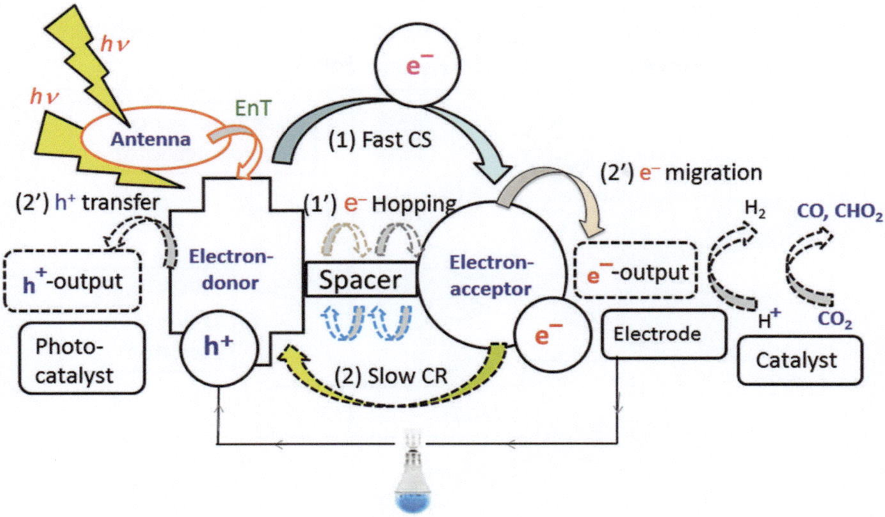

Fig. 2 Electron transfer in D–A molecular system; charge separation occurs from the direct D-photoexcitation or via energy transfer (EnT) process from antenna. (*1*) Fast CS process occurs through space or through bond, generating $D^{\bullet+}–A^{\bullet-}$. (*2*) Migrations of electron (e^-) and hole (h^+) subsequently occur, competitively with CR process. Then, the e^- and h^+ can be taken out into electrode for photovoltaic cell or as appropriate catalysts for H_2 evolution and CO_2 conversion

longer times. This would provide sufficient time to mediate the electron and hole outside of the $D^{\bullet+}–A^{\bullet-}$ radical ion pair for solar light-energy conversion devices, as depicted in Fig. 2. In such simple D–A model systems, well-adjusted energies of the donor and acceptor molecules and finely tuned electronic coupling between them are important to control the electron transfer rates on the basis of Marcus theory [5]. For example, small overall reorganization energy has been a significant factor to increase the rate of forward electron transfer and to slow down the rate of charge recombination. Subsequently, long-lived electron and hole can be used in building photovoltaic cells and also for photocatalytic schemes to evolve H_2 from water or to convert CO_2 to conventional fuels.

Recently, new nanocarbon architectures such as fullerenes, single-walled carbon nanotubes (SWCNTs), carbon nanohorns (CNHs), graphene, and graphene oxide linked with photosensitizers have received considerable attention for their distinguishing electron transfer ability suitable to develop devices that operate under visible-light illumination [6, 7]. Since these new carbons have larger π-electron surfaces than the usual aromatic compounds, modest functionalizations of these carbon surfaces can be considered as slight surface modifications of the wide curved π-electron systems (fullerenes, SWCNTs, CNHs) as well as flat π-electron graphene systems. The roles of these nanocarbons are pictorically summarized in Fig. 3, where zinc porphyrin (ZnP) is taken as an example for the visible-light-absorbing electron donor.

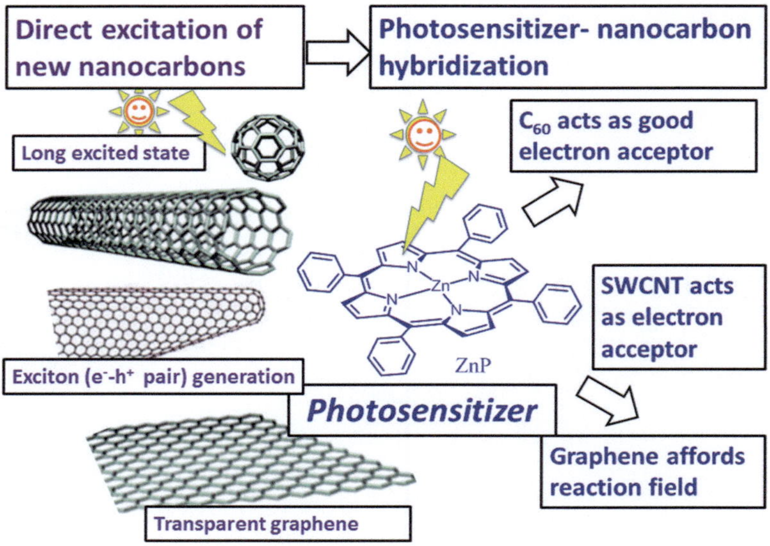

Fig. 3 Photophysical characteristics of carbon nanomaterials (C_{60}, SWCNT, CNH, and graphene) (*left side*) and photoinduced processes of carbon nanomaterials in the presence of zinc porphyrin (ZnP) (*right side*)

Fullerenes act as electron acceptors (A) under direct light excitation in the presence of electron donors, generating long-lived radical ion pairs (RIPs), although the absorption intensity of fullerenes is not high in the visible region. Functionalized with the visible-light-absorbing electron donors such as ZnP on the fullerenes, higher solar-energy conversion could be realized [8].

In the case of SWCNTs, on the other hand, direct photoexcitation generates short-lived exciton (electron–hole pair), which may be difficult to take out from the SWCNTs, because of the short lifetime. Thus, functionalization of SWCNTs with the photosensitizers must be crucial to construct light-energy-harvesting systems. Similar situation would be anticipated for sensitizer-functionalized CNHs, which are one of the new nanocarbons. In the case of a few-layer graphene, which is mostly transparent in the visible light region, some modifications with photoactive molecules are necessary to convert them into the light-harvesting systems [9].

As for the method of chemical modifications of these carbon materials, covalent functionalization is considered to be useful as a target-oriented strategy; however, currently employed synthetic procedures are elaborate due to the multistep synthetic procedures. This issue becomes more problematic for nanocarbons due to their poor solubility in organic solvents used in syntheses. In the case of open-edge SWCNTs and graphene with dangling bonds, covalent bond formations are easy at the terminal carbons, keeping their main π-electron structure intact. Covalent functionalization of the sidewalls of SWCNTs and CNHs can be performed in the same ways to fullerenes using method such as Prato method [10], although the product yields are low because of difficulty in handling. However, such scanty covalent functionalizations would largely retain the π-electron structures of nanocarbons. In the case of the functionalized fullerenes, positions and number of the substituents can be identified and counted using traditional analytical/spectroscopic methods such as NMR and mass spectroscopies. However, in the case of SWCNTs, CNHs, and graphene, such information cannot be obtained easily; the numbers of substituents must be estimated from conventional chemical analysis (elemental analysis) and thermal analysis etc. [11].

On the other hand, non-covalent self-assemblies of photosensitizers with intermolecular interactions retain completely the π-networks of these nanocarbons, making them useful for a wide range of applications including photo-devices and optoelectronics. Among the non-covalent functionalization approaches, the simplest method is the direct π–π stacking of porphyrins (MP) and phthalocyanines (MPc) onto the SWCNT, CNH, and graphene surfaces. However, such direct π–π interactions promote close contact between the entities, which often induce very fast photoinduced ET followed by equally fast back ET process. Although such phenomena are suitable for fast photoresponsible devices, they are not always advantageous to produce the long-lived radical ion pair (RIP). Thus, it is necessary to insert spacers avoiding such direct interaction. For this purpose, appropriate linkages are necessary to connect the sensitizers to nanocarbons to achieve controlled CS process. A popular methodology of this strategy includes usage of small aromatic compounds as adsorbents onto the nanocarbon surface via π–π stacking, i.e., aromatic compounds such as pyrene are useful. For connecting visible-light-absorbing photosensitizers to such pyrene molecule, metal–ligand coordination, inclusion complexation, and ion pairing and/or hydrogen-bonding interactions are employed in addition to traditional covalent bonds. To increase the photoinduced ET efficiencies, additional photosensitizers must be introduced.

This chapter documents the recent progress on supramolecular constructions of nano-architectures using photosensitizing D and A molecules with well-defined nanocarbons, which are key findings in the areas of photochemistry, photoelectrochemistry, and photocatalyst. In the case of SWCNTs, the chirality- and size-sorted SWCNTs can be employed, which makes each SWCNT to consider as a "molecule," although mixtures of various SWCNTs must be treated as materials. Thus, it has been possible to introduce the molecular orbital considerations to treat these SWCNT hybrids similar to D-fullerene molecules, whereas other nanocarbons such as CNHs and graphene (and graphene oxide) can be considered as materials of semiconductor or metallic character.

2 Photoinduced Electron Transfer of Fullerene-Donor Systems

2.1 Molecular Orbital Description of Photoinduced Electron Transfer

Fundamental events of photoinduced ET process between D and A can be inferred by the schematic molecular orbital (MO) energy levels as illustrated in Fig. 4, where ZnP and C_{60} are adopted as a component of weakly associated D–A complex [12].

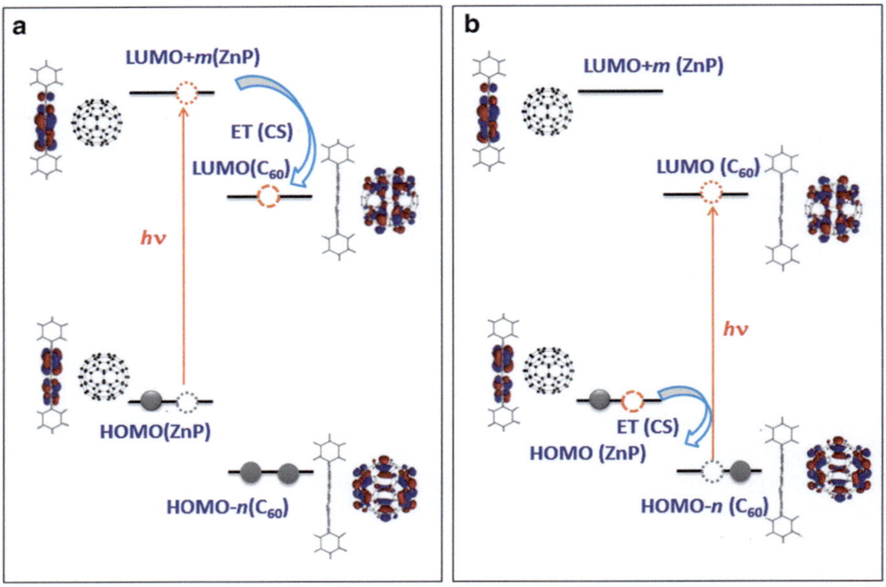

Fig. 4 MO energy level diagrams showing photoinduced ET between ZnP and C_{60} in ZnP–C_{60} hybrids from their excited states; the HOMO and LUMO are for a whole complex molecule (HOMO–*n* is one of the lower occupied MOs and LUMO+*m* is one of the lower unoccupied MOs). (**a**) ZnP excitation and (**b**) C_{60} excitation. *Dotted orange circle*, moving electron; *dotted gray circle*, hole. In this chapter, ET is mainly used for intermolecular events, whereas charge separation (CS) is used for intramolecular events. Reconstructed using data from reference [12]

In the dyad, the HOMO is localized within ZnP, whereas the LUMO is localized on C_{60}, which shows no appreciable charge–transfer (CT) interaction in the ground state. This is also a model of ET process occurring in a collisional encountered mixture of D and A. In both cases, photoinduced ET could occur both by ZnP excitation and C_{60} excitation in appropriate polar solvents. However, for the sake of differentiation, in this chapter, ET is mainly used for intermolecular events, whereas charge separation (CS) is used for intramolecular events.

As shown in Fig. 4a, light excitation of ZnP elevates an electron on the HOMO(ZnP) to the LUMO+*m*(ZnP), from which the electron jumps down to the LUMO(C_{60}), giving ($ZnP^{•+}$,$C_{60}^{•-}$) or $ZnP^{•+}$–$C_{60}^{•-}$. When D and A are close enough, through-space ET would occur. For D–A, in general, the free-energy change for ET process via the excited state of D (=$\Delta G^{o}_{ET(D*)}$) can be evaluated from energy differences, [LUMO + *m*(D) – LUMO(A)] and [HOMO(D) – LUMO(A)], after taking the excitation energy $E_{EX(D)}$ (=[LUMO + *m*(D) – HOMO(D)]) into consideration.

On the other hand, excitation of C_{60} rises an electron from the HOMO – *n*(C_{60}) to the LUMO(C_{60}) (=$E_{EX(C60)}$) as shown in Fig. 4b; then, one of the two electrons of the HOMO(ZnP) falls down to the half-vacant HOMO – *n*(C_{60}), giving $ZnP^{•+}$–$C_{60}^{•-}$. Although the electron configuration of the radical ion pair (RIP) via this ET process is the same as that obtained from ZnP excitation, trail of the excess electron in

$C_{60}^{\bullet-}$ is different. That is, for the ZnP excitation, the extra electron of $C_{60}^{\bullet-}$ on the LUMO comes from LUMO + m(ZnP) originated from HOMO(ZnP). For C_{60} excitation, on the other hand, the extra electron of $C_{60}^{\bullet-}$ on LUMO comes from the HOMO − m(C_{60}); successively, an electron comes from the HOMO(ZnP) to the half-vacant HOMO − m(C_{60}).

When D and A are connected with short spacer (sp), CS process can take place via the through-bond mechanism. If the electron does not stay on the sp moiety with higher LUMO(sp) level, the MO diagram in Fig. 4 might be enough to interpret the CS process. On the other hand, if the electron stays for a distinct time on the sp moiety, $D^{\bullet+}$–($sp^{\bullet-}$)–A or D–($sp^{\bullet+}$)–$A^{\bullet-}$ must be presented as an intermediate, which implies the so-called stepwise mechanism. If the energy level of the LUMO + n(sp) is as low as LUMO(A), such $D^{\bullet+}$– ($sp^{\bullet-}$)–A would be formed via D*. For high energy LUMO + n(sp) such as σ*-bonds, the tunneling mechanism must be taken into consideration. In the case of D–($sp^{\bullet+}$)–$A^{\bullet-}$ via A*, a stepwise process may be possible when the energy level of HOMO − n(sp) is as high as HOMO(D) [13].

2.2 *Donor and Acceptor Mixture in Polar Solvents*

In the presence of high concentrations of zinc phthalocyanine (ZnPc) and C_{60}, photoinduced ET processes via the singlet excited state of either ZnPc or C_{60} in polar solvents is suggested by the fluorescence quenching of ZnPc or C_{60} in the picosecond time regions. Occurrence of ET processes is confirmed by observing $ZnPc^{\bullet+}$ and $C_{60}^{\bullet-}$ in the transient absorption spectra with a short laser pulse. The lifetimes of $ZnPc^{\bullet+}$ and $C_{60}^{\bullet-}$ are usually not long enough to persist beyond 1,000 ps [14, 15]. Under the conditions of low concentrations of ZnPc and C_{60}, long-living $ZnPc^{\bullet+}$ and $C_{60}^{\bullet-}$ are observable in ns–μs region. Such an example for a mixture of ZnP and C_{60} is shown in the nanosecond transient spectra (Fig. 5a). The $^3ZnP^*$ (840 nm) and $^3C_{60}^*$ (740 nm) are initially produced; then, they decay with concurrent occurrence of $C_{60}^{\bullet-}$ (1,080 nm) and $ZnP^{\bullet+}$ (at 620 nm, which is not shown). Energy diagram supports that the energy level of the radical ion pair ($ZnPc^{\bullet+}$, $C_{60}^{\bullet-}$) is lower than the energy levels of either $^3ZnP^*$ or $^3C_{60}^*$ in polar solvents. The $^3C_{60}^*$ may be generated either by direct excitation or by subsequent energy transfer (EnT) from $^1ZnP^*$ to C_{60}, since the excitation energy of the $^1ZnP^*$ is higher than $^1C_{60}^*$. Thus, the observed absorption intensity of $C_{60}^{\bullet-}$ is a total of the ET processes via $^3ZnP^*$ and via $^3C_{60}^*$. In highly polar solvents such as benzonitrile (PhCN), from the dependence of rise rates of $C_{60}^{\bullet-}$ and decay rates of $^3ZnP^*$ on the concentration of added C_{60}, the second-order ET rate constant via $^3ZnP^*$ ($k_{ET(2nd\text{-}order)}$ via $^3ZnP^*$) is evaluated to be ca. $10^9\ M^{-1}\ s^{-1}$, near to the diffusion-controlled limit. In nonpolar solvents, the energy level of the radical ion pair is higher than $^3ZnP^*$ and $^3C_{60}^*$ energy; thus, the decays of $^3ZnP^*$ and $^3C_{60}^*$ are not observed.

In the case of a mixture of C_{60} and ZnPc (Fig. 5b), selective excitation of C_{60} generates $^3C_{60}^*$ (740-nm band), which decays by accepting an electron from ZnPc; subsequently, $ZnPc^{\bullet+}$ (860 nm) and $C_{60}^{\bullet-}$ (1,080 nm) are produced as ET products.

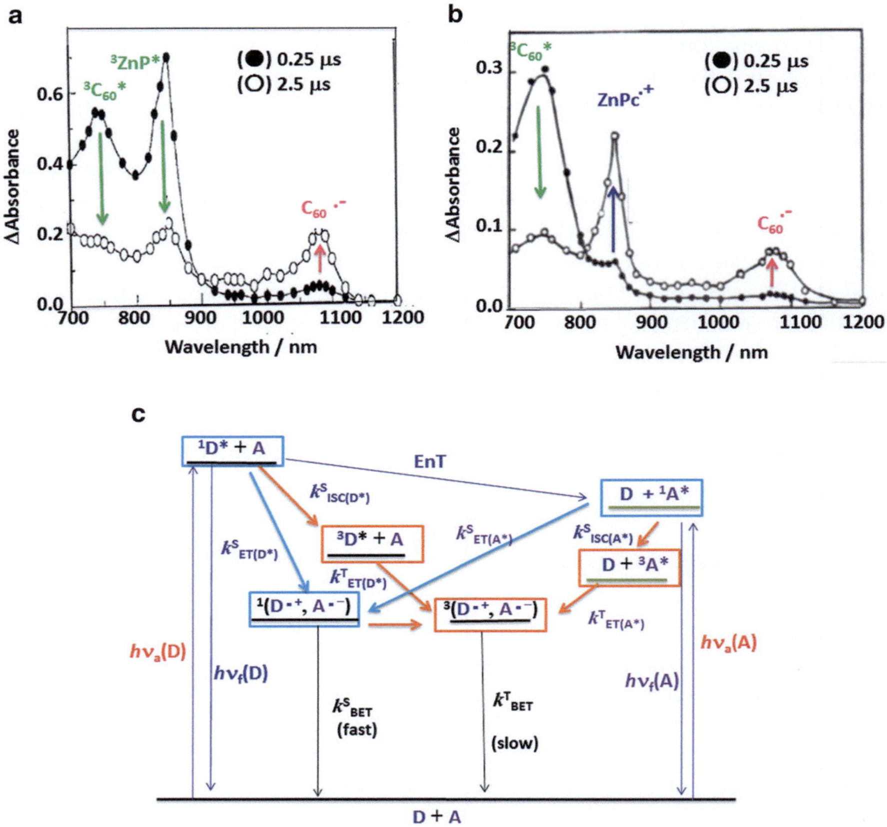

Fig. 5 Nanosecond transient absorption spectra confirming intermolecular photoinduced ET in PhCN; (**a**) a mixture of ZnP and C_{60} with 532-nm excitation, which excites both components; (**b**) a mixture of ZnPc and C_{60} with predominant excitation of C_{60}. (**c**) Energy diagram for overall photoinduced ET processes. BET and CR are used to differentiate the intermolecular process from intramolecular process. Reconstructed from the data in reference [14–16]

The energy level of ($ZnPc^{\bullet+}$,$C_{60}^{\bullet-}$) is lower than $^3C_{60}$* energy in polar solvents. In highly polar solvents, from dependences of the rise rates of $ZnPc^{\bullet+}$ and $C_{60}^{\bullet-}$ and the decay rate of $^3C_{60}$* on the concentration of added ZnPc, the ET rate constant via $^3C_{60}$* ($k^T_{ET(2nd\text{-}order)}$) is evaluated to be 10^8 M^{-1} s^{-1}, which is an order smaller than the diffusion-controlled rate constant (k_{dif}).

These observations are summarized in an energy diagram as shown in Fig. 5c [16]. The ET via ^{1}D* may be competitive to intersystem crossing (ISC) to produce ^{3}D* with $k_{ISC} \approx 10^8$ s^{-1}. To increase the ET contribution via ^{1}D*, concentration of A [A] must be high enough to avoid the competitive ISC. The ET process becomes predominant, when $k^S_{ET(Pseudo\text{-}1st\text{-}order)} = k^S_{ET(2nd\text{-}order)}[A] \gg k^S_{ISC(D)}$; that is, [A] must be higher than 10^{-1} M for $k^S_{ET(2nd\text{-}order)} \approx k_{dif} \approx 10^9$ M^{-1} s^{-1}. On the other hand, when $[A] \ll 10^{-1}$ M, ET process via ^{3}D* would be predominant.

Similarly, when [D] ≫ 10^{-1} M ($k^{S}_{ET(2nd\text{-}order)}$[D] ≫ $k^{S}_{ISC(A)}$), ET process via $^{1}A^{*}$ would occur, whereas for [D] ≪ 10^{-1} M, ET process via $^{3}A^{*}$ would be anticipated to occur. When the generated $D^{\bullet+}$ and $A^{\bullet-}$ interact in solution and exist as a radical ion pair ($D^{\bullet+}$,$A^{\bullet-}$) (RIP), the RIP would gain a triplet spin character. For $^{1}(D^{\bullet+},A^{\bullet-})$, back ET (BET) process to the singlet ground state is fast, whereas back ET process from the $^{3}(D^{\bullet+},A^{\bullet-})$ to the singlet ground state is relatively slow, because of the spin-forbidden process.

In the case of covalently connected D–A system, spin multiplicity of the ET process (or CS process) depends on the distance between D and A (or bonds). For short distance, CS process takes place via the singlet excited states, giving short-lived $^{1}(D^{\bullet+}–A^{\bullet-})$. On the other hand, for longer distance, CS process takes place via the triplet excited states, giving long-lived $^{3}(D^{\bullet+}–A^{\bullet-})$. Even for $^{1}(D^{\bullet+}–A^{\bullet-})$, the lifetime is long enough for hyperfine interaction (usually 1 ns) to promote the spin conversion from $^{1}(D^{\bullet+}–A^{\bullet-})$ to generate $^{3}(D^{\bullet+}–A^{\bullet-})$.

2.3 Charge Separation via Acceptor Excitation in Covalently Connected C_{60}–Phenothiazine System

For covalently connected systems such as C_{60}–phenothiazine (C_{60}–PTZ), long-living radical ion pair is observed by selective excitation of the C_{60} moiety. The structure of C_{60}–PTZ is shown in Fig. 6a [17], in which C_{60} and PTZ moieties are separated with several single bonds, suggesting absence of direct π-interactions. Fluorescence of the C_{60} moiety is strongly quenched upon connecting PTZ; with an increase of solvent polarity, the fluorescence quenching rates increase (Fig. 6b). From shortening of the fluorescence lifetimes, k^{S}_{CS} values via $^{1}C_{60}^{*}$ are evaluated as 1.0×10^{9} s^{-1} in toluene, 2.6×10^{9} s^{-1} in DCB, 4.5×10^{9} s^{-1} in PhCN, and 6.2×10^{9} s^{-1} in DMF. The HOMO and LUMO in Fig. 6c suggest formation of the distinct radical anion ($C_{60}^{\bullet-}$) and radical cation ($PTZ^{\bullet+}$) as a CS product. Nanosecond transient absorption spectra support the formation of the $PTZ^{\bullet+}$ at 600 nm and $C_{60}^{\bullet-}$ at 1,020 nm. The formed $C_{60}^{\bullet-}$–$PTZ^{\bullet+}$ decays with the CR process, giving a lifetime for RIP of about 350 ns (Fig. 6d). The k_{CR} values of $C_{60}^{\bullet-}$–$PTZ^{\bullet+}$ decrease with solvent polarity (1.1×10^{6} s^{-1} in DCB, 2.9×10^{6} s^{-1} in PhCN, 17.0×10^{6} s^{-1} in DMF), which is an opposite trend to that of k^{S}_{CS}.

The slow CR rate constant (k_{CR}) compared to the fast CS rate constant (k_{CS}) in addition to the observed solvent polarity effect could be reasonably interpreted by the Marcus theory [18]. The free-energy changes of the CR process (ΔG_{CR}) are usually more negative (−1.1 to −1.7 eV) than that of the CS process (ΔG_{CS} = −0.1 to −0.6 eV) as shown in Fig. 7a for C_{60}–PTZ in the studied organic solvents. Since the reorganization energy (λ_{reorg}) is in the region of 0.4–0.7 eV for C_{60} derivatives [19], the CS process falls in the uphill region near the summit, whereas the CR process is in the deep inverted region, that is, ($|\Delta G_{CS}| \ll \lambda_{reorg} \ll |\Delta G_{CR}|$) as shown in Fig. 7b. Furthermore, as the $|\Delta G_{CS}|$ values become larger with the solvent polarity, the CS process approaches the top region, resulting in an increase in the k_{CS} values as

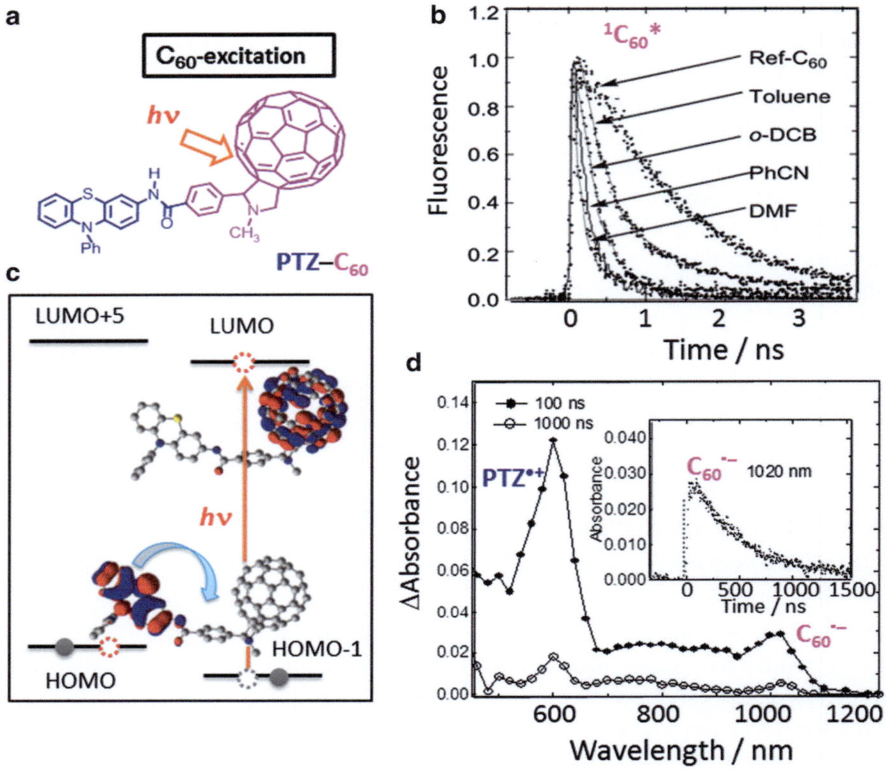

Fig. 6 (**a**) Molecular structure of C_{60}–phenothiazine (C_{60}–PTZ), (**b**) MO description of CS process via C_{60} excitation, (**c**) solvent polarity effect on the fluorescence quenching (fluorescence lifetimes), and (**d**) nanosecond transient absorption spectra in PhCN. *Inset*: time profile of $C_{60}^{\bullet-}$ at 1,020 nm. Reconstructed using data from reference [17]

observed from the fluorescence lifetimes. On the other hand, as the $|\Delta G_{CR}|$ values become larger with decreased solvent polarity, the k_{CR} values decrease as observed from the decay time of the $C_{60}^{\bullet-}$–$PTZ^{\bullet+}$ in the ns-transient absorption bands. Thus, the solvent polarity effect on k_{CS} values is opposite to that of k_{CR} values.

Recently, a multi-modular donor–sensitizer–acceptor system with three PTZ moieties appending to subphthalocyanine (SubPc) with axial C_{60} (PTZ_3–SubPc–C_{60}) is reported as shown in Fig. 8a [20]. The visible light excites mainly the central SubPc having strong absorption in the visible region. On the basis of the MO energy diagram, visible-light excitation of the central SubPc induces CS process from SubPc* to C_{60}, which corresponds to an electron transfer from the LUMO+1 of SubPc to the LUMO of C_{60}, generating PTZ_3–$SubPc^{\bullet+}$–$C_{60}^{\bullet-}$. Subsequently, an electron from the HOMO of PTZ unit transfers to the half-vacant HOMO−1 of $SubPc^{\bullet+}$, generating $(PTZ)_2PTZ^{\bullet+}$–SubPc–$C_{60}^{\bullet-}$, in which the hole (radical cation) is localized on one of the

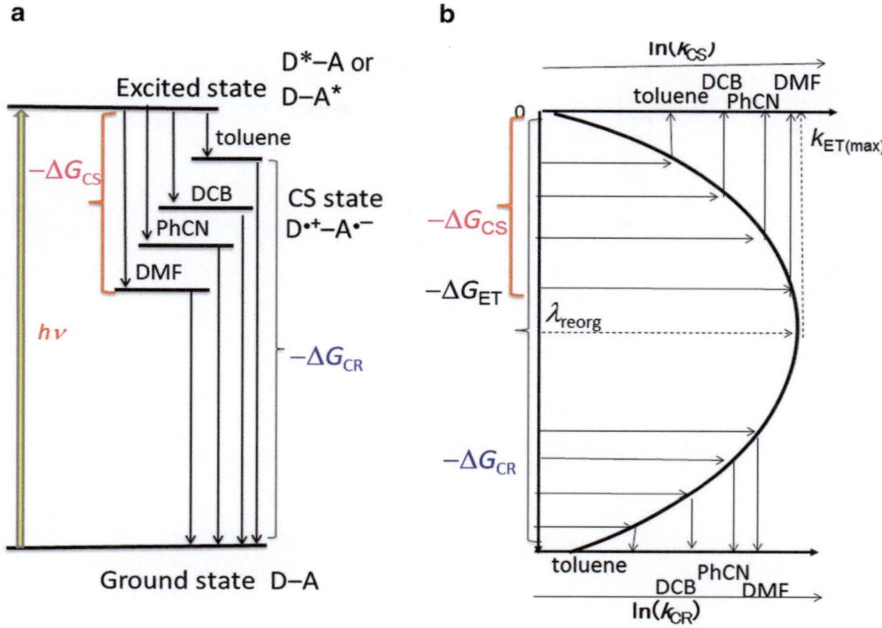

Fig. 7 (**a**) Solvent polarity effect on the energy level of the RIP, (**b**) the Marcus parabola and solvent polarity effect on the k_{CS} and k_{CR} values, which show the opposite trends. Reconstructed using data from reference [17]

three PTZs as seen from the HOMO in Fig. 8a. In $(PTZ)_2PTZ^{\bullet+}$–SubPc–$C_{60}^{\bullet-}$, the electron and hole are separated in longer distance over central SubPc.

Since the fluorescence of SubPc is efficiently quenched by the connection of C_{60} in PhCN, the initial CS from ^{1}SubPc* occurs to C_{60}. In the picosecond transient spectra shown in Fig. 8b, the absorption band of the $C_{60}^{\bullet-}$ clearly appears at 1,000 nm after 35 ps. In addition, a broad absorption band is found at 900 nm, which could be attributed to $SubPc^{\bullet+}$ [21], suggesting the formation of PTZ_3–$SubPc^{\bullet+}$–$C_{60}^{\bullet-}$. At 500 ps, the 900-nm band decreases, leaving the 650-nm band due to $PTZ^{\bullet+}$, suggesting the formation of $(PTZ)_2PTZ^{\bullet+}$–SubPc–$C_{60}^{\bullet-}$. Electron transfer from PTZ to $SubPc^{\bullet+}$ occurs within 500 ps; that is, the hole transfer (HT) process occurs in the opposite direction. This process is equivalent to electron migration from the HOMO on PTZ to the half-vacant HOMO − 1 of $SubPc^{\bullet+}$ (Fig. 8a).

In the nanosecond transient absorption spectra (Fig. 8c), the broad band due to $SubPc^{\bullet+}$ is replaced by the 700-nm band of $PTZ^{\bullet+}$ at 50 ns, suggesting the formation of $(PTZ)_2PTZ^{\bullet+}$–SubPc–$C_{60}^{\bullet-}$, that is, a hole transfer (HT) from $SubPc^{\bullet+}$ to PTZ. The lifetime of the final $(PTZ)_2PTZ^{\bullet+}$–SubPc–$C_{60}^{\bullet-}$ is as long as ca. 150 ns. It is rather surprising that the lifetime $(PTZ)_2PTZ^{\bullet+}$–SubPc–$C_{60}^{\bullet-}$ is shorter than that of $C_{60}^{\bullet-}$–$PTZ^{\bullet+}$ (350 ns). Some conformational changes in PTZ_3–SubPc–C_{60} may proximate the $PTZ^{\bullet+}$ and $C_{60}^{\bullet-}$ units, promoting through-space CR process.

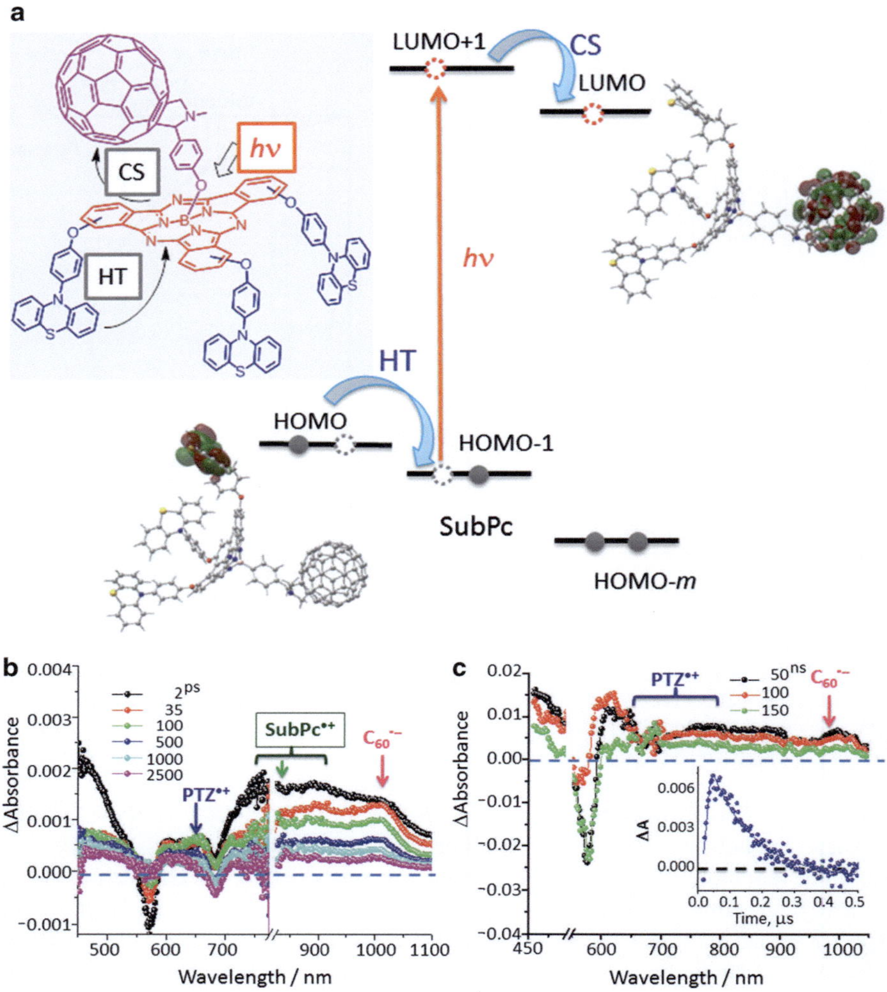

Fig. 8 (**a**) MO diagram for CS and hole transfer (HT) after light excitation of SubPc moiety of PTZ_3–SubPc–C_{60}; the HOMO is localized on one of the three PTZs; the HOMO−1 and LUMO+1 are localized on SubPc; the LUMO and HOMO−*m* are localized on C_{60}. (**b**) Picosecond transient absorption spectra of PTZ_3–SubPc–C_{60} in PhCN. (**c**) Nanosecond transient absorption spectra of PTZ_3–SubPc–C_{60} in PhCN. *Inset*: time profile of $C_{60}^{\bullet-}$ at 1020 nm. Reconstructed using data from reference [20]

2.4 *Charge Separation via Donor Excitation for Substituted Porphyrin Connected to C_{60} Molecule*

For covalently connected systems, a new concept of charge stabilization via extensive delocalization of the unpaired electron over the porphyrin substituents has been reported in a porphyrin–C_{60} dyad shown in Fig. 9a [22, 23]. This newly synthesized

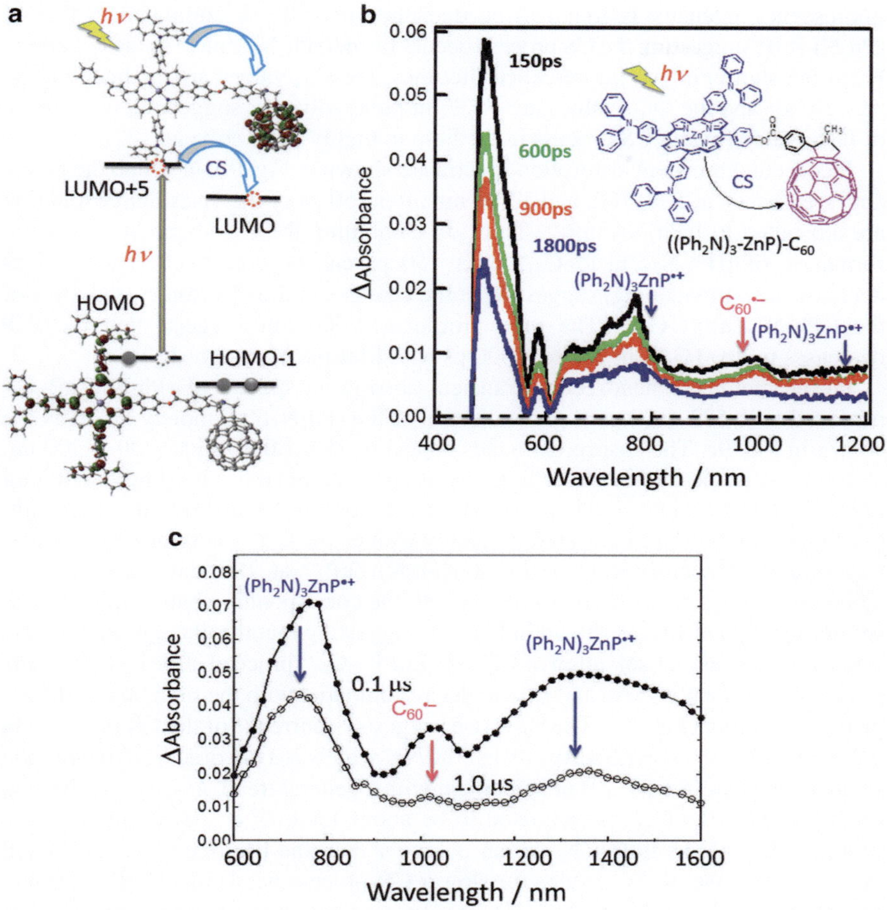

Fig. 9 (**a**) Molecular structure of $((Ph_2N)_3ZnP)–C_{60}$, MO description of CS process using the HOMO and LUMO calculated by B3LYP/3-21G method. (**b**) Picosecond transient absorption spectra of $((Ph_2N)_3ZnP)–C_{60}$ in Ar-saturated PhCN obtained by 400-nm laser-light excitation. (**c**) Nanosecond transient absorption spectra obtained by 532-nm laser-light irradiation (6-ns laser pulse). *Inset*: absorption time profile at 1,300 nm. Reconstructed using data from reference [22, 23]

multimolecular system is composed of a porphyrin ring with three covalently linked diphenylamine (Ph_2N) entities at the *meso*-phenyl positions of the porphyrin ring ($(Ph_2N)_3ZnP$). Delocalization of the π-system over the *meso*-substituents and porphyrin is anticipated from the HOMO in Fig. 9a, predicting that ($(Ph_2N)_3ZnP$) acts as an electron donor. When a fulleropyrrolidine is connected at the fourth *meso-phenyl* position with single bonds, $(Ph_2N)_3ZnP–C_{60}$ is synthesized. The HOMO and LUMO suggest $[(Ph_2N)_3ZnP]^{•+}–C_{60}^{•-}$ as the most stable CS state.

Experimentally, CS process via $^1[(Ph_2N)_3ZnP]^*$ is confirmed by the quenching of fluorescence located around 650 nm in $((Ph_2N)_3ZnP)–C_{60}$. In polar solvents, the fluorescence intensity is found to be quenched over 85 % compared to that of $(Ph_2N)_3ZnP$, suggesting the CS process occurs from $^1[(Ph_2N)_3ZnP]^*$ to the C_{60} entity. From the shortening of fluorescence lifetimes, the $k^S{}_{CS}$ values are evaluated to be $>5\times10^9$ s^{-1} and the $\Phi^S{}_{CS}$ values are >0.90 in polar solvents, suggesting occurrence of faster and efficient CS process in medium to highly polar solvents [22].

Picosecond transient absorption spectra are shown in Fig. 9b, in which the newly appeared bands at 500, 780, and 1,200 nm after 150 ps of the laser-light excitation are attributed to $[(Ph_2N)_3ZnP]^{\bullet+}$. Thus, on combining the $C_{60}{}^{\bullet-}$ band at 1,000 nm, formation of $[(Ph_2N)_3ZnP]^{\bullet+}$–$C_{60}{}^{\bullet-}$ after 150 ps can be confirmed. These bands decrease with broadening, suggesting some changes in the electronic properties of the $[(Ph_2N)_3ZnP]^{\bullet+}$–$C_{60}{}^{\bullet-}$. The time profile at 1,000 nm suggests that the RIP decreases by ca. 1/3 from 150 to 3,000 ps with lifetime of ca. 2 ns [23].

Furthermore, the nanosecond transient absorption spectra of $(Ph_2N)_3ZnP$–C_{60} observed with 532-nm laser-light excitation of the $(Ph_2N)_3ZnP$ moiety in PhCN are shown in Fig. 9c. The absorption bands appear at 750, 1,000, and 1,200–1,300 nm at 100 ns after the 532-nm laser light, exciting predominantly the $(Ph_2N)_3ZnP$ unit [22]. Similar transient absorption spectra are observed with 355-nm laser-light excitation; the 355-nm light predominantly excites the C_{60} moiety of $(Ph_2N)_3ZnP$–C_{60}. Importantly, absorption peaks of $[(Ph_2N)_3ZnP]^{\bullet+}$ at 780 and 1,300 nm and 1,000 nm of $C_{60}{}^{\bullet-}$ are broader compared with the corresponding bands appearing at shorter times than 1 ns as shown in Fig. 9b, suggesting that the changes in the electronic property and/or solvation of $[(Ph_2N)_3ZnP]^{\bullet+}$–$C_{60}{}^{\bullet-}$ proceed after 1 ns. The time profile shows the slow rise and slower decay compared with the time profile during picosecond order (Fig. 9b). The slow rise suggests occurrence of the CR process via $^3[(Ph_2N)_3ZnP]^*$ ($k^T{}_{CS}=1.3\times10^7$ s^{-1} in PhCN), which is two orders of magnitude smaller than $k^S{}_{CS}$ (5×10^9 s^{-1}) in agreement with a general trend, $k^T{}_{CS}<k^S{}_{CS}$. The $k^T{}_{CR}$ for $[(Ph_2N)_3ZnP]^{\bullet+}$–$C_{60}{}^{\bullet-}$ is evaluated to be about 1.0×10^6 s^{-1} corresponding to a lifetime of 1,000 ns in PhCN, which is longer than the lifetime = 2 ns evaluated in Fig. 9b. As one of the reasons for slower CR process persisting 1,000 ns, triplet spin character of $[(Ph_2N)_3ZnP]^{\bullet+}$–$C_{60}{}^{\bullet-}$ may be considered. Furthermore, the changes in the solvation and electronic properties may also contribute to such prolonging $[(Ph_2N)_3ZnP]^{\bullet+}$–$C_{60}{}^{\bullet-}$.

In a comparison of the lifetimes of radical ion pairs (RIPs) of simple dyads with similar linkage but without Ph_3N entities ($ZnP^{\bullet+}$–(sp)–$C_{60}{}^{\bullet-}$) the $[(Ph_2N)_3ZnP]^{\bullet+}$–$C_{60}{}^{\bullet-}$ has longer lifetime, suggesting that the delocalization of radical cation in the wider region of the larger donor entity. Such highly stable radical ions in the RIP are one of the factors to achieve long-living CS state. In addition, contribution of the triplet excited states of the D and/or A to the CS process increases the triplet spin character of RIP resulting in prolonged lifetime. Even if the CS takes place via the singlet excited state, when the energy of the RIP is close to the triplet excited state, the RIP may gain triplet spin character. To realize the spin conversion, the initial RIP with the singlet spin character must have a certain long lifetime. To achieve such long lifetime of RIP ($D^{\bullet+}$–$A^{\bullet-}$) with short linkage, the Marcus theory must be considered.

2.5 *Cascade Electron and Hole Transfer After Photoinduced Charge Separation*

Another strategy to fulfill the tendency of $k_{CS} \gg k_{CR}$ is to sequentially move the electron and hole from the RIPs by appending electron mediator or hole transfer units, that is, D_2–D_1–A, in which the electron donor ability of D_2 is higher than D_1, and D–A_1–A_2, in which the electron acceptor ability of A_2 is higher than A_1.

An example is shown in Fig. 10a for Fc–($(Ph_2N)_2$ZnP)–C_{60}, in which one of the Ph_3N units in Fig. 9a is replaced by a ferrocene (Fc) entity [23]. Here, Fc is selected, because it has higher electron donor ability than $(Ph_2N)_2$ZnP. Immediately after photoexcitation of the central ($(Ph_2N)_2$ZnP) unit, an electron transfers from $^1[(Ph_2N)_2ZnP]^*$ to the C_{60} unit, generating $[(Ph_2N)_2ZnP]^{•+}$ and $C_{60}^{•-}$. Then, an electron on the HOMO(Fc) of the vicinal Fc transfers to $((Ph_2N)_2ZnP)^{•+}$ to the half-vacant HOMO−1, since the HOMO(Fc) is higher than the HOMO−1 (Fig. 9b). This hole transfer process must be competitive with the short distant CR process between the vicinal $[(Ph_2N)_2ZnP]^{•+}$ and $C_{60}^{•-}$, in which the distance between the hole (center of $[(Ph_2N)_2ZnP]^{•+}$) and electron (center of $C_{60}^{•-}$) is ca. 18 Å providing an estimate of

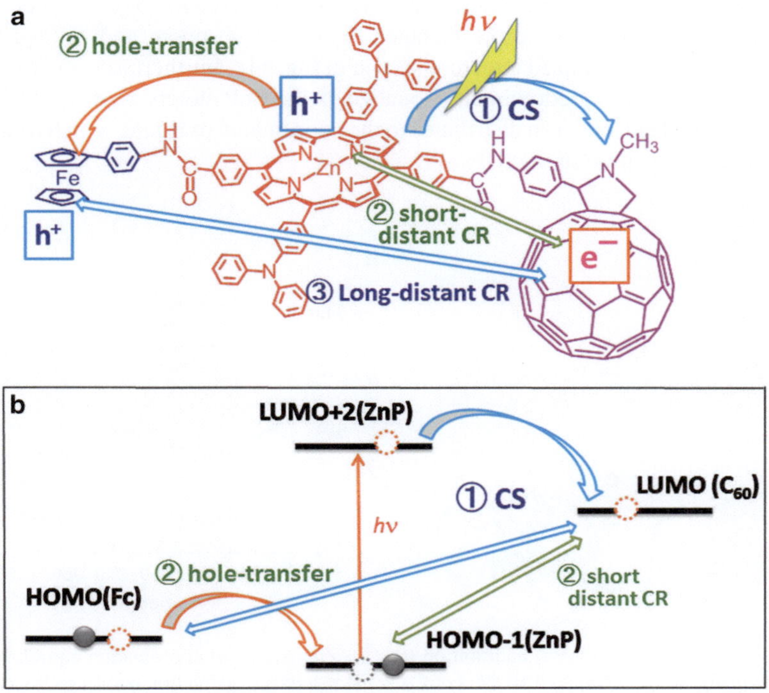

Fig. 10 (**a**) Cascade hole transfer following initial photoinduced CS of Fc–($(Ph_2N)_2$ZnP)–(sp)–C_{60} to produce long-living $Fc^{•+}$–($(Ph_2N)_2$ZnP)–(sp)–$C_{60}^{•-}$, (**b**) MO description. Reconstructed using data from references [22] and [23]

CR rate of ca. 10^6 s^{-1}. Thus, the hole transfer rate constant must be larger than ca. 1×10^6 s^{-1}. Then, the final $Fc^{\bullet+}$–($(Ph_2N)_2$ZnP)–(sp)–$C_{60}^{\bullet-}$ would be produced, in which the distance between $Fc^{\bullet+}$ and $C_{60}^{\bullet-}$ units is as long as ca. 35 Å, achieving long-living CS state. Indeed, experimentally, lifetime of the long distant CS state is 8.5 μs, which is longer than that observed for ($(Ph_2N)_3$ZnP)–C_{60} being ca. 1 μs.

2.6 *Fullerene–Porphyrin Molecule with Flexible Connector*

In another study, pyridine coordination effect on the CS and CR processes of covalently linked C_{60}~ZnP dyads is investigated (Fig. 11a) [24, 25]. The MO calculations of C_{60}~ZnP at the B3LYP/3-21G level are shown in Fig. 11b, in which the optimized structure has a short distance between the C_{60} sphere and ZnP surface of C_{60}~ZnP. In the HOMO, although the electron mainly localized on ZnP moiety, a small amount of electron is distributed to the nearside of C_{60} sphere, suggesting that charge–transfer (CT) interaction exists in the ground state. Therefore, the observed transient species must be short lived, because the excited state of the CT state is considered to be delocalized RIP over the entire C_{60}~ZnP.

On the other hand, coordination of pyridine to ZnP from the opposite side to the C_{60} sphere tends to increase the distance between C_{60} sphere and ZnP plane as revealed from the optimized structure shown in Fig. 11c. Furthermore, such coordination makes HOMO predominantly localize on the ZnP moiety, destroying the CT interaction. Instead, electron distributes to the coordinated pyridine, which improves the hole delocalization in the RIP.

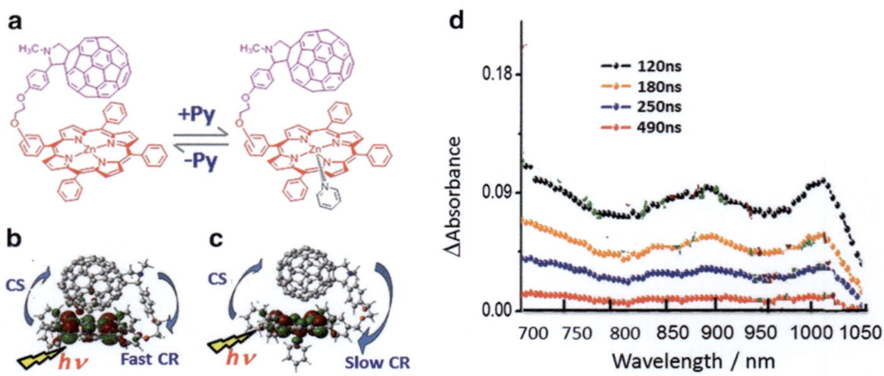

Fig. 11 Effect of pyridine (Py) addition on the C_{60}~ZnP dyad; (**a**) coordination equilibrium on addition of Py; (**b**) structure and HOMO of C_{60}~ZnP, where CT interaction between closed structure of C_{60}~ZnP is seen; (**c**) HOMO of C_{60}~ZnP coordinated to Py, which eliminates CT by increasing distance between C_{60} and ZnP; (**d**) nanosecond transient absorption spectra of C_{60}~ZnP in the presence of excess Py in *o*-DCB, which show broadening of the bands with slow decay. Reconstructed using data from references [24, 25]

In o-DCB, photoexcitation yields $C_{60}^{\bullet-}$~$ZnP^{\bullet+}$ via C_{60}~$^1ZnP^*$ and/or $^1C_{60}^*$~ZnP in ca. $5\times10^9\ s^{-1}$ as revealed by the fluorescence lifetime and picosecond transient absorption spectra. The $C_{60}^{\bullet-}$~$ZnP^{\bullet+}$ displays relatively short lifetimes similar to the coordination complex (<20 ns), because of close vicinity of $C_{60}^{\bullet-}$ and $ZnP^{\bullet+}$. However, on addition of a small amount of pyridine to coordinate to the Zn ion of ZnP moiety in o-DCB solution, the transient absorption spectra change to Fig. 11d [24, 25]. After 10 ns laser-light pulse, the 1,000-nm peak reaches to a maximum at 100 ns. Before 100 ns, sharp peaks appear at 700 nm and 800–900 nm, which can be attributed to $^3C_{60}^*$ and $^3ZnP^*$, respectively. A sharp peak is also observed at 1,000 nm, which can be ascribed to $C_{60}^{\bullet-}$. After 100 ns, broad bands are found at 1,000 nm and 880 nm, which are also attributed to $C_{60}^{\bullet-}$ and $ZnP^{\bullet+}$, respectively, although solvation of the RIP ($C_{60}^{\bullet-}$ and $ZnP^{\bullet+}$) may change before 100 ns. Lifetime of the RIP (τ_{RIP}) of $C_{60}^{\bullet-}$ and $ZnP^{\bullet+}$ prolongs up to 200 ns. These characters of the HOMO result in prolonging of the RIP as observed in Fig. 11d.

Coordination of additional pyridine to the ZnP plane from the same side to the C_{60} sphere can increase the distance between both entities; then, the photoexcitation produces pure RIP ($C_{60}^{\bullet-}$~(Py→)$ZnP^{\bullet+}$) and its CR rate is slowed down, which may be one of the reasons for long-lived $C_{60}^{\bullet-}$~(Py→)$ZnP^{\bullet+}$.

In addition, other electronic factors can be considered to be responsible for this observation, because the broadening of transient bands after 100 ns suggests the characteristic interactions between pyridine molecules and C_{60} molecule and sometimes induces aggregations of C_{60}~ZnP molecules resulting into fine particles.

2.7 Fullerene–Zn Porphyrin Coordination Systems

When C_{60} is functionalized with phenylimidazole (C_{60}Im) or pyridine (C_{60}Py) entities, the supramolecular coordination complex (C_{60}Im → ZnP in Fig. 12a) exists as major equilibrium species in o-DCB, giving $K_{complex} \approx 1\times10^5\ M^{-1}$ from the Benesi–Hildebrand plots for the 1:1 complex. When ZnP is photoexcited, HOMO electron of the ZnP (Fig. 12b) is promoted to the upper LUMO+5 localizing on the ZnP, from which the excited electron transfers to the LUMO localizing on the C_{60} moiety (Fig. 12c), leaving a hole on HOMO localized on ZnP, which gives $C_{60}^{\bullet-}$Im → $ZnP^{\bullet+}$ as a stable RIP.

When C_{60} moiety of the dyad is photoexcited, electron of the C_{60} moiety in the ground state (HOMO–3) is promoted to the excited state C_{60} (LUMO); subsequently, the half-vacant HOMO–3 abstracts an electron from the HOMO on ZnP moiety, giving the CS state, $C_{60}^{\bullet-}$Im→$ZnP^{\bullet+}$.

In the coordination system, transient absorption band due $C_{60}^{\bullet-}$Im is observed at 1,020 nm as shown in Fig. 12d [26]. The absorption bands of $C_{60}{\bullet}^{-}$Im at 1,020 nm and $ZnP^{\bullet+}$ in the 600–680-nm region show rapid rise within 100 ps, suggesting that the CS process takes place via $^1ZnP^*$ entity within the coordination complex with a CS rate constant (k_{CS}) of $>10^{10}\ s^{-1}$. From the decay of $C_{60}^{\bullet-}$Im within ca. 10 ns, rate constant of CR process, k_{CR}, in the studied dyads was obtained to be ca. $3\times10^8\ s^{-1}$,

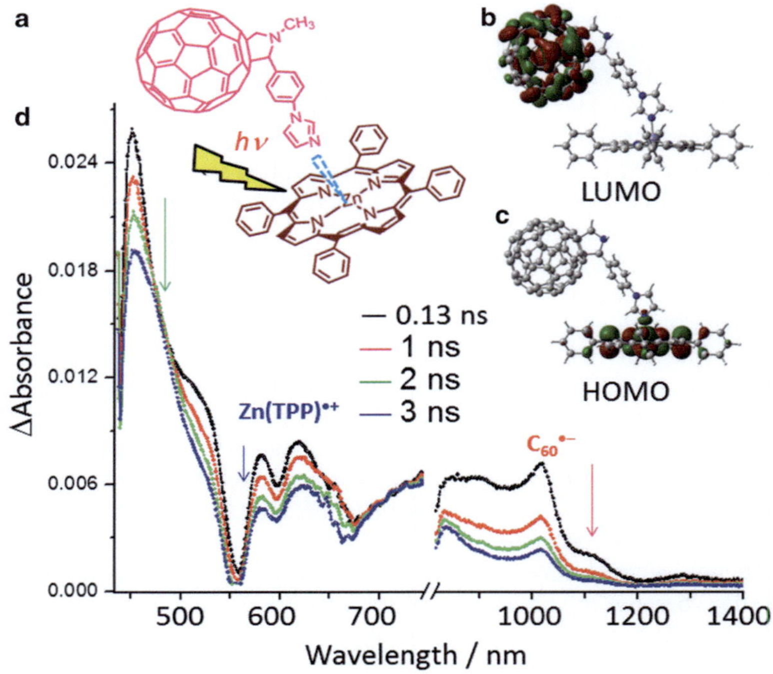

Fig. 12 (**a**) Structure of $C_{60}Im{\rightarrow}ZnP$ complex; (**b** and **c**) the LUMO and HOMO, respectively; (**d**) picosecond transient absorption spectra of $C_{60}Im{\rightarrow}ZnP$ complex in *o*-DCB solution. Reconstructed using data from reference [26]

which is also reasonable due to the close distance between the electron and hole in $C_{60}^{\bullet-}Im{\rightarrow}ZnP^{\bullet+}$ and also due to almost face-to-face (parachute) configuration.

Bis(pyridine)C_{60} derivative is capable of coordinating both ZnP and ZnPc. An example is shown for Zn naphthalocyanine (ZnNc) [27]; the formations of coordination complexes, $NcZn{\rightarrow}C_{60}$ and $NcZn{\rightarrow}C_{60}{\leftarrow}ZnNc$, are confirmed by the absorption spectral changes and fluorescence quenching phenomena. The k^{S}_{CS} values evaluated from fluorescence lifetimes indicate that $NcZn{\rightarrow}C_{60}{\leftarrow}ZnNc$ is more efficient in the CS process than $NcZn{\rightarrow}C_{60}$. Nanosecond transient spectra support the formation of RIPs, $NcZn^{\bullet+}{\rightarrow}C_{60}^{\bullet-}$ and $NcZn^{\bullet+}{\rightarrow}C_{60}^{\bullet-}{\leftarrow}ZnNc$. The k_{CR} values evaluated from the 1,000-nm band decay indicate that the lifetime of $NcZn^{\bullet+}{\rightarrow}C_{60}^{\bullet-}{\leftarrow}ZnNc$ is longer than that of $NcZn^{\bullet+}{\rightarrow}C_{60}^{\bullet-}$, probably due to the conformational reason (i.e., looser complex of $NcZn{\rightarrow}C_{60}{\leftarrow}ZnNc$) or electronic reason (i.e., delocalization of the cation radical between the two ZnNc moieties).

A further addition of ZnP to the ZnP–ZnPc dyad forms an interesting coordination complex as shown in Fig. 13a, in which one of the ZnP acts as an antenna, whereas the other ZnP along with the ZnPc coordinates to bis(pyridine)C_{60}. These coordinated ZnP and ZnPc act as electron donors with respect to the coordinated C_{60}, that is, (ZnP)–($PZn{\rightarrow}C_{60}{\leftarrow}ZnPc$) [28]. Other coordination structure ((ZnPc)–($PZn{\rightarrow}C_{60}{\leftarrow}ZnP$)) is also possible. Experimentally, the picosecond transient

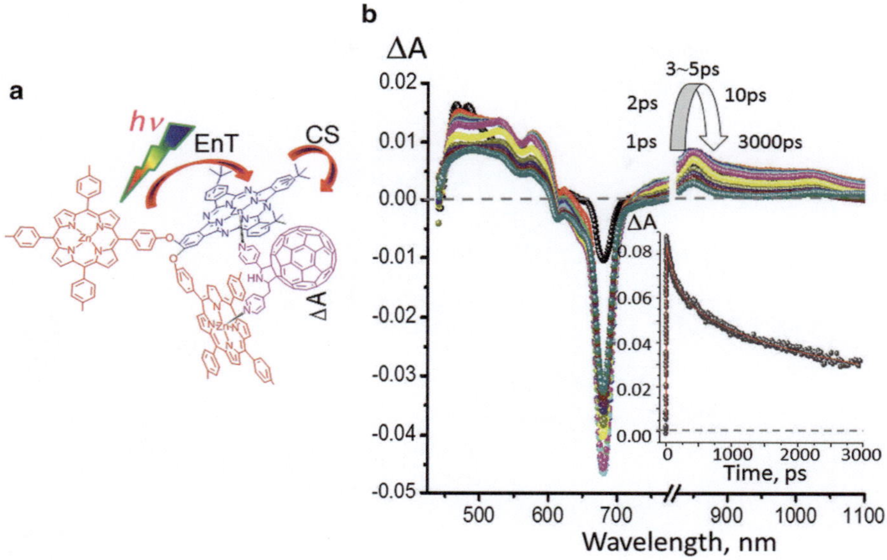

Fig. 13 (**a**) Proposed coordination complex of (ZnP)–(PZn→C_{60}←ZnPc) and (**b**) picosecond transient absorption spectra in toluene solution. Reconstructed from the data in reference [28]

spectra exhibit the 880-nm band due to $ZnPc^{•+}$ and the 1,000-nm band due to $C_{60}^{•-}$ with strong depletion at 680 nm due to the absorption of ZnPc as shown in Fig. 13b. Thus, the main RIP is considered as (ZnP)–(PZn→$C_{60}^{•-}$←$ZnPc^{•+}$), suggesting the presence of (ZnP)–(PZn→C_{60}←ZnPc) in solution. The maximum intensity of the transient absorption appears at 3–5 ps, suggesting fast CS process. The time profile at 880 nm in the inset shows the initial fast decay until 500 ps and slow decay after 1,000 ps continuing even after 3,000 ps. From the latter part, the lifetime of (ZnP)–(PZn→$C_{60}^{•-}$←$ZnPc^{•+}$) is evaluated ca. 10 ns.

2.8 Successive Energy Transfer and Electron Transfer

The axial ligation can be extended to construct a working model of the photosynthetic antenna–reaction center complex. In particular, a supramolecular triad (Fig. 14a) comprises a covalently linked BF_2-chelated dipyrromethene (BDP) and zinc porphyrin (BDP~ZnP), which is coordinated by ImC_{60}, forming BDP~ZnP→ImC_{60} [29]. In this triad, BDP acts as an energy-funneling antenna (^{1}BDP*~ZnP→ImC_{60}) with respect to the covalently linked ZnP to receive photoenergy from the excited BDP via Förster (resonance) mechanism to the vicinal ZnP to form BDP~^{1}ZnP*→ImC_{60}. The ^{1}ZnP* thus formed donates an electron to the coordinate ImC_{60}, finally generating BDP~$ZnP^{•+}$→ $ImC_{60}^{•-}$. Experimentally, fluorescence lifetime measurements initiated with selective excitation of the BDP moiety

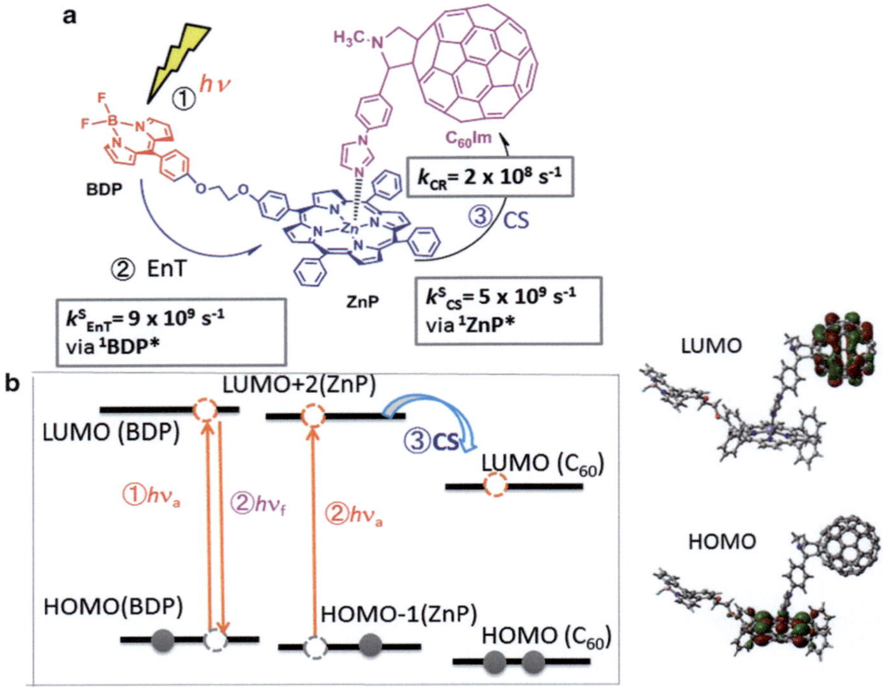

Fig. 14 Supramolecular triad with coordination bonding (BDP~ZnP→ImC_{60}) as a model of antenna–reaction center complex. The rate constants were evaluated from the fluorescence lifetime and transient absorption measurements. Energy diagram explaining the EnT-CS processes. ① Light absorption of BDP, ② EnT via Forster (resonance) mechanism from ^{1}BDP* to ZnP, ③ CS via ^{1}ZnP* to give BDP~$ZnP^{\bullet+}$→Im$C_{60}^{\bullet-}$ [29]

support efficient intramolecular singlet excited energy transfer from ^{1}BDP to the vicinal ZnP, creating singlet excited ZnP with a large rate constant ($k^S_{EnT}=9\times10^9$ s^{-1}) and high quantum yield of ca. 0.83. Then, the indirectly excited ZnP participates in efficient CS process with the coordinated vicinal ImC_{60} ($k^S_{CS}=9\times10^9$ s^{-1}), resulting in the transient absorption of Im$C_{60}^{\bullet-}$ as an evidence of the CS state with a lifetime of ~5 ns. These energy transfer and electron transfer events after the photoexcitation are summarized in Fig. 14b.

This approach to build a model of antenna–reaction center mimicking with BDP, ZnP, and C_{60} entities can be further extended by constructing a model, employing the inclusion complex between crown ether and alkyl ammonium cation [30]. For the construction, BDP was initially covalently attached to a zinc porphyrin entity bearing a benzo-18-crown-6 host segment at the opposite end of the porphyrin ring (BDP~ZnPCrown). Next, an alkyl ammonium functionalized fullerene, ($NH_3^+C_{60}$), is used to self-assemble the crown-ether entity via ion–dipole interactions, finally to form the triad, BDP~ZnPCrown:$NH_3^+C_{60}$. The newly formed supramolecular triad is fully characterized by the spectroscopic, computational, and electrochemical methods. Selective excitation of the BDP moiety resulted in efficient singlet EnT over 97 % efficiency creating the singlet excited ZnP (BDP~^{1}ZnP*Crown:$NH_3^{+}C_{60}$). The rate of EnT from the decay measurements of the time-correlated singlet photon counting (TCSPC) and up-conversion techniques agrees well with that obtained by pump–probe technique. Then, it is revealed that efficient photoinduced energy transfer (time constant in the order of 10–60 ps depending upon the conformer) takes place. Then, the excited ZnP resulted in charge separation to the coordinated C_{60} yielding a CS state (BDP~$ZnP^{•+}$Crown:$NH_3^{+}C_{60}^{•-}$), thus mimicking the antenna–reaction center functionalities of photosynthesis. Nanosecond transient absorption studies yielded a relatively long-lived lifetime of the CS state suggesting the presence of remote conformers separating the $ZnP^{•+}$ from $C_{60}^{•-}$ apart. These supramolecular systems represent successful model to mimic the rather complex "combined antenna–reaction center" events of photosynthesis.

2.9 Photoinduced Charge Separation Through Long Spacers

The spacer (sp) connecting D and A entities plays several important roles in creating effective electronic communication between the two entities. When D and A are connected with π-conjugative spacer, MO of the sp must be taken into consideration for CS and CR processes through the spacer with π-conjugation ability. Such π-conjugative spacer is usually called as molecular wire and molecular bridge.

An example is shown for ZnP–(thienylene-vinylene)*n*–C_{60} system in Fig. 15a [31]. The MOs participating in the photoinduced ET process are shown in Fig. 15b, c. The computed HOMO and LUMO suggest no appreciable π-conjugation between the ZnP moiety and the adjacent thienylene-vinylene moiety, due to almost perpendicular orientation. Since the HOMO and LUMO are localized on the ZnP and C_{60}, respectively, formation of $ZnP^{•+}$–(thienylene-vinylene)*n*–$C_{60}^{•-}$ is envisioned as a most stable CS state. By the photoexcitation of ZnP moiety, an electron from the HOMO is elevated to the LUMO+6 (for $n=3$) and LUMO+7 (for $n=8$), as a local electronic transition within the LUMO(ZnP). Then, there may be two possible routes. The first one is the direct CS mechanism, in which an electron transfers to the LUMO localized on the C_{60} through the (thienylene-vinylene)*n* bridge. The second one is the stepwise CS process; for short spacer, CS takes place by promoting an electron from the ZnP (LUMO+6) to the higher LUMO+*m* ($m>6$) on the (thienylene-vinylene)$_3$ (CS-1). Then, the electron propagates through the (thienylene-vinylene)*n* bridge (CS-2), as an intermediate (or phantom) RIP ($ZnP^{•+}$–(thienylene-vinylene)$n^{•-}$–C_{60}). Finally, electron drops to the LUMO of C_{60} (CS-3), which generates stable $ZnP^{•+}$–(thienylene-vinylene)*n*–$C_{60}^{•-}$. For longer one (ZnP–(thienylene-vinylene)$_8$–C_{60}), the CS-1 is exothermic, because the energy of the LUMO+2 on (thienylene-vinylene)$_8$ is lower than that of the LUMO+7 on the ZnP. Thus, $ZnP^{•+}$–(thienylene-vinylene)$n^{•-}$–C_{60} might be real. Similar results are obtained for ZnP–(thiophene)*n*–C_{60} systems ($n=4$–12) [32].

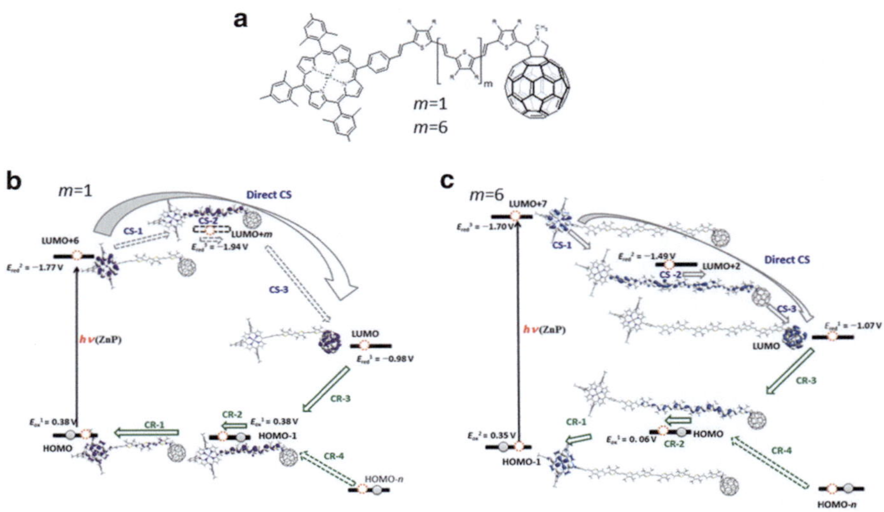

Fig. 15 (**a**) Molecular structure of ZnP–(thienylene-vinylene)*n*–C_{60}; $n=m+2$; (**b**) Schematic MO representations calculated by DFT B3LYP/3-21G level for photoinduced CS processes via ZnP excitation of ZnP–(thienylene-vinylene)$_3$–C_{60}; (**c**) For ZnP–(thienylene-vinylene)$_8$–C_{60}. Reconstructed using data from reference [32]

For the CR process involving a short bridge (ZnP–(thienylene-vinylene)$_3$–C_{60}), the electron jumps from the thienylene-vinylene chain (HOMO−1) to the $ZnP^{•+}$ (half-vacant HOMO) denoted as CR-1. Then, the electron on the HOMO−2 moves to the edge of C_{60} (CR-2); the excess electron of $C_{60}^{•-}$ (LUMO) moves down to the half-vacant HOMO−1 (CR-3), returning to the original neutral molecule.

From the fluorescence lifetime measurements and sub-picosecond transient spectra, the measured $k^S{}_{CS}$ values are in 10^9–10^{10} s^{-1} range depending on the length of thienylene-vinylene π-systems, whereas the measured $k^S{}_{CR}$ values from the nanosecond transient absorption measurements are in 10^6–10^5 s^{-1} range. As reasons for smaller $k_{CR}{}^S$ values than $k^S{}_{CS}$, electron conductivity of the LUMO+6 on the thienylene-vinylene chain (CS-2) is higher than hole conductivity of the HOMO (CR-2) could be considered as one of the factors.

Experimentally, for the dependence of $k^S{}_{CS}$ values on the length of bridges, the attenuation factors (β_{CS}) are evaluated [33]. For (phenylene-vinylene)*n*, (thienylene-vinylene)*n*, and (thiophene)*n* systems, β_{CS}=ca. 0.03 Å^{-1}, which is corresponding to the decrease in $k^S{}_{CS}$ values by ca. 1/10 with an increase in the bridge length of 100 Å. In the case of ZnP–(thiophene)*n*–C_{60}, a larger β_{CS} value in less polar solvents (β_{CS}=0.15 Å^{-1} in *o*-DCB) is reported [32], suggesting that long distance CS process via the oligothiophenes is advantageous in polar solvent. Here, the attenuation factor for the CR process (β_{CR}) determined by transient absorption measurements is almost the same as the β_{CS} values. Depending on the amplitude of the β_{ET} (β_{CS} and β_{CR}) values, the spacer is called as bridge (intermediate β_{ET} values) and molecular wire (small β_{ET} values). The k_{ET} values (k_{CS} and k_{CR}) and β_{ET} values are important to design and evaluate organic polymer film photo-solar cells, in which the exciton (electron–hole pair) migrates along the π-conjugated polymer chains.

2.10 Output Systems of Electron and Hole from the Charge-Separated States

To realize artificial photosynthesis for fuel or electricity generation, it is necessary to extract electron and hole of the RIP after the photoinduced CS state, before it returns to the starting neutral molecule in the ground state. For this aim, an electron from $C_{60}^{\bullet-}$ in $D^{\bullet+}$–(sp)–$C_{60}^{\bullet-}$ can be extracted with addition of electron mediator such as alkyl viologen dication (RV^{2+}). As shown in Fig. 16, $RV^{\bullet+}$ with the absorption peak at 620 nm can be accumulated with the visible-light irradiation in the presence of appropriate hole-trapping reagents [18]. The electron mediator efficiency increases, when the rate of electron-mediating process (R_{md}) must be faster than the CR rate (R_{CR}). The intermolecular R_{md} can be increased with increasing the RV^{2+} concentration ($k_{md}^{2nd}[RV^{2+}]$), whereas the intramolecular R_{CR} is kept constant. For example, k_{CR}^{1st} for $D^{\bullet+}$–$A^{\bullet-}$ is 1×10^6 s^{-1} and $k_{med}^{pseudo\text{-}1st}$ ($=k_{med}^{2nd}[RV^{2+}]$) must be larger than 1×10^6 s^{-1}; namely, if k_{med}^{2nd} is as fast as diffusion limit ($k_{dif}\approx5\times10^9$ M^{-1} s^{-1}),

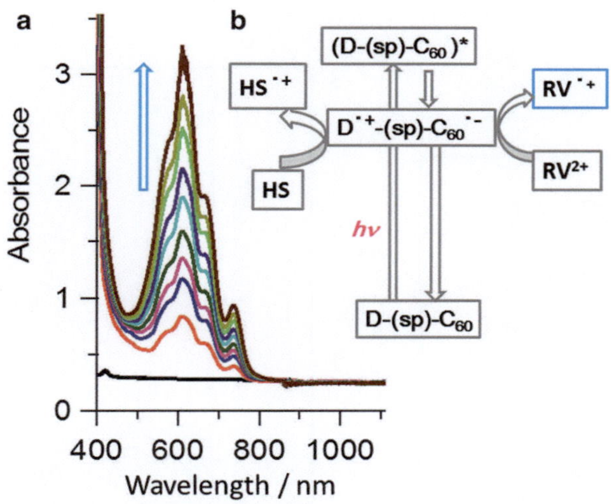

Fig. 16 Steady-state absorption spectra obtained by photolysis of D–(sp)–C_{60} with light longer than 430 nm in the presence of RV^{2+} (with hole-shifting reagent (HS)). As D–(sp)–C_{60}, phenothiazine–C_{60} (0.05 mM), as RV^{2+}, hexyl viologen (0.1 mM), as HS, BNAH (benzyl nicotinamide dihydride; 1 mM) are employed in PhCN for 0–60 min (time interval = 10 min). *Inset*: electron transfer mechanism. Electron accumulation as $RV^{\bullet+}$ at 620 nm during the visible-light irradiation time, in which D–(sp)–C_{60} can be repeatedly used. Reconstructed using figures from reference [18]

$k_{med}^{pseudo\text{-}1st} \approx 5 \times 10^6\ s^{-1}$ when $[RV^{2+}] \approx 1$ mM. On further addition of hole-shifting reagent ([HS] ≈ 1 mM, $k_{HS}^{2nd} \approx k_{dif}$), the hole can be extracted from $D^{•+}$–(sp)–$C_{60}^{•-}$ to retard the R_{CR}. When the hole-shifting reagent is spontaneously converted into a closed ion and molecule, the electron trapped in RV^{2+} showing the blue color due to $RV^{•+}$ can be accumulated. In the case of semiconducting electrodes, photocurrent can be taken out from $C_{60}^{•-}$ or $RV^{•+}$. On addition of H_2-active catalysts such as Pt nanoparticles, electron injection into H^+ promotes the hydrogen evolution.

2.11 Application to Solar Cell

A photoelectrochemical solar cell was constructed using porphyrin–C_{60} dyads having efficient CS ability; an example is shown for aluminum porphyrin (AlP) directly connected with C_{60} via Al–C covalent bond (abbreviated as $AlP^{•+}$–C_{60}), which is adsorbed onto the SnO_2 nanoparticle surface pasted on ITO (Fig. 17a) [34]. In polar solution, $AlP^{•+}$–$C_{60}^{•-}$ is confirmed by nanosecond transient absorption and ESR measurements in polar solvent. Thus, the solar cell is constructed by expecting that the photogenerated electron transfers from $AlP^{•+}$–$C_{60}^{•-}$ to SnO_2/ITO, generating electric current. This new type of solar cell is denoted as a charge-separation-type solar cell, which is a kind of modification of the dye-sensitized Grätzel solar cell, where an electron is directly injected from the photoexcited dye to TiO_2/ITO [35].

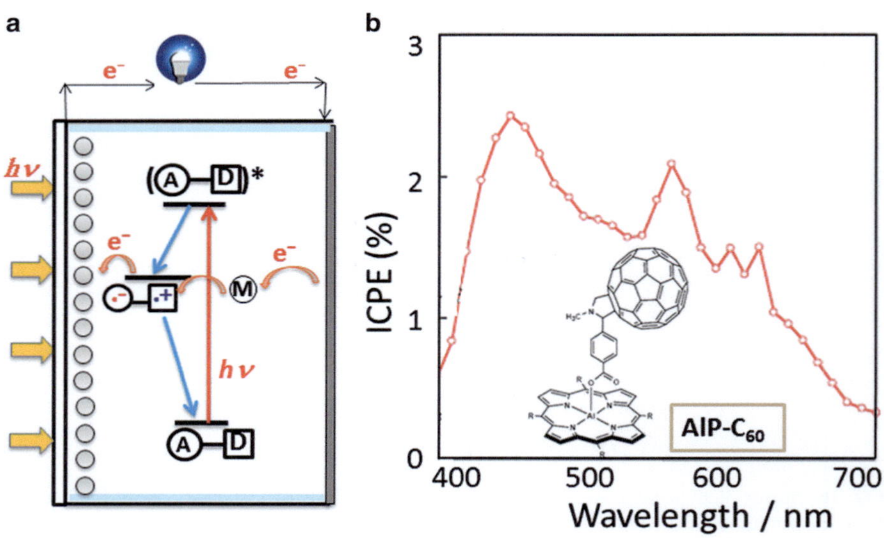

Fig. 17 (**a**) A new type of photovoltaic cell (charge-separation-type solar cell). (**b**) Incident photon-to-photocurrent efficiency (IPCE) vs. wavelength of the illumination light of the FTO/SnO_2 electrode coated with AlP–C_{60}. Reconstructed using figures from reference [34]

The photocurrent measurements show good photovoltaic performance of the cell (Fig. 17b for AlP–C_{60}); the maximal incident photon-to-current conversion efficiency (IPCE) is 25 % at 450 nm, which is higher than those of individual components. In addition to the $AlP^{•+}$–$C_{60}^{•-}$ formation, morphology is also an important factor, because the AlP–C_{60} dyad forms a chained nanoparticle, which may promote efficient electron hopping. TEM studies suggest that high IPCE is achieved, when the AlP–C_{60} molecules are arranged as a chain on the electrodes (SnO_2/ITO); that is, the morphology is an important factor. In this respect, the highly efficient solar cells based on the meso-superstructured organometal halide perovskites are being developed in this field for wider practical applications [36].

In the case of organic polymer film photo-solar cells, the organic π-conjugated polymers are mixed with fullerene derivatives such as [6,6]-phenyl-C_{61}-butyric acid methyl ester (PCBM) as an electron acceptor [37]. Here, the adsorbed light excites mainly the colored organic π-conjugated polymers, generating an exciton (electron–hole pair). When this exciton travels near the electron-accepting fullerene, the electron jumps into the fullerene phase, leaving the hole in the polymer phase. The electron can move toward the electrode via hopping through the chained fullerene particles (phase) to reach the anode, whereas the holes transfer along the π-conjugated polymer chains to reach the cathode, generating electricity. Here, the rate constants gathered from D–(π-conjugated polymer chains)–C_{60} are useful to design efficient film-type solar cells. Indeed, polythiophenes with small β_{ET} values are quite suitable to the exciton migration and hole transfer in polymers.

3 Photosensitizer–SWCNT Systems

3.1 Visible-Light Photosensitizing Molecule on SWCNT

It is essential to employ chirality-sorted SWCNT for application to photoactive devices and also to establish meaningful structure–reactivity relationships. Separation of semiconductive SWCNTs from metallic ones is very important, because they have quite different electronic properties. Several methods including gel chromatography have been successfully applied for the separation [38, 39]. Although mixture of the semiconductive and metallic SWCNTs is difficult to handle as materials for electro-devices, selective use increases the quality of each SWCNT as materials. Furthermore, the separation of semiconducting SWCNTs according to their chirality (diameter) makes one to recognize each SWCNT as a molecule, although length could have some distributions. Furthermore, enriched samples of SWCNT(6,7) and SWCNT(7,6) are now commercially available. For such chirality-sorted SWCNT, MO descriptions for the CS processes are possible, which helps us to understand photochemical and photophysical events of each SWCNT. Electron distribution of the HOMO of one of the semiconductive SWCNTs shows helical structure along the π-orbitals surrounding the SWCNT, whereas that of the metallic SWCNT shows linearly distributed π-orbitals along the

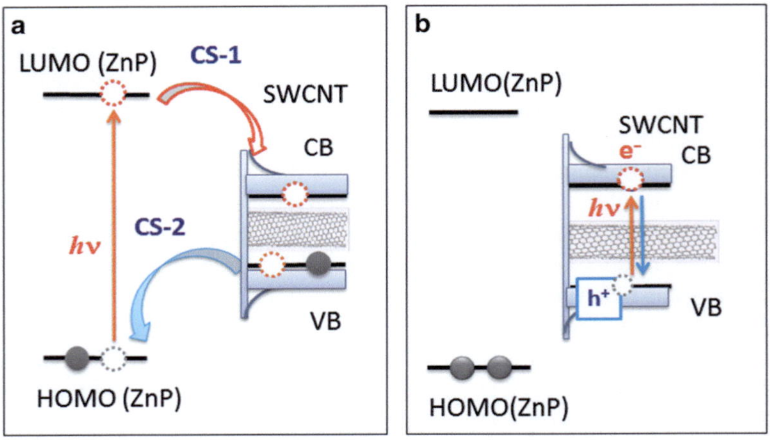

Fig. 18 Energy diagrams showing photoinduced electron transfer between photosensitizing molecule (ZnP) and SWCNT; (**a**) ZnP excitation and (**b**) SWCNT excitation. The CB and VB denote the conduction band and valence band of semiconducting SWCNT, respectively

nanotube, supporting its high electric conductivity. In the cases of functionalized SWCNTs, MO distribution including the sensitizing molecule can be obtained.

At first, energy level diagrams for the possible photoinduced processes of sensitizer molecule (i.e., ZnP) linked to semiconductive SWCNT are schematically illustrated in Fig. 18a, b. In the case of SWCNTs, band gaps between the valence band (VB) and the conduction band (CB) are reported around 1 eV. In the band electronic structures, the highest and lowest energy levels correspond to the HOMO and LUMO in the MO concept, respectively. The band gap decreases with an increase in the diameter of SWCNT [40, 41]. Both the VB and CB energy levels position within the wider HOMO–LUMO gap of ZnP. Consequently, two types of photoinduced CS processes could be envisioned, when ZnP is selectively excited (Fig. 18a). In the first case, an electron transfers from the LUMO(ZnP) to the CB of SWCNT, generating $ZnP^{\bullet+}$/$SWCNT^{\bullet-}$ ion pair (denoted as CS-1). In the second case, an electron transfers from the VB of SWCNT to the half-filled HOMO(ZnP) level of the excited ZnP* generating $ZnP^{\bullet-}$/$SWCNT^{\bullet+}$ ion pair (denoted as CS-2). When both CS-1 and CS-2 simultaneously occur, excited energy transfers from ZnP*/SWCNT to yield ZnP/SWCNT*. In the case of SWCNTs, this point can be experimentally confirmed by observing increase in the fluorescence of SWCNT* in the near-IR region by selective excitation of ZnP. Although the measurements are preliminary, such appreciable increase in fluorescence of SWCNT* is not observed by selective excitation of ZnP in ZnP/SWCNTs, although strict selective excitation of ZnP is difficult, because of the overlap of the second absorption bands of SWCNT with that of ZnP.

Direct excitation of SWCNTs usually results in the formation of the exciton (electron–hole pair). However, owing to shorter lifetime of the exciton being in the range of 1–10 ps, the electron–hole pair recombines rapidly, relaxing to the ground states of SWCNTs before an electron or hole could transfer to nearby ZnP (Fig. 18b) [42]. Hence, no long-lived ET products could be expected via direct excitation of SWCNTs. However, it is possible to generate solvated electron via photo-ejection from SWCNTs due to high VB levels. Such electron can be captured by ZnP, finally generating $ZnP^{\bullet-}/SWCNT^{\bullet+}$, which is denoted as indirect ET or trivial ET process. Furthermore, it is not advantageous to employ such a process as a main strategy to achieve an efficient photoinduced electron transfer process.

3.2 Covalently Linked Porphyrin–SWCNT

Recently, photophysical studies of covalently functionalized SWCNT(*n*,*m*) (where (*n*,*m*) = (7,6) or (6,5)) with ZnP are performed to establish diameter-dependent CS efficiencies [43]. In order to visualize the geometry of the covalently linked porphyrin–nanotube hybrids (Fig. 19a), DFT energy optimization using B3LYP/3-21Gbasis set on these computationally challenging systems is performed. Over 300 carbons for building the nanotube with one porphyrin unit are taken into the calculations with hydrogen atoms bonded to the terminal carbons of the open SWCNT(*n*,*m*). Figure 19b shows the optimized structure of ZnP–(sp)–SWCNT(6,5) hybrid. The calculated diameter and length are 7.6 and 34.3 nm for SWCNT(6,5), in which the diameters are in good agreement with the earlier reported STM and AFM images.

In the optimized structure, a close association of ZnP to the surface of SWCNT is observed, which is possible from the bent ester bond connecting the ZnP moiety and pyrrolidine ring. The distance between Zn atom to the closest carbon of SWCNT was found to be ~3.4 Å for the optimized structures. The alkoxy chain of pyrrolidine ring was also close to the nanotube surface revealing some interactions. As shown in Fig. 19b, the ZnP positioning on the SWCNT surface was slightly off due to the smaller diameter of SWCNT(6,5).

The energy levels and electron distributions of representative MOs are obtained by DFT–MO calculations. Some of the HOMOs and LUMOs are shown in Fig. 19b. The HOMO and HOMO−1 have electron densities mainly on SWCNT(6,5), corresponding to the valence band of SWCNT(6,5). Although the HOMO and HOMO−1 are similar, the HOMO−1 has higher electron densities on the middle part of SWCNT(6,5). In addition, some electron distributions are also found on the ZnP moiety, suggesting a weak charge–transfer interaction. On the other hand, HOMO−2 has the electron density predominantly localized on the ZnP moiety, corresponding to the HOMO of unbounded ZnP moiety. Then, the HOMO−3 has again the electron density localized on SWCNT(6,5). The LUMO and LUMO+1 have the electron density localized on the whole SWCNT(6,5) corresponding to the conduction band of SWCNT(6,5).

These LUMOs are also typical for LUMO+*n*, although the ranges of the distributions vary with the LUMO+*n*. At the LUMO+*n*, electron density is mainly spread on the ZnP moiety. By photoexcitation of the ZnP moiety, an electron on the HOMO−3 is elevated to the LUMO+*n* within the ZnP moiety, from which electron

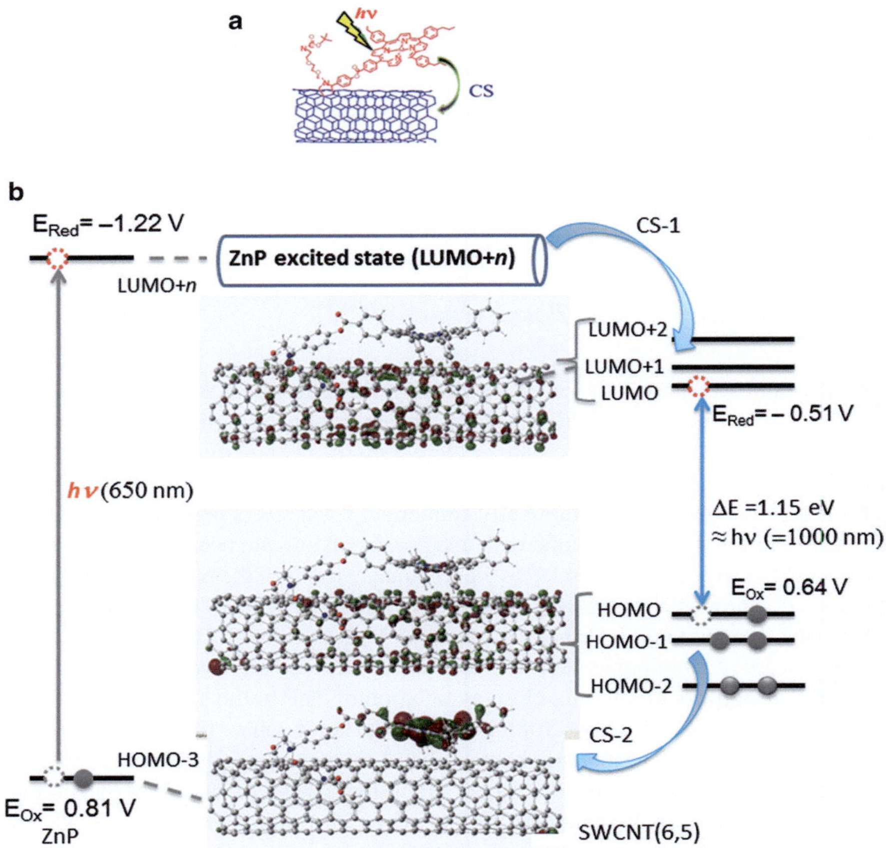

Fig. 19 (**a**) ZnP–(sp)–SWCNT(6,5). (**b**) Energy diagrams with MOs for photoinduced CS and CR processes; although the LUMO+*n* is not shown, electron distributes mainly on the ZnP similar to HOMO−3. The E_{Ox} and E_{Red} values of ZnP and SWCNT(6,5) vs. Ag/AgCl. Reconstructed using figures from reference [43]

can transfer to the LUMO mainly localized on SWCNT(6,5); then, $ZnP^{\bullet+}$–(sp)–$SWCNT^{\bullet-}$ can be obtained via CS-1 process (Fig. 19b). Similar MO distributions and energy levels are obtained for other ZnP–SWCNT(*n*,*m*). Using E_{Red} of SWCNT(*n*,*m*) and E_{Ox} of ZnP, free-energy change of CS ($\Delta G_{CS\text{-}1}$) via $^1ZnP^*$ (2.03 eV) can be evaluated to be −0.71 eV and −0.72 eV for SWCNT(6,5) and SWCNT(7,6), respectively, indicating thermodynamic feasibility.

From the energy level diagram in Fig. 19b, another CS process from the HOMO levels of SWCNT(*n*,*m*) to the half-vacant HOMO of $^1MP^*$ is also possible as denoted as CS-2 process, which gives the radical ion pair of opposite charges, $ZnP^{\bullet-}$–(sp)–$SWCNT^{\bullet+}$. The free-energy estimations for this process yielded $\Delta G_{CS\text{-}2}$ values of −0.17 eV and −0.31 eV via $^1ZnP^*$ for SWCNT(6,5) and SWCNT(7,6), respectively. Considering lower exothermicity of CS-2 processes, CS-1 can be considered being the predominant route for the occurrence of photoinduced CS.

Fluorescence intensity quenching and lifetime shortening of ZnP are observed in SWCNT covalently connected systems, which could be either due to the occurrence of photoinduced CS or EnT processes. From the transient absorption spectra showing 600–700-nm band due to $ZnP^{•+}$ and 900–1,300-nm band of $SWCNT^{•-}$, formation of $ZnP^{•+}$–(sp)–SWCNT$(n,m)^{•-}$ with relatively long lifetimes of ca. 100 ns is witnessed. As a reason for the preference of CS over EnT via $^{1}ZnP^{*}$, the HOMO electrons spreading widely along the SWCNT may be difficult to inject to small size $ZnP^{•+}$ with the half-vacant HOMO. Relatively long lifetimes (ca. 100 ns) of $ZnP^{•+}$–(sp)–$SWCNT^{•-}$ are observed likely due to the spreading of the injected electron across the entire π-system of SWCNT, avoiding return path to $ZnP^{•+}$ for charge recombination.

When dialkyl viologen dication (RV^{2+}) is added as an electron-accepting mediator ($RV^{•+}$) with an appropriate sacrificial hole shifter, $RV^{•+}$ can be accumulated as a blue solution by the visible-light irradiation [44]. It is rather interesting to see that the extra electron on SWCNT, stabilized by the low LUMO and low CB (small E_{Red} value (≈ -0.5 V vs. Ag/AgCl) in Fig. 15), prevents efficient electron mediating, that is, the electron-mediating rate is not fast enough to compete with CR rate (10^{7}–10^{8} s^{-1}). Thus, higher concentrations of RV^{2+} are necessary to observe the electron pooling. On the other hand, RV^{2+} also tends to adsorb on the SWCNT walls thus easily accepting electron from the $SWCNT^{•-}$. Furthermore, the generated electron can be used for photovoltaic solar cell and H_2-evolution applications (IPCE ≈ 20 %). One example is shown for H_2 evolution by the visible-light irradiation of ZnP–SWCNT in the presence of Pt-nanoparticle catalysts as shown in Fig. 20, in which the ZnP is covalently connected to SWCNT with short flexible bond. From the observed pH dependence, the electron flow can be evaluated.

3.3 *Supramolecular Porphyrin–(Pyrene)/SWCNT Hybrid*

In addition to the covalent bonding, utilization of various intermolecular binding strategies has been widely used to construct ZnP/SWCNT systems. Polycyclic aromatic compounds with highly conjugated π-electrons such as pyrene, to which the visible-light-absorbing ZnP are appended, can be employed as adsorbent on the SWCNT surface via π–π stacking. Figure 21a shows an example of the π–π stacking between the surface of SWCNTs and pyrene (Pyr) with ZnP; that is, the ZnP is functionalized with four π-stacking pyrene entities ($ZnP(Pyr)_4$) at the macrocycle periphery [45].

Photophysical properties are monitored by fluorescence measurements of the ZnP moiety in the 600–700-nm region. As shown in Fig. 21b, ZnP-fluorescence intensity quenching and the lifetime shortening in the presence of SWCNT(n,m) are observed due to fast events originating from $^{1}ZnP^{*}$ in $ZnP(Pyr)_4$/SWCNT(n,m). Possible quenching mechanisms involve either CS process or EnT process, in which the CS process is suggested by the appearance of broad transient absorption bands in the 600–700 nm and near-IR regions with lifetimes of ca. 50 ns, although their

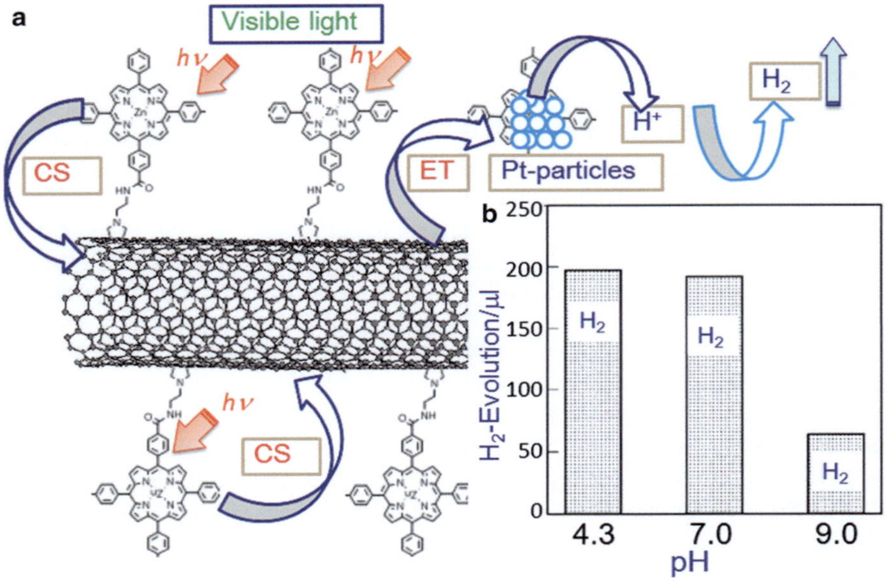

Fig. 20 (**a**) H_2 evolution by the visible-light irradiation of ZnP–SWCNT hybrid in the presence of Pt-nanoparticle catalyst in aqueous solution; arrows are electron flows. (**b**) pH dependence of H_2 evolution. Reconstructed using figures from reference [44]

assignments are still tentative [46]. The possible CS processes are supported by accumulation of the reduced chemical product ($HV^{•+}$) under visible-light irradiation of ZnP(Pyr)$_4$/SWCNT(*n*,*m*) in the presence of hexyl viologen (HV^{2+}) and hole shifter. Yields of the accumulated $HV^{•+}$ are 40–60 % for the ZnP(Pyr)$_4$/SWCNT(*n*,*m*). Furthermore, photoelectrochemical cells composed of ZnP(Pyr)$_4$/SWCNT/ITO/SnO_2 electrodes show a distinct photocurrent, although the IPCE is not high. Returning to the ZnP-fluorescence lifetime shortening in Fig. 21c, the k^S_{CR} values via $^1ZnP^*$ are evaluated to be (5–10) × 10^9 s^{-1}. In these cases, the hybrid of SWCNT(7,6) tends to show higher CS efficiency than SWCNT(6,5), which may be related to easily reducible large-bore tube of SWCNT(7,6) (−0.41 V vs. Ag/AgCl and 1.1 nm) compared to that of lower reduce ability of small size of SWCNT(6,5) (−0.43 V and 0.8 nm).

Metal–ligand coordination binding strategy can be applied to assemble ZnP/SWCNT, i.e., the pyrene (Pyr) is connected with pyridine (PyrPy) or imidazole (PyrIm), which is adsorbed onto SWCNT surface via π–π interactions using Pyr part and SWCNT (SWCNT/PyrPy or SWCNT/PyrIm). The appended Py and Im can coordinate to ZnP, forming SWCNT/PyrPy→ZnP or SWCNT/PyrIm→ZnP hybrids, or can coordinate to ZnPc forming SWCNT/PyrPy→ZnPc or SWCNT/PyrIm→ZnPc hybrids [46]. The transient absorption peak at 900–1,000 nm (lifetime ≈ ca. 100 ns) observed for SWCNT/PyrIm→ZnPc is a clear evidence of the formation of $ZnPc^{•+}$; on combining with the near-IR transient broad bands, the formation of $SWCNT^{•-}$/PyrPy→$ZnPc^{•+}$ with lifetime of ca. 80 ns ($k_{CR} \approx 10^7$ s^{-1}) is

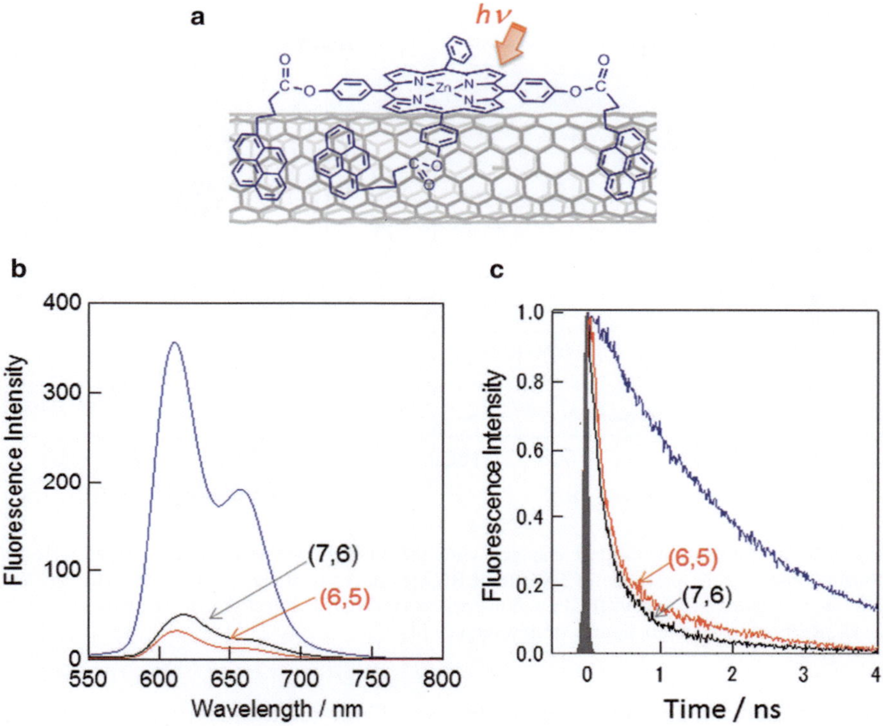

Fig. 21 Supramolecular hybrid using π–π interaction between SWCNT and pyrene moiety appending porphyrin (ZnP(Pyr)$_4$/SWCNT(*n*,*m*)). Fluorescence quenching of the ZnP moiety proving an evidence of hybrid formation and occurrence of the excited-state events; *blue*, ZnP; *red*, ZnP(Pyr)$_4$/SWCNT(6,5); *black*, ZnP(Pyr)$_4$/SWCNT(7,6) in DMF. Reconstructed using figures from reference [45]

supported. By utilizing $k^S_{CS} \approx 10^9$ s^{-1} from the fluorescence lifetime shortening, the k_{CS}/k_{CR} value, which is an index of CS stabilization, is ca. 100.

Photocatalytic role of the photosensitizers (ZnP, ZnPc, etc.) in the pooling of the reduced product in solution is clearly observed for naphthalocyanine-based (ZnNc) SWCNT hybrids (SWCNT/PyrIm→ZnNc) in the presence of RV^{2+} and BNAH in Fig. 22a. During the increases of $RV^{\bullet+}$ (620-nm band) by the predominant excitation of the ZnNc unit, the absorption peak intensity of ZnNc at 780 nm is constant, confirming regeneration of ZnNc after repeated use of ZnNc as photosensitizer, electron donor toward SWCNT, and hole mediator to BNAH, returning to neutral ZnNc as observed in Fig. 22a. Indeed, the transient appearance of the 1,000-nm band due to $ZnNc^{\bullet+}$ is observed in the transient absorption spectra. Compared with apparent evidences of ZnNc and $ZnNc^{\bullet+}$, the role of SWCNT is rather scarcely envisioned; however, as a pair of $ZnNc^{\bullet+}$, the formation of $SWCNT^{\bullet-}$/PyrIm→$ZnNc^{\bullet+}$can be suggested; indeed, the near-IR absorption broad bands might be a trace of an excess electron on SWCNT, which may act as electron acceptor from ZnNc,

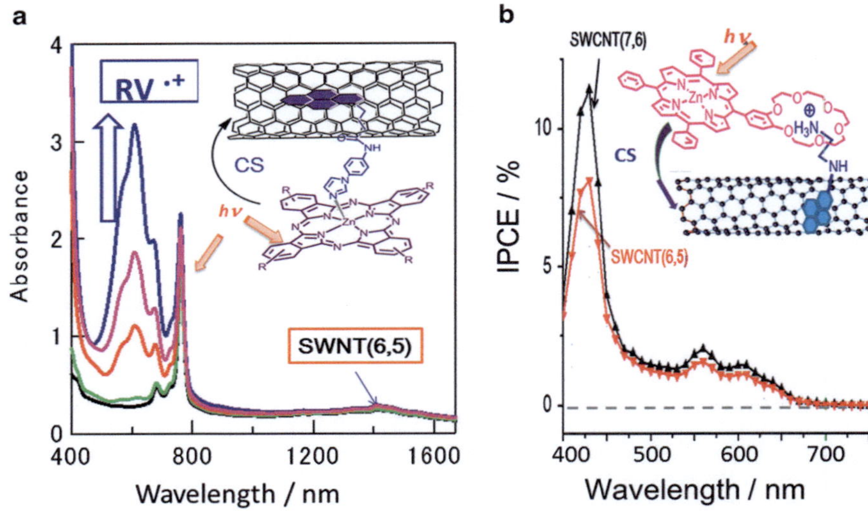

Fig. 22 (**a**) Absorption spectral changes with the visible-light irradiation of SWCNT(6,5)/PyrIm→ZnNc in the presence of RV^{2+} (and BNAH) in Ar-saturated *o*-DCB [45]. (**b**) Incident photon conversion efficiency (IPCE) action spectra of SWCNT(*n*,*m*)/$PyrNH_3^+$:(Crown)ZnP/FTO/SnO_2. Reconstructed using figures from reference [47]

electron mediator to RV^{2+}, to generate $RV^{•+}$. These photocatalytic procedures are the basis of photocurrent generation and H_2-gas evolution in the presence of nanoparticles. Furthermore, it is notable that higher $RV^{•+}$ formation is observed for SWCNT(7,6) hybrids than SWCNT(6,5) hybrids, which may be one of references for using size-sorted SWCNT.

Crown ether–ammonium ion inclusion complex strategy is also employed for SWCNT. One example is shown in the inset of Fig. 22b, in which the ZnP with a crown ether ((Crown)ZnP) is employed using Pyr as a glue. The (Crown)ZnP is attached on the SWCNT via $PyrNH_3^+$, constructing SWCNT/$PyrNH_3^+$:(Crown)ZnP hybrid [47]. The CS process producing $SWCNT^{•-}$/$PyrNH_3^+$:(Crown)$ZnP^{•+}$ is confirmed by the transient absorption spectral measurements. Photoelectrochemical solar cells of SWCNT/$PyrNH_3^+$:(Crown)ZnP on SnO_2/FTO show IPCE action spectra as depicted in Fig. 22b. The spectra for SWCNT(6,5) and SWCNT(7,6) trace absorption band of ZnP (main band at 420 nm and sub-band in 550–620 nm), supporting occurrence of CS process via $^1ZnP^*$. The higher IPCE value is obtained for SWCNT(7,6) than for SWCNT(6,5); as affecting factors, their band gaps, electric conductivities, radii, and surface area can be additional contributing factors.

As a more simple strategy, ion-pairing construction can also be applied for SWCNTs using charged tetrapyrroles as photosensitizers and Pyr derivatives with opposite charges. For example, a positively charged ($TMPyP^+$)Zn (tetrakis(4-N-methylpyridyl)porphyrin zinc) can be ion paired with a negatively charged $PyrCOO^-$

adsorbed on surface of SWCNT [47]. Similarly, SWCNT(n,m)/pyrNH_3^+/($TPPS^-$)Zn hybrids are formed by ion pairing a negatively charged ($TPPS^-$)Zn (tetrakis-4-sulfonatophenylporphyrinato zinc) with a positively charged PyrNH_3^+ on the SWCNT. The hybrid formations are confirmed by fluorescence quenching and TEM images. The formation of $SWCNT^{\bullet-}$/PyrNH_3^+/($TPPS^-$)$Zn^{\bullet+}$ ion pair via ^{1}($TPPS^-$)Zn* would be anticipated in CS-1 route in Fig. 18a, since ($TPPS^-$)Zn has high LUMO due to the electron-rich porphyrin ring. As for SWCNT/PyrCOO^-/($TMPyP^+$)Zn, the electron-deficient ($TMPyP^+$)M is capable of accepting an electron from SWCNT, generating $SWCNT^{\bullet+}$/PyrNH_3^+/($TMPyP^+$)$Zn^{\bullet-}$ by an electron transfer from the valence band to the half-vacant HOMO of ^{1}($TMPyP^+$)Zn* (CS-2 route in Fig. 18a). High $HV^{\bullet+}$ accumulation (>40 % vs. [HV^{2+}]) and IPCE (>8 %) are observed [48].

3.4 Supramolecular Fullerene–(Pyrene)/SWCNT Hybrid

The MO descriptions for CS process between the photoexcited state of C_{60} moiety and SWCNTs are different from the porphyrin–SWCNT systems, because the LUMO(C_{60}) is usually lower than the conduction band of SWCNT as shown in Fig. 23a [49]. When C_{60} is attached to SWCNTs via the Pyr moiety, CS process from the valence band of SWCNT to the half-vacant HOMO of photoexcited C_{60}

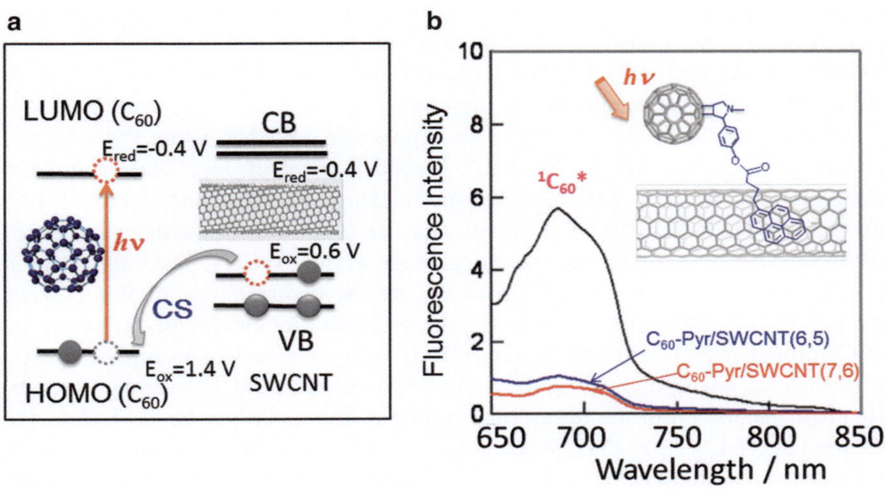

Fig. 23 (**a**) Energy diagram for CS process of SWCNT–C_{60}, (**b**) fluorescence spectra of C_{60} in SWCNT(7,6)/PyrC_{60} hybrid induced by photoexcitation of the C_{60} moiety. Reconstructed using figures from references [49, 50]

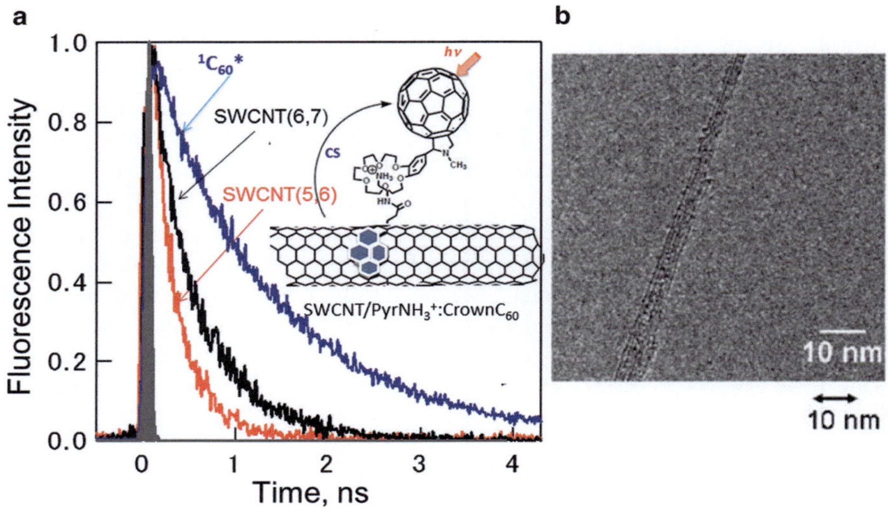

Fig. 24 (**a**) Fluorescence decay time profiles observed by selective excitation of C_{60} in SWCNT(*n*,*m*)/PyrCrown:$NH_3^+C_{60}$, (**b**) TEM image of SWCNT(6,5)/PyrCrown:$NH_3^+C_{60}$. Reconstructed using figures from reference [51]

would be possible. Fluorescence quenching of the C_{60} moiety with SWCNTs suggests formation of the SWCNT/PyrC_{60} complex (Fig. 23b) [50] in which the CS process occurs from the SWCNT to the C_{60} moiety. In this case, EnT with electron-exchange mechanism is difficult to occur, because electron exchange from LUMO(C_{60}) to CB(SWCNT) is energetically unfavorable. Thus, generation of $SWCNT^{•+}$/Pyr$C_{60}^{•-}$ is unequivocally suggested.

Crown ether inclusion strategy is also applied to SWCNT, using PyrNH_3^+ and CrownC_{60}, forming SWCNT/PyrNH_3^+:CrownC_{60} (inset of Fig. 24a). The TEM image is shown in Fig. 24b, in which the thickness of nanotubes is about 3 nm; probably C_{60} moieties with ca. 1 nm are attached both sides of the outer surface of SWCNT with 1 nm, rather than inserting inside of the SWCNT. The photoinduced events originating from photoexcited C_{60} can be elucidated from fluorescence quenching (Fig. 24a). Shortening of C_{60}-fluorescence lifetimes suggests occurrence of CS process with rate constants of $(2–3) \times 10^9\ s^{-1}$. The transient absorption spectra of SWCNT/PyrNH_3^+:CrownC_{60} show the 1,000-nm band of $C_{60}^{•-}$, among the near-IR broad bands. From the decay time profile of the 1,000-nm band of $C_{60}^{•-}$, the lifetime of $SWCNT^{•+}$/PyrNH_3^+:Crown$C_{60}^{•-}$ is evaluated to be 70 ns. On addition of HV^{2+}, accumulation of $HV^{•+}$ is observed as a blue solution by visible-light irradiation, suggesting that the extra electron on the C_{60} moiety mediates to HV^{2+}. Higher $HV^{•+}$ formation and IPCE are observed for SWCNT(7,6) and SWCNT(6,5), in which the former has higher efficiency.

3.5 *Supramolecular Fullerene–Porphyrin–SWCNT "Triple" Hybrid*

When ZnP is attached to SWCNT in a way other than the coordination method, ZnP has the capability to bind one more secondary donor or acceptor functionalized with a coordination ligand. Although the absorption intensity of the C_{60} moiety in the visible region is not high enough to utilize it as a sole solar light sensitizer, the combination with porphyrins or phthalocyanines gives higher photoinduced electron transfer efficiency. Thus, "trilevel" supramolecular hybrids, C_{60}–porphyrin–SWCNT(n,m), are constructed via self-assembly method [52]. At first, ZnP(alkyl)$_4$/SWCNT(n,m) hybrids are formed via intermolecular alkyl–π and π–π interactions between the ZnP(alkyl)$_4$ and SWCNT(n,m). Subsequently, phenylimidazole-functionalized fulleropyrrolidine, C_{60}Im, is coordinated to ZnP entity resulting in C_{60}Im→ZnP(alkyl)$_4$/SWCNT(n,m) "trilevel" systems as shown in Fig. 25. Formation of ZnP(alkyl)$_4$/SWCNT(n,m) hybrids is confirmed by the ZnP-fluorescence quenching in organic solution, suggesting occurrence of some photophysical events via ^{1}ZnP*. On addition of C_{60}Im to ZnP(alkyl)$_4$/SWCNT(n,m), further fluorescence quenching is observed, indicating that the C_{60}Im→ZnP(alkyl)$_4$/SWCNT(n,m) is formed via the coordination of C_{60}Im to Zn ion (Fig. 25), confirming formation of the triple hybrid.

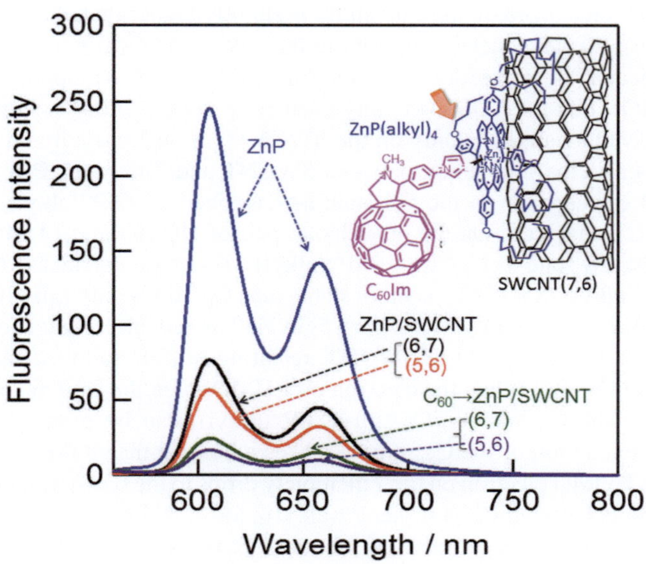

Fig. 25 Steady-state fluorescence spectra of ZnP(alkyl)$_4$ on addition of SWCNT(n,m) and further addition of C_{60}Im and structure of C_{60}Im→ZnP(alkyl)$_4$/SWCNT(7,6). Reconstructed using figures from reference [52]

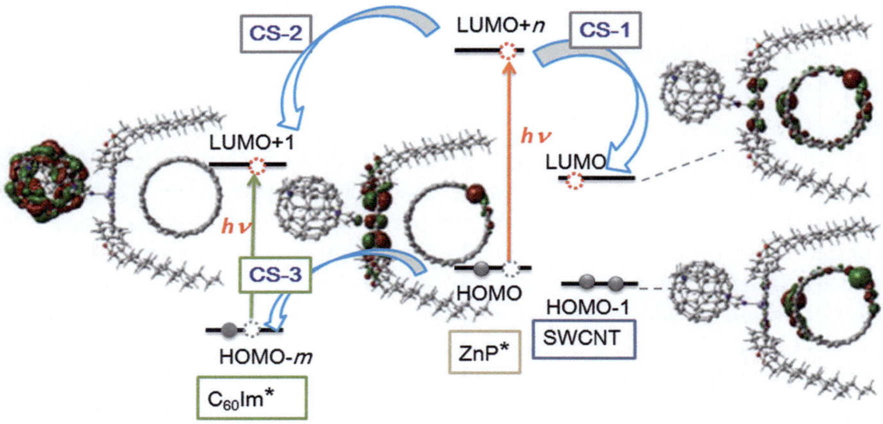

Fig. 26 Illustration of C_{60}Im→ZnP(alkyl)$_4$/SWCNT(7,6) donor–acceptor nanohybrids. Edge-on view of the optimized structure, energy diagram, and electron distributions for C_{60}Im→ZnP(alkyl)$_4$/SWCNT(7,6) calculated at the B3LYP/3-21Glevel. Reconstructed using figures from reference [52]

The DFT–MO calculations of C_{60}Im→ZnP(alkyl)$_4$/SWCNT(7,6) show an optimized structure, in which the alkyl chains get their arms around SWCNT(7,6), as shown in Fig. 26. The frontier molecular orbitals of C_{60}Im→ZnP(alkyl)$_4$/SWCNT(7,6) are also shown in Fig. 26.

The HOMO localized on the ZnP entity is slightly higher than the HOMO−1 on SWCNT(7,6); this tendency is opposite to the electrochemically obtained trend in Fig. 23a, because of interaction between ZnP and SWCNT. An opposite tendency in the LUMOs, that the LUMO+1 fully located on the C_{60} entity is slightly higher than the LUMO localized mainly on the SWCNT(7,6) and partially on the ZnP, is also attributed to the interaction between SWCNT and ZnP parts. Pair of HOMO and LUMO corresponds to the CS state as described in $ZnP^{•+}$(alkyl)$_4$/$SWCNT^{•-}$ with slight CT character. On the other hand, pair of HOMO and LUMO+1 corresponds to the CS state as $C_{60}^{•-}$Im→$ZnP^{•+}$(alkyl)$_4$. From the calculated MO energy levels, $ZnP^{•+}$(alkyl)$_4$/$SWCNT^{•-}$ is more stable than $C_{60}^{•-}$Im→$ZnP^{•+}$(alkyl)$_4$. When an electron is elevated to the LUMO+n ($n>5$) of ZnP moiety after light excitation, the electron drops to the LUMO of SWCNT, resulting in $ZnP^{•+}$(alkyl)$_4$/$SWCNT^{•-}$ via CS-1 route in Fig. 26. When the electron on LUMO+n on the ZnP moiety drops to LUMO+1 on the C_{60} moiety, $C_{60}^{•-}$Im→$ZnP^{•+}$(alkyl)$_4$ can be produced via CS-2 route. Photoexcitation of the C_{60} moiety makes a hole on the HOMO − m ($m>10$), to which the HOMO electron on the ZnP moiety drops to the half-vacant HOMO−m of the C_{60} moiety, giving $C_{60}^{•-}$Im→$ZnP^{•+}$(alkyl)$_4$, too (CS-3 route).

Information on the occurrence of CS and CR processes is obtained from the time-resolved spectroscopies. From the shortening of fluorescence lifetimes of the ZnP moiety, $k^{S}_{CS\text{-}1} = 6 \times 10^9$ s^{-1} and $k^{S}_{CS\text{-}2} = 2 \times 10^9$ s^{-1} are evaluated in o-DCB. The transient absorption spectra showing the broad bands in 650–700-nm region and near-IR region suggest the formations of $C_{60}^{•-}$Im→$ZnP^{•+}$(alkyl)$_4$/SWCNT and

$C_{60}Im \rightarrow ZnP^{\bullet +}(alkyl)_4/SWCNT(7,6)^{\bullet -}$. Decays of these transient absorption bands indicate that the CR processes occur within about 100 ns. These observations prompted us to perform photoelectrochemical studies to visualize their ability to convert light energy into electricity. The photocurrent action spectra of $ZnP(alkyl)_4$/SWCNT(*n*,*m*)/SnO_2/FTO and $C_{60}Im \rightarrow ZnP(alkyl)_4$/SWCNT(*n*,*m*)/$SnO_2$/FTO show IPCE values in a wide wavelength region covering up to 800 nm with a maximum near 420 nm, which tracks the absorption spectra of the ZnP moiety. The light on–off switching shows quick response and the high light stability during the repetition. Higher IPCE values at 420 nm are obtained for $C_{60}Im \rightarrow ZnP(alkyl)_4$/SWCNT than those for $ZnP(alkyl)_4$/SWCNT by a factor of ca. 2, suggesting that the $C_{60}Im$ coordination contributes to additional CS process via $^1C_{60}^*Im$ and via $^1ZnP^*$. In the case of $C_{60}Im \rightarrow ZnP(alkyl)_4$/SWCNT(*n*,*m*) hybrids, SWCNT(7,6) performs slightly better than SWCNT(6,5).

3.6 Supramolecular Porphyrin–ssDNA–SWCNT Triple Hybrids

Another example for the "triple" hybrids is to employ ssDNA to wrap around the SWCNT and attach a porphyrin outside of the wrapped ssDNA by Coulomb attractions. Thus, ssDNA acts as an electron mediator from SWCNT to the outside-bound porphyrin [53]. To accomplish the three-component hybrids, chirality-sorted semiconducting SWCNT(*n*,*m*)s and zinc porphyrin bearing peripheral positive charges (($TMPyP^+$)Zn (tetrakis(4-*N*-methylpyridyl)porphyrin)Zn) serving as light-absorbing photoactive materials are utilized. The "triple" hybrids are held by ion pairing between the negatively charged phosphate groups of ssDNA on the surface of SWCNT and the positively charges at the periphery porphyrin macrocycle. The newly assembled bio-nano donor–acceptor hybrids have been characterized by TEM and spectroscopic methods. From the optimized structure in Fig. 27, distance between ($TMPyP^+$)Zn and ssDNA bases (blue dots) is almost the same as the distance between ssDNA bases and SWCNT surface.

Photoinduced electron transfer from $^1ZnP^*$ to SWCNTs directly and/or via ssDNA as an electron mediator has been established by performing systematic studies. In a mixture of ($TMPyP^+$)M and ssDNA in aqueous solution, slight shortening of the fluorescence lifetime of ZnP is observed as shown in Fig. 28a, suggesting occurrence of CS-2 in Fig. 28b; $k_{CS\text{-}2} \approx 5 \times 10^8\ s^{-1}$. Furthermore, additional shortening of fluorescence lifetime of the ZnP is observed in ($TMPyP^+$)Zn/ssDNA/SWCNT, suggesting that CS-1 occurs from the $^1ZnP^*$ to SWCNT as shown in Fig. 28b; $k_{cs\text{-}1} \approx 4 \times 10^9\ s^{-1}$. Energetically, combination of CS-2 and CS-3 is possible, where ssDNA acts as an electron mediator. Therefore, higher charge-separation efficiency has been successfully demonstrated by the selection of appropriate semiconductive SWCNTs with right band gap, with the help of ssDNA as an electron mediator.

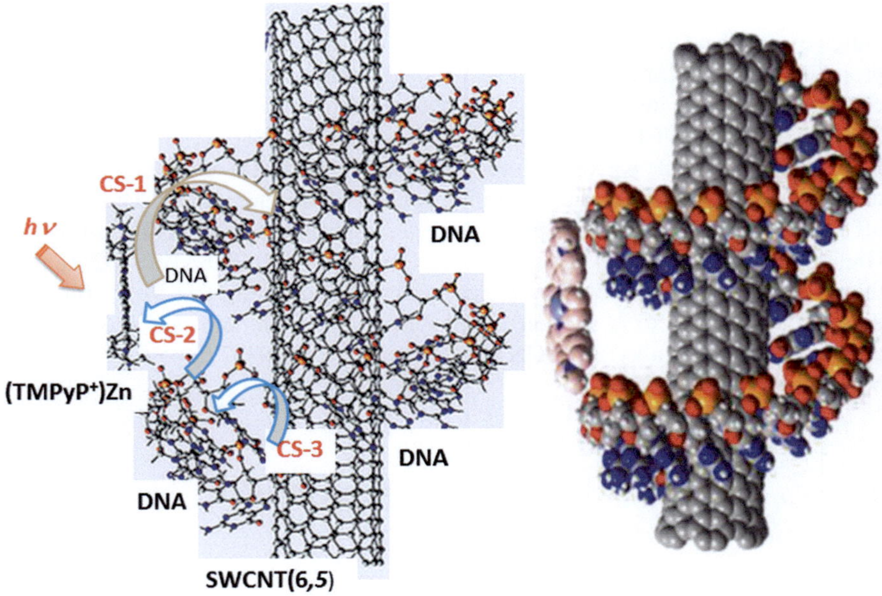

Fig. 27 Triple hybrids, (TMPyP⁺)Zn/ssDNA/SWCNT(6,5) formed by a combination of ssDNA wrapping and ion-pairing binding protocols. Outside of ssDNA, negatively charged phosphate anions; inside, C=O; middle of ssDNA, DNA bases. In ssDNA: *brown ball* P, *red ball* O, *blue ball* N. In (TMPyP⁺)Zn: *pink*, (TMPyP⁺), *sky*; Zn. Reconstructed using figures from reference [53]

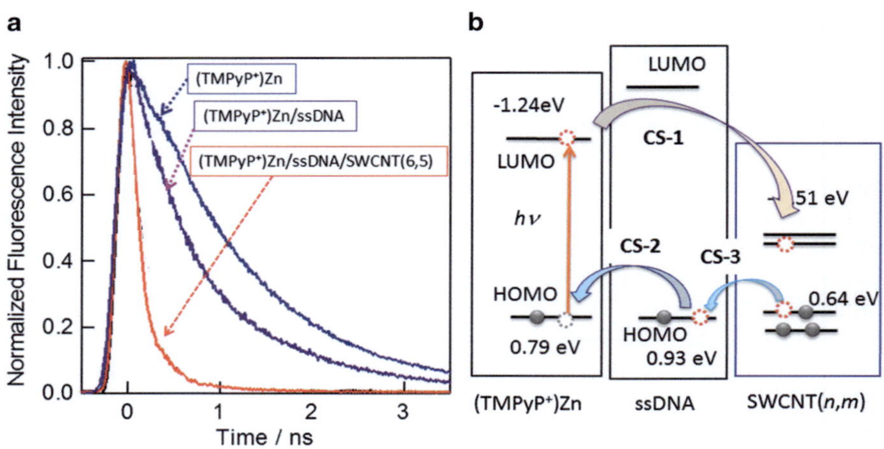

Fig. 28 (**a**) Fluorescence decays in water (λ_{ex}=408 nm). (**b**) Schematic representation of the energy level diagram for photoinduced electron transfer processes of (TMPyP⁺)Zn/ssDNA/SWCNT. *Red dotted circle* is moving electron. Reconstructed using figures from reference [53]

4 Photosensitizer–Carbon Nanohorn Systems

The research field with carbon nanohorns (CNHs), which are prepared by laser ablation of pure graphite, is developing, because they hold the advantages for the absence of any metal catalyst. They have structural characteristics as surface curvature with high stretched five-membered rings present at the conical tips and rough secondary spherical superstructure preventing further contact and aggregation. In addition, pristine CNHs showed semiconducting properties [54].

In order to use CNHs in wider applications, chemical functionalization with photosensitizers is essential. Various strategies are available to introduce organic addends covalently attached onto the outer surface of CNHs, thus yielding soluble CNH-based materials [55]. Moreover, some novel CNH-based hybrid materials, in which the characteristic properties of CNHs are combined with photoactive components such as porphyrins, have been prepared. An example is shown in Fig. 29a, and the TEM images are shown in Fig. 29b, suggesting that (A) each aggregate is a dahlia like, (B) each horn is single walled, and (C) attached porphyrin molecules are not aggregated on the CNH, thus transparent (invisible).

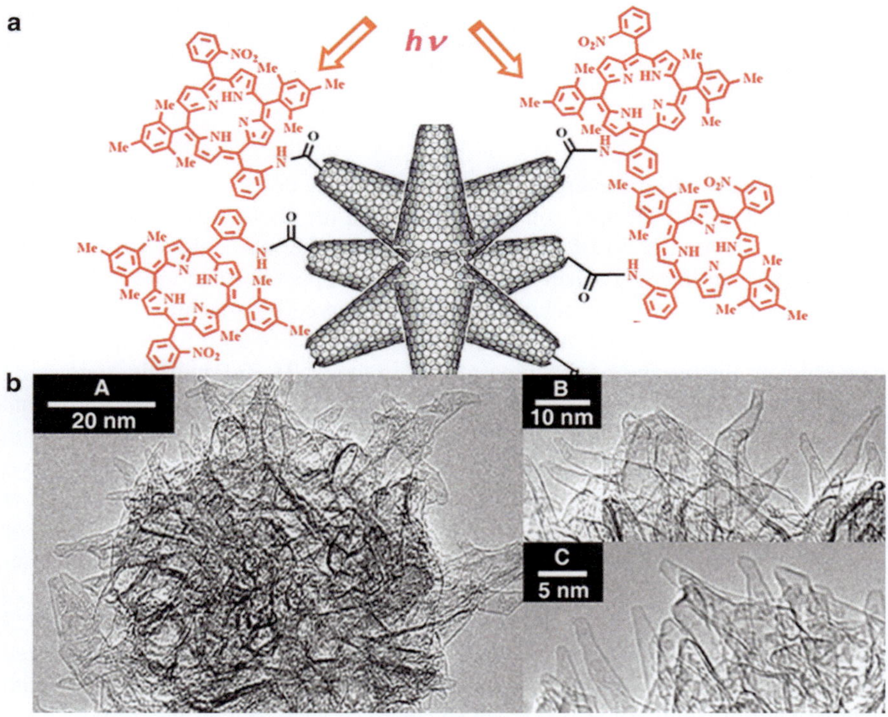

Fig. 29 (**a**) Porphyrin connected CNH pretreated with H_2O_2 to make holes near top of horns. (**b**) TEM image of CNH aggregates forming a nanoparticle (*A*). TEM images near the top after chemical bonding with porphyrins (*B* and *C*). Reconstructed from figures in reference [55]

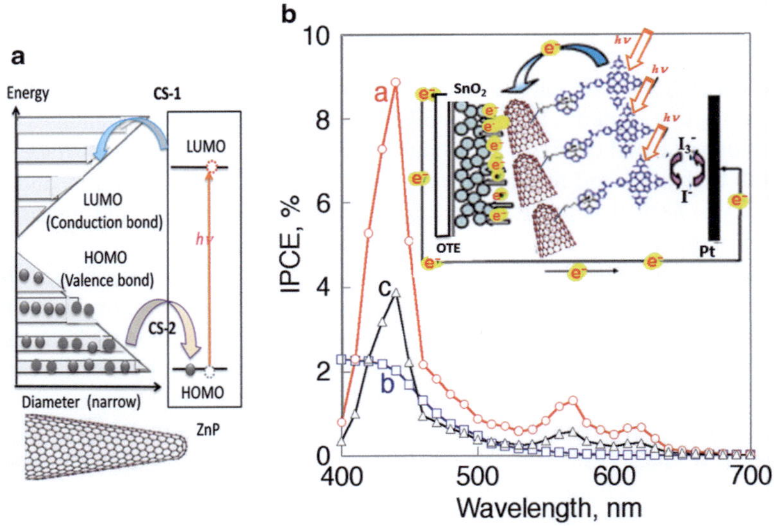

Fig. 30 (**a**) Electronic energy states of the CNH and CS process via the photoexcited ZnP; (**b**) (*a*) IPCE spectra of photoelectrochemical solar cell of CNH–NH_3^+:Crown–ZnP on SnO_2/OET (*inset*), compared with (*b*) ZnP/SnO_2/OET and (*c*) CNH/SnO_2/OET. *Inset* is photoelectrochemical cell design. Reconstructed using figures from reference [56]

In Fig. 30a, electronic energy states of the CNH are described as semiconductor with gradient radius. The electron of the LUMO of the photoexcited ZnP can inject into the conduction band of the CNH (CS-1). Furthermore, the electron in the valence band can enter to the half-vacant HOMO after photoexcitation of ZnP (CS-2). Indeed, the fluorescence quenching of ZnP is observed by attachment on CNH in polar solvents. Steady-state and lifetime measurements of ZnP fluorescence indicate efficient fluorescence quenching of ZnP in the CNH–ZnP hybrids, suggesting that photoinduced events occur from the photoexcited ZnP to CNHs. Additional proof for the photoinduced charge-separated state $CNH^{•-}$–$ZnP^{•+}$ is derived from electron-pooling studies in the presence of HV^{2+} and a hole trap. Figure 30b shows the IPCE spectrum of CNH–NH_3^+:Crown–ZnP on SnO_2/OET, which has a peak at 420 nm corresponding to ZnP absorption band, whereas the component (ZnP or CHN) has only low IPCE value. Inset of Fig. 30b explains the flow of electron after photoexcitation of ZnP on CNH in a photoelectrochemical cell.

Similarly, porphyrin dimer covalently bonded to CNHs shows spectroscopic CS evidences and also accumulation of $HV^{•+}$ [57]. The hybrid materials are integrated into a photoactive electrode of an optical transparent electrode (OTE) cast with nanostructured SnO_2 films (OTE/SnO_2), which reveal stable and reproducible photocurrent and photovoltage responses, with an incident photon-to-current conversion efficiency value as large as 9 %, without any applied bias voltage.

5 Photosensitizer–Graphene Systems

Single-layered graphene is almost a flat material; thus, compared with the spherical fullerenes and SWCNTs, the flat graphene has the highest electron delocalization, although the energy levels of a graphene layer must be represented by solid-state theory and not by MO description. Thus, the optical property of the single-layered graphene and electron transfer ability cannot be explained in similar way of fullerenes and SWCNTs. It is also notable that the electronic states and the electron transfer ability depend strongly on the presence of the dangling bonds at the terminal carbons and defects within the graphene sheet. Alternatively, graphene oxide (GO) can more easily be chemically treated, and it is possible to consider their electronic states in an analogy to the carboxylated aromatic hydrocarbons, thus to describe the MOs.

The graphene oxide (GO) is covalently functionalized with H_2P as shown in Fig. 31a by Tagmatarchis's group [58]. The resulting GO–H_2P hybrids form stable dispersions in DMF and have been characterized by spectroscopic method (UV–Vis, ATR-IR, Raman, etc.) and thermal analysis (TGA) methods. The UV–Vis absorption spectrum of GO–H_2P exhibits the main peak at 420 nm and weak ones in the region of 500–650 nm due to the H_2P moiety, while the GO sheet shows broad continuous absorption extending from the 800 nm to the UV region with increasing intensity, suggesting that various sizes of π-conjugative islands are included in the GO sheet.

In general, electronic states of GO can be considered as aromatic hydrocarbons as shown in Fig. 31b. For photoinduced CS process for GO–H_2P, an electron at the LUMO(H_2P) level elevated by photoexcitation of the H_2P unit transfers to the LUMO(GO), generating $GO^{\bullet-}$–$H_2P^{\bullet+}$ as a result of CS-1 route. Furthermore, EnT process must be also considered by the simultaneous occurrence of CS-1 and CS-2, when the HOMO–LUMO gap of the GO decreases with an increase in the π-area of GO. The transient absorption spectra show main absorption at 650 nm, suggesting the formation of $GO^{\bullet-}$–$H_2P^{\bullet+}$ with relatively short lifetime (ca. 50 ns), although quantum yield for the CS process is not high, because H_2P-fluorescence quenching is not appreciable. In a standard photoelectrochemical cell constructed by electrophoretic deposition onto SnO_2 electrodes, the IPCE of 1.3 % at 430 nm is observed for OTE/SnO_2/GO–H_2P (Fig. 31c), which is still low, requiring better device optimization.

When both donor and acceptor molecules are attached to a single-layer GO (SLGO), the SLGO may afford a good reaction field. An example is shown for ZnPc–(SLGO)–C_{60} in Fig. 32, where the C_{60} and ZnPc moieties are separately (independently) attached to the SLGO with short linkages [59]. During steady-light excitation of the ZnPc part and/or the C_{60} part of ZnPc–(SLGO)–C_{60} in the presence of RV^{2+} and BNAH, accumulation of $RV^{\bullet+}$ is recognized as a characteristic 620-nm band as shown in Fig. 32a, exhibiting a blue solution [60]. The 620-nm band intensity increases with the added amounts of BNAH and also irradiation time, showing a tendency to saturate at appropriate concentration of BNAH and light-illumination time.

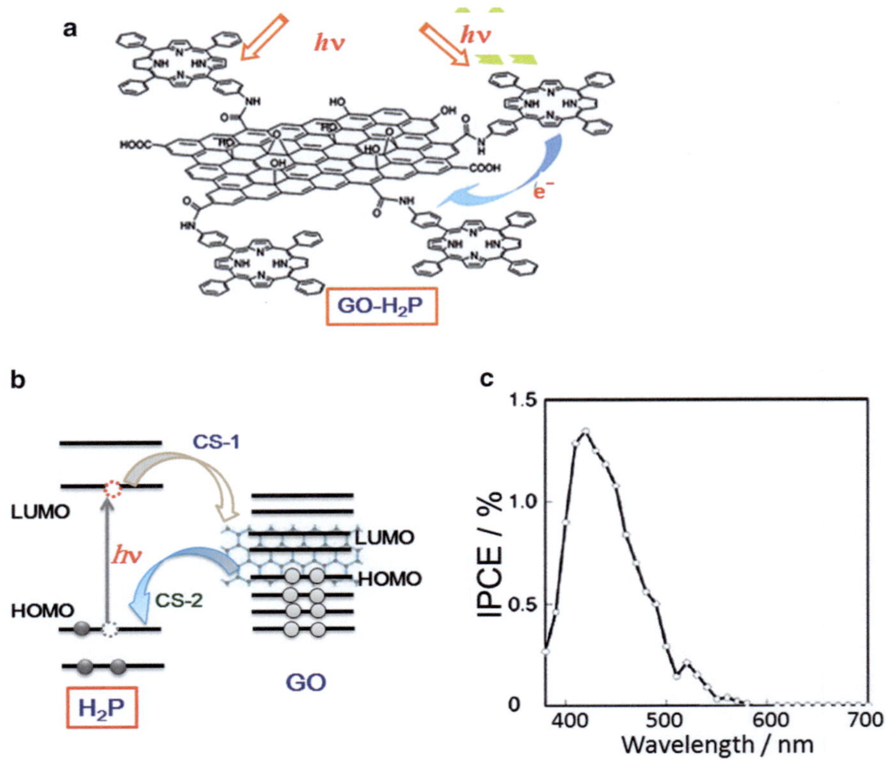

Fig. 31 (**a**) Porphyrin-functionalized graphene oxide (GO), (**b**) the energy diagram, (**c**) IPCE spectrum of a H_2P-functionalized graphene oxide on SnO_2/FTO in Ar-saturated acetonitrile containing 0.5 M LiI and 0.1 M I_2 (I^-/I_3^-) mediator. Reconstructed using figures from reference [58]

The maximal 620-nm band intensity is plotted against BNAH concentration in Fig. 32b. The 620-nm band intensity obtained from ZnPc–SLGO–C_{60} is extremely higher than those of individual ones (ZnPc–SLGO and SLGO–C_{60}), suggesting that through-space ET takes place between the photoexcited ZnPc and C_{60} part or between the photoexcited C_{60} and ZnPc, generating directly $ZnPc^{•+}$–(SLGO)–$C_{60}^{•-}$. Furthermore, the photoinduced ET (CS) between the photoexcited ZnPc and SLGO, generating $ZnPc^{•+}$–$(SLGO)^{•-}$, and between the photoexcited C_{60} and SLGO, generating $(SLGO)^{•+}$–$C_{60}^{•-}$, is considerable. Finally, coupling between the electron on $(SLGO)^{•-}$ and the hole on $(SLGO)^{•+}$ occurs on SLGO, prolonging $ZnP^{•+}$–(SLGO)–$C_{60}^{•-}$, as a tandem mechanism. Thus, higher $RV^{•+}$ accumulation has been achieved.

Nanosecond transient spectra of ZnP–(SLGO)–C_{60} are shown in Fig. 33a, in which among the broad absorption peak chain, the 830-nm band due to $ZnPc^{•+}$ and the 1,000-nm band of $C_{60}^{•-}$ are recognized [59, 60]. Time profiles at 830 and 1,000 nm decay within 200 ns with lifetimes of ca. 80 ns as shown in Fig. 33b, c. Such relatively long-lived $ZnPc^{•+}$ and $C_{60}^{•-}$ can be used as electron source for output.

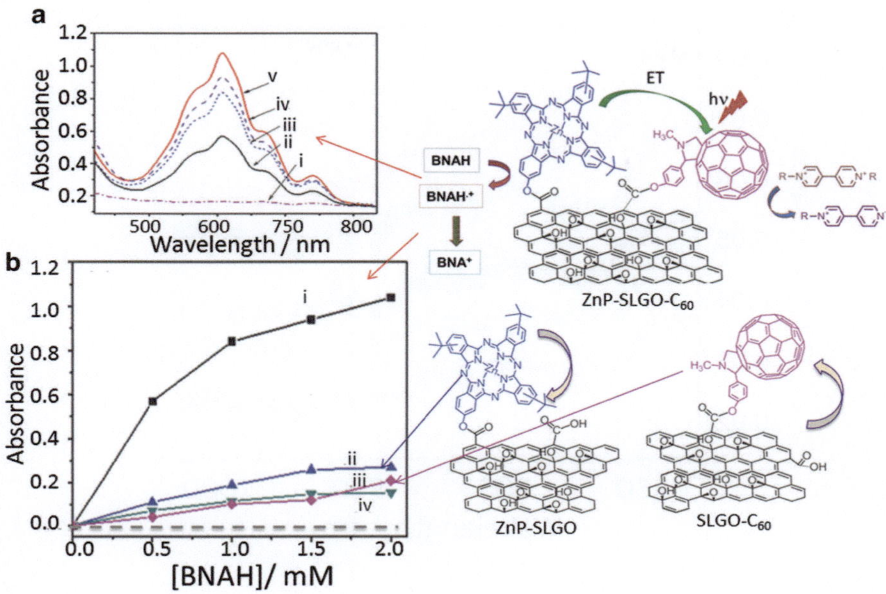

Fig. 32 Accumulation of $RV^{•+}$; (**a**) steady-state absorption spectral observation of the $RV^{•+}$ by the light illumination of ZnPc (and also C_{60}) in the presence of RV^{2+} and BNAH (*i–iv*, increase in [BNAH]; *v*, [BNAH] = 0). (**b**) *black*, ZnP–(SLGO)–C_{60}; *blue*, ZnPc–(SLGO); *violet*, (SLGO)–C_{60}; *green*, SLGO. Reconstructed using figures from reference [60]

On the other hand, the larger size graphene (reduced graphene) can also be functionalized by the light-adsorbing photosensitizing electron donor via the adsorption methods using the pyrene moieties. An example using $ZnPc(Pyr)_4$ is shown for $ZnPc(Pyr)_4/G$ in Fig. 34 [61]. The large area graphene is considered as a zero-gap semiconductor with low optical density, because of its thin sheet. In the case of such graphene, energy level diagrams depicting the possible photoinduced processes for sensitizer molecule (ZnPc) linked to graphene (abbreviated as G). In the case of graphene nano-ribbon and double-layered graphene, their electronic structures are described as the band structure because of their semiconducting character.

With the ZnPc excitation (Fig. 34), the CS process is possible, in addition to EnT process. That is, an electron transfers from the LUMO of ZnPc* to the upper cone (CB-band) of G generates the $ZnPc^{•+}(Pyr)_4/G^{•-}$ ion pair (denoted as CS-1). Furthermore, an electron transfers from the lower cone of G (VB-band) to the half-filled HOMO(Zn) level of the excited ZnPc* generating $M^{•-}(Pyr)_4/G^{•+}$ ion pair (denoted as CS-2) is also possible. When CS-1 and CS-2 simultaneously occur, excited energy transfer (EnT) from $^*ZnPc(Pyr)_4/G$ results in $ZnPc(Pyr)_4/G^*$, although the excited-state property (G*) is not clarified yet. Experimentally, traces of formation of short-lived $M^{•-}$–$G^{•+}$ is gathered from femtosecond transient studies [61].

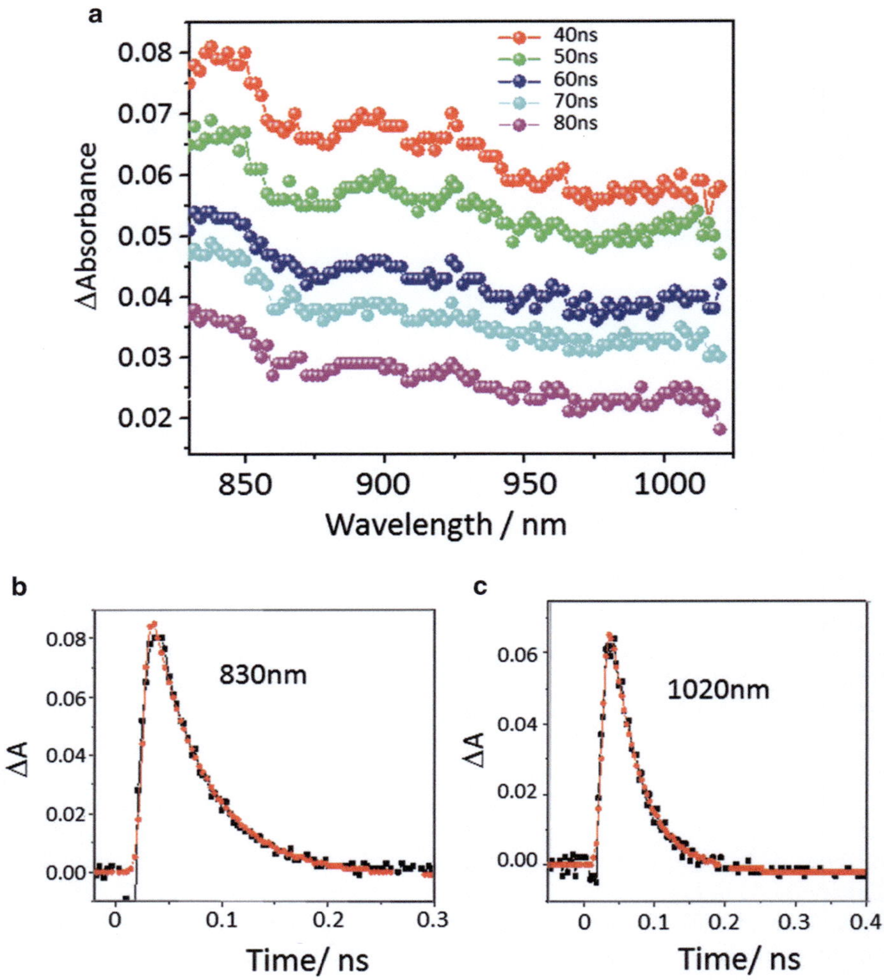

Fig. 33 (**a**) Nanosecond transient spectra of ZnP–(SLGO)–C_{60} in PhCN, (**b**) time profiles at 830 nm, (**c**) time profiles at 1,000 nm. Reconstructed using figures from reference [59, 60]

Graphene functionalization achieved by the light-adsorbing thiophene oligomers appended with the ZnPc moieties as photosensitizing electron donor and also adsorbent is also reported by Guldi's group [62]. The reported transient spectra with short-pulse laser technique show the formations of short-lived species as photophysical intermediates. It is demonstrated that such novel nanohybrid materials are suitable for new electronic materials, especially for printed electronics. Non-covalent functionalization of graphene gives rise to cost-efficient novel materials. In addition, the functional groups mediate exfoliation and solubility of graphene sheets. Devices constructed by tuning the electronic structure of graphene by chemical

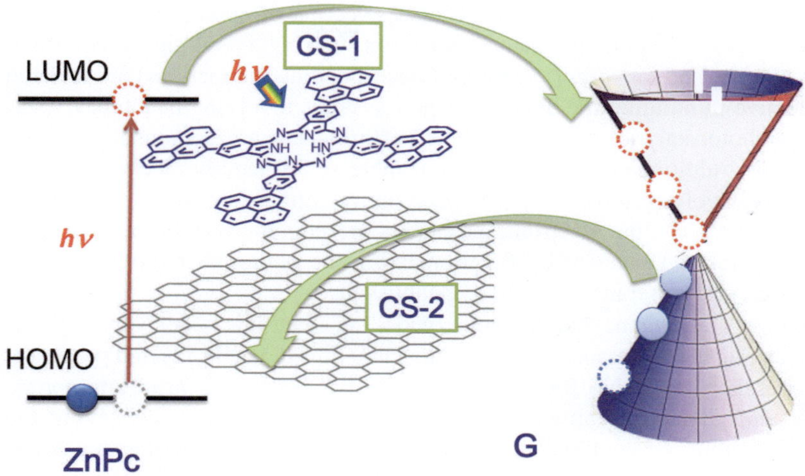

Fig. 34 Nanohybrid consists of ZnPc with four pyrene molecules adsorbed on graphene surface ($ZnPc(Pyr)_4/G$). Reconstructed using figures from reference [61]. Schematic energy diagram of the electron transfer pathway

functionalization in an appropriate manner are thus conceivable. The resulting functionalized graphene hybrids are employed as photoactive components in photoelectrochemical cells, which are challenging in easy applications.

6 Summary and Perspective

To achieve efficient photoinduced processes aiming at their potential applications in the fields of photovoltaics and photocatalysis, various nanocarbon materials have emerged out to be superior candidates to construct donor–acceptor hybrids. The fullerenes belong to the class of usual organic molecules; hence, the MOs provide useful information to understand the photosensitized electron transfer processes, on combining with the spectral and electrochemical data. Consequently, many elegant donor–acceptor hybrids using fullerenes are designed and studied to establish formation of long-lived charge-separated states. The role of spacer connecting the donor and acceptor on the long-range electron transfer processes have also been revealed, resulting in several interesting features relevant to applications in the field of photovoltaics and optoelectronics.

The band gaps of the SWCNTs are narrower than the HOMO–LUMO gaps of typical photosensitizers. Consequently, photoinduced electron transfer process can only be induced via the appended photoexcitable electron donor or electron acceptor entities on the nanotube surface. Photochemical behavior of the donor–acceptor hybrids comprised of SWCNT also depends upon the chirality (diameter (size)) of the tubes, which controls their band gaps and redox potentials. Studies performed by using enriched single-size nanotubes show that higher efficiency of the photoinduced processes is established by the selection of appropriate size-sorted SWCNTs. Supramolecular approaches using π–π stacking, metal–ligand coordination, ion pairing, and ion–dipole (with hydrogen-bonding interactions) are realized to be effective in building the donor–acceptor systems involving nanotubes. Size-dependent photocatalytic and photovoltaic performances are demonstrated using these hybrids. Additionally, by using two or more self-assembly mechanisms, the triple hybrids involving three redox and photoactive entities have been constructed. In the case of CNHs, the electronic properties are complicated, because of their gradient structures. However, by functionalizing the photosensitizers on CNHs, photoinduced electron transfer has been established.

In the case of graphene and graphene oxide, the electronic states vary with the area of π-conjugation. Because of lack of large amounts of single-layered graphene, the commercially available few-layer graphene has been used after appropriate chemical and physical treatments. In addition, the graphene oxide having the aromatic π-regions is found to be a useful material for photochemical applications. It is evident that the expansion of supramolecular chemistry concepts to graphene materials needs further work for fine-tuning of the electronic properties of graphene materials to develop a new generation of materials for light-energy harvesting and optoelectronics devices.

As a whole, the combination of several kinds of photosensitizers with different donor–acceptor properties on the large surface of the nanocarbons may open a new perspective. Such ambitious attempts have been tested for fullerenes, SWCNTs, and CNHs; now their usefulness has been confirmed. Then, these attempts have been extended to the graphene. So far, the solar voltaic cells involving dye-sensitized cell configuration and thin-film OPV type have revealed modest light-energy conversion efficiencies. Solar H_2-evolution techniques employing the combination of photocatalysts similar to the natural photosynthetic Z-scheme are also promising [62, 63]. It is also reported that the photocatalytic efficiency can be increased on combination of nanocarbons with semiconducting photocatalysis for H_2 evolution from water [64, 65]. On the other hand, artificial CO_2 conversion systems using solar light and sensitizer–nanocarbons are another appealing research field [66]. For such purposes, the carbon-based materials discussed in this chapter have the high potentials so as to say "nanocarbon materials to combat environmental carbon problems."

Acknowledgment The authors are thankful to their collaborators at the Wichita State University, University of North Texas, Tohoku University and JAIST, Keio University, NIMS for their useful scientific discussions. Work described here was supported by grants from the National Science Foundation.

References

1. G. I. Likhtenshtein: *Solar Energy Conversion: Chemical Aspects* (Wiley-VCH, Weinheim, 2012)
2. G. Renger: *Primary Processes of Photosynthesis: Principles and Apparatus* (RSC Publishing, Cambridge, 2008)

3. R. Razeghifard: *Natural and Artificial Photosynthesis*: *Solar Power as an Energy Source* (John Wiley & Sons, New York, 2013)
4. S. Fukuzumi and T. Kojima: J. Mater. Chem. **18**, 1427 (2008)
5. R. M. Marcus: J. Electroanal. Chem. **438**, 251 (1997)
6. F. D'Souza and O. Ito: Sci. Prog. **96**, 369 (2013)
7. M. Bernardi, J. Lohrman, P. V. Kumar, A. Kirkeminde, N. Ferralis, J. C. Grossman, and S. Ren: ACS Nano **6**, 8896 (2012)
8. T. Umeyama and H. Imahori: J. Phys. Chem. C **117**, 3195 (2013).
9. K. Dirian, M. A. Herranz, G. Katsukis, J. Maling, L. Rodrigues-Perez, C. Romero-Nieto, V. Strauss, N. Marin, and D. Guldi: Chem. Sci. **4**, 4335 (2013).
10. M. Quintana, E. Vazquez, M. Prato, Acc. Chem. Res. 46, 138 (2013)
11. F. D'Souza, A. S. D. Sandanayaka and O. Ito: in *Organic Nanomaterials*; *Synthesis, Characterization, and Device Application*, edited by T. Torres and G. Bottari (John Wiley & Sons, Hoboken, NJ, 2013) chapter 9
12. M.-S. Liao, J. D. Watts, and M.-J. Huang: Phys. Chem. Chem. Phys. **11**, 4365 (2009)
13. A. B. Ricks, K. E. Brown, M. Wenninger, S. D. Karlen, Y. A. Berlin, D. T. Co, and M. R. Wasielewski: J. Am. Chem. Soc. **134**, 4581 (2012)
14. T. Nojiri, M. M. Alam, H. Konami, A. Watanabe, and O. Ito: J. Phys. Chem. A **101**, 7943 (1997)
15. T. Nojiri, M.M. Alam, H. Konami, A. Watanabe, and O. Ito: J. Phys. Chem. A, **102** (1998)
16. A. S. D. Sandanayaka, Y. Araki, C. Luo, M. Fujitsuka, and O. Ito: Bull. Chem. Soc. Jpn. **77**, 1313 (2004)
17. H. Kawauchi, S. Suzuki, M. Kozaki, K. Okada, D.-M.S. Islam, Y. Araki, and O. Ito, K. Yamanaka: J. Phys. Chem. A **112**, 5878 (2008)
18. P. Schere and S. F. Fischer: *Theoretical Molecular Biophysics* (Springer Berlin-Heidelberg, 2010)
19. Y. Araki and O. Ito: J. Photochem. Photobiol. C: Photochem. Rev. **9**, 93 (2008)
20. C. B. Kc, G. N. Lim, M. E. Zandler, and F. D'Souza: Org. Lett. **15**, 4612 (2013)
21. M. E. El-Khouly: Phys. Chem. Chem. Phys. **12**, 12746 (2010)
22. F. D'Souza, S. Gadde, D.-M. S. Islam, C. A. Wijesinghe, A. L. Schumacher, M. E. Zandler, Y. Araki, and O. Ito, J. Phys. Chem. A **111**, 8552 (2007)
23. C. A. Wijesinghe, M. E. El-Khouly, M. E. Zandler, S. Fukuzumi, and F. D'Souza: Chem. Eur. J. **19**, 9629 (2013)
24. F. D'Souza, G. R. Deviprasad, M. E. Zandler, M. E. El-Khouly, M. Fujitsuka, and O. Ito: J. Phys. Chem. A **106**, 4952 (2002)
25. O. Ito and F. D'Souza: Molecules **17**, 5816 (2012)
26. S. K. Das, B. Song, A. Mahler, V. Nesterov, A. K. Wilson, O. Ito, and F. D'Souza: J. Phys. Chem. A **118**, 3994 (2014)
27. F. D'Souza, S. Gadde, M. E. El-Khouly, M. E. Zandler, Y. Araki, and O. Ito: J. Porphyrins Phthalocyanines **9**, 698 (2005)
28. C. B. Kc, G. N. Lim, P. A. Karr, and F. D'Souza: Chem. Eur. J. **20**, 7725 (2014)
29. F. D'Souza, P. M. Smith, M. E. Zandler, A. L. McCarty, M. Itou, Y. Araki, and O. Ito: J. Am. Chem. Soc. **126**, 7898 (2004)
30. E. Maligaspe, N. V. Tkachenko, N. K. Subbaiyan, R. Chitta, M. E. Zandler, H. Lemmetyinen, F. D'Souza: J. Phys. Chem. A **113**, 8478 (2009).
31. F. Oswald, D.-M. S. Islam, M. E. El-Khouly, Y. Araki, R. Caballero, P. de la Cruz, O. Ito, and F. Langa: Phys. Chem. Chem. Phys. 16, 2443 (2014)
32. T. Nakamura, H. Kanato, Y. Araki, O. Ito, K. Takimiya, T. Otsubo, and Y. Aso: J. Phys. Chem. A **110**, 3471 (2006)
33. O. Ito and K. Yamanaka: Bull. Chem. Soc. Jpn. 82, 316 (2009).
34. P. K. Poddutoori, A. S. D. Sandanayaka, T. Hasobe, O. Ito, and A. van der Est: J. Phys. Chem. B **114**, 14348 (2010)
35. M. Grätzel: Nature, **414**, 338 (2001)
36. A. Kojima, K. Teshima, Y. Shirai, and T. Miyasaka: J. Am. Chem. Soc. **131**, 6050 (2009)

37. H. Hoppe and N. S. Sariciftci: in *Photoresponsive Polymers* II, *ed. by* S. R. Marder and K.-S. Lee: Springer, Berlin-Heidelberg, 2008
38. H. Liu, D. Nishide, T. Tanaka, and H. Kataura: Nature Commun. **2**, 1313 (2011)
39. G. D. Scholes, S. Tretiak, T. J. McDonald, W. K. Metzger, C. Engtrakul, G. Rumbles, and M.J. Heben: J. Phys. Chem. C **111**, 11139 (2007)
40. H. Kataura, Y. Kumazwa, Y. Maniwa, I. Umezu, Y. Ohtsuka, and Y. Achiba: Synth. Met. **103**, 2555 (1999)
41. Y. Tanaka, Y. Hirana, Y. Niidome, K. Kato, S. Saito, and N. Nakashima: Angew. Chem. Int. Ed. **48**, 7655 (2009)
42. F. Wang, G. Dukovic, L. E. Brus, and T. F. Heiz: Phys. Rev. Letts. **92**, 177401 (2007)
43. S. K. Das, A. S. D. Sandanayaka, N. K. Subbaiyan, M. E. Zandler, O. Ito, and F. D'Souza: Chem. Eur. J. **18**, 11388 (2012)
44. T. Arai, S. Nobukuni, A. D. S. Sandanayaka, and O. Ito: J. Phys. Chem. C., **113**, 14493 (2009)
45. E. Maligaspe, A. S. D. Sandanayaka, T. Hasobe, O. Ito, and F. D'Souza: J. Am. Chem. Soc. **113**, 8158 (2010)
46. R. Chitta, A. S. D. Sandanayaka, A. L. Schumacher, L. D'Souza, Y. Araki, O. Ito, and F. D'Souza: J. Phys. Chem. C **111**, 6947 (2007)
47. A. S. D. Sandanayaka, N. K. Subbaiyan, R. Chitta, E. Maligaspe, T. Hasobe, O. Ito, and F. D'Souza: ChemPhysChem **12**, 2266 (2011)
48. F. D'Souza R. Chitta, A. S. D. Sandanayaka, N. K. Subbaiyan, L. D'Souza, Y. Araki and O. Ito: Chem. Eur. J. **13**, 8277 (2007)
49. A. S. D. Sandanayaka, E. Maligaspe, T. Hasobe, O. Ito and F. D'Souza: Chem. Commun. **46**, 8749 (2010)
50. J. L. Delgado, M. Á. Herranz and N. Martín: J. Mater. Chem. **18**, 1417 (2008)
51. F. D'Souza, R. Chitta, A. S. D. Sandanayaka, N. K. Subbaiyan, L. D'Souza, Y. Araki, and O. Ito: J. Am. Chem. Soc. **129**, 15865 (2007)
52. F. D'Souza, S. K. Das, A. S. D. Sandanayaka, N. K. Subbaiyan, D. R. Gollapalli, M. E. Zandler, T. Wakahara, and O. Ito: Phys. Chem. Chem. Phys. **14**, 2940 (2012)
53. F. D'Souza, S. K. Das, M. E. Zandler, A. D. S. Sandanayaka, O. Ito: J. Am. Chem. Soc. **133**, 19922 (2011)
54. S. Iijima, M. Yudasaka, R. Yamada, S. Bandow, K. Suenaga, F. Kokai, and K. Takahashi: Chem. Phys. Lett. **309**, 165 (1999)
55. G. Pagona, A. S. D. Sandanayaka, Y. Araki, J. Fan, N. Tagmatarchis, G. Charalambidis, P. Trohopoulos, A. G. Coutsolelos, B. Boitrel, M. Yudasaka, S. Iijima, and O. Ito: Adv. Funct. Mater. **7**, 1705 (2007)
56. M. Vizuete, M. José Gómez-Escalonilla, J. Luis G. Fierro, A. S. D. Sandanayaka, T. Hasobe, M. Yudasaka, S. Iijima, O. Ito, and F. Langa: Chem. Eur. J. **16**, 10752 (2010).
57. G. Pagona, G. E. Zervaki, A. S. D. Sandanayaka, O. Ito, G. Charalambidis, T. Hasobe, A. G. Coutsolelos, and N. Tagmatarchis: J. Phys. Chem. C, **116**, 9439 (2012)
58. N. Karousis, A. S. D. Sandanayaka, T. Hasobe, S.P. Economopoulos, E. Sarantopoulou, and N. Tagmatarchis: J. Mater. Chem. **21**, 109 (2011)
59. S. K. Das, C. B. Kc, K. Ohkubo, Y. Yamada, S. Fukuzumi, F. D'Souza: Chem. Commun. **49**, 2013 (2013)
60. S. K. Das, C. B. Kc, and F. D'Souza: Fuller. Nanotub. Carbon Nanostruct. **22**, 116 (2014)
61. C. B. Kc, S. K. Das, K. Ohkubo, S. Fukuzumi, and F. D'Souza: Chem. Commun. **48**, 11859 (2012)
62. J. Malig, N. Jux, D. Kiessling, J.-J. Cid, P. Vázquez, T. Torres, and D. M. Guldi, Angew. Chem. Inter. Ed. **50**, 3561 (2011).
63. Y. Yachibana, L. Vayssierres, and J. R. Currant: Nature Photonics 6, 511 (2012)
64. Y. K. Kim and H. Park: Energy Environ. Sci. **4**, 685 (2011)
65. D. D. Chronopoulos, N. Karousis, S. Zhao, Q. Wang, H. Shinohara, and N. Tagmatarchis: J. Chem. Soc. Dalton Trans. **43, 7429** (2014)
66. S. Sato, T. Arai, T. Morikawa, K. Uemura, T. M. Suzuki, H. Tanaka, and T. Koji: J. Am. Chem. Soc. **133**, 15240 (2011)

Plasmonic Photocatalysts with Wide Light Absorption Spectra and High Charge Separation Efficiencies

Zeyan Wang, Baibiao Huang, Peng Wang, Hefeng Cheng, Zhaoke Zheng, Zaizhu Lou, and Ying Dai

Abstract Plasmonic photocatalysts have recently attracted increasing interests due to its superior photocatalytic performances and potential applications on solving energy and environmental problems. Due to the unique localized surface plasmon resonance effects (LSPRs), plasmonic photocatalysts exhibit wide light absorption spectra and high charge carriers separation efficiencies. In this chapter, we briefly introduced the basic principles of LSPRs, and reviewed recent development of plasmonic photocatalysts.

Keywords Plasmonic • Photocatalysis • Light absorption • Charge separation

1 Introduction

Plasmonic photocatalysts, typically composed of noble metal nanoparticles and semiconductor photocatalysts, have recently attracted increasing interests due to their superior photocatalytic performances and potential applications. The predominant feature for plasmonic photocatalyst is the localized surface plasmon resonance effects (LSPRs) originated from noble metal nanoparticles. Due to the coherent oscillation of the conduction electrons in noble metal upon interaction with incident light radiation, plasmonic photocatalysts usually exhibit strong and broad light absorption spectra that cover a range from the ultraviolet region to near-infrared region [1–6]. In addition, the LSPRs of noble metal nanoparticles can also be easily adjusted by tuning the size, shape, and surrounding medium [7–12]. Another benefit of plasmonic photocatalyst is the formation of the Schottky junction that results from the contact of a noble metal and a semiconductor. It can build up an internal electric field in the region near the metal–semiconductor interface (the space-charge region)

Z. Wang • B. Huang (✉) • P. Wang • H. Cheng • Z. Zheng • Z. Lou • Y. Dai
State Key Laboratory of Crystal Materials, Shandong University, Jinan 250100, China
e-mail: bbhuang@sdu.edu.cn

E.A. Rozhkova, K. Ariga (eds.), *From Molecules to Materials*,
DOI 10.1007/978-3-319-13800-8_9

and force the electrons and holes to move in different directions, thus stimulating the charge separation [13, 14]. The above two features allow plasmonic photocatalysts to not only exhibit intense and broad light absorption spectra but also suppress the electron–hole recombination, thus leading to higher photocatalytic efficiencies.

2 Basic Principles of LSPRs

The LSPRs of noble metal NPs refer to the phenomenon in which the conducting electrons on the NPs undergo a collective oscillation (excitation) induced by the oscillating electric field of the impinging light. The oscillating charges increase an electric field near the surface. When the frequency of the incident light satisfies the resonance conditions of the noble metal NPs, the LSPRs occur with the associated light absorption (also including light scattering) [15, 16]. The LSPRs offer an exciting way of inducing light absorption in visible and near-IR regions [17]. The control of LSPRs is a key feature determining the performance of nanoscale optical and photonics devices. The LSPRs of a spherical metal NP are dominated by the dipolar mode and are described by the polarizability α:

$$\alpha = 3\varepsilon_0 V(\varepsilon - \varepsilon_{\mathrm{m}}) / (\varepsilon + 2\varepsilon_{\mathrm{m}})$$

where V is the volume of the nanoparticle, ε_0 the vacuum permittivity, ε_{m} the dielectric constant of the surrounding medium, and $\varepsilon(\omega) = \varepsilon_{\mathrm{r}}(\omega) + \mathrm{i}\varepsilon_{\mathrm{i}}(\omega)$ the frequency-dependent complex dielectric function of the metal with $\varepsilon_{\mathrm{r}}(\omega)$ and $\varepsilon_{\mathrm{i}}(\omega)$ as the real and imaginary components of $\varepsilon_{\mathrm{r}}(\omega)$, respectively. At the electromagnetic frequency ω for which $\varepsilon_{\mathrm{r}}(\omega) = -2\varepsilon_{\mathrm{m}}$, a strong resonance occurs, determining the LSPR frequency. The latter depends on variables such as the nature of the metal, the size and shape of the metal NP, and the dielectric property of the surrounding medium. These variables allow us to tune the NP optical properties [18]. The electric polarization associated with the LSPRs results in a large local electric-field enhancement and strongly enhances light absorption by the NPs. Noble metal NPs' LSPRs can be tuned to locate within the visible-light region and hence are useful for applications requiring visible-light absorption [19]. The Fermi level of noble metals lies between the valence band maximum (VBM) and conduction band minimum (CBM) of a polar semiconductor forming a plasmonic photocatalyst.

The noble metal NPs deposited on the surface of polar semiconductors are most likely charge-polarized and work as light trapping, scattering, and concentration centers. It can excite the conduction electrons at the interface between a metal and a dielectric, and by proper tuning of these noble metal NP structures, light can be concentrated and "folded" into a semiconductor layer, thereby enhancing the absorption [17]. Furthermore, the SPR of noble metal NPs can induce a fast electron transfer from the photoexcited noble metal NP to the polar semiconductor to which the NP is attached [20–22]. The presence of noble metal NPs in contact with the semiconductor surface can also accelerate the redox reaction between the semiconductor and H_2O, CO_2, or organic pollutants [23–25]. The latter facilitates the transfer of the photogenerated electrons and holes from the plasmonic photocatalyst to such molecular species on the surface of the photocatalyst.

3 Proposed Mechanisms on Charge Transfer

The incorporation of plasmonic metal nanoparticles into semiconductors can effectively improve the photocatalytic efficiencies by extending the light absorption spectra, increasing light scattering and exciting charge carriers in semiconductors by the charge transfer from metal to semiconductors [27]. Among these three processes, the first two mainly focus on the absorption of photon energy, and the last process focuses on the charge separation of charge carriers and utilization of photon energies. However, at present, the exact mechanism on LSPR-induced charge separation still remains unclear. Based on the experimental results reported recently, three hypothesized energy transfer mechanisms have been proposed, namely, direct electron transfer (DET) mechanism, local electromagnetic field (LEMF) enhancement, and resonant energy transfer (RET) mechanism. Details about the three mechanisms are discussed below.

3.1 Direct Electron Transfer (DET) from Metal to Semiconductor

When the metal is in direct contact with semiconductors, LSPR-mediated “hot electrons” can be injected to the conduction band of semiconductors from metal, which is similar as the dye sensitization process. The metallic plasmonic nanoparticles act as a dye sensitizer by absorbing photon energy and transferring the energetic electrons to the nearby semiconductors. Compared to conventional dyes, plasmonic metals exhibit an excellent mobility of charge carriers and higher absorption cross-sections. Furthermore, the LSPR light absorption spectra can be adjusted to the entire solar spectrum by tuning the size or shape of the metal nanoparticles, which are likely to become even more ideal sensitizers than a typical dye sensitizer. However, the DET material selection is limited by the electronic band structure alignment of the semiconductor and metal.

3.2 Local Electromagnetic Field (LEMF) Enhancement

As metal and semiconductor are separated by nonconductive spacers, direct charge transfer from the metal to the semiconductor would be blocked. However, experimental results illustrate that the photocatalysis performances were still enhanced after an insulating interlayer is added between the metal and the semiconductor [28]. This mechanism is called LSPR-mediated local electromagnetic field enhancement

(LEMF), which can contribute to the generation of electron–hole pairs radiatively. The local electric fields excited by plasmonic nanostructures are usually orders of magnitude higher than the incident photons, which are spatially nonhomogeneous, and decreasing exponentially with distance from the surface of plasmonic metals within tens of nanometers. When a semiconductor is near enough to the plasmonic metals, strong plasmonic electric fields will be able to excite electron–hole pairs in the semiconductor. Due to the enhancement of the LSPR-mediated electromagnetic fields, the rate of electron–hole formation in the semiconductors will also be increased by a few orders of magnitude. One requirement needed in the LEMF mechanism is that the LSPR energies should be above the bandgaps of the semiconductors.

3.3 Resonant Energy Transfer (RET)

Besides the abovementioned two mechanisms in plasmonic photocatalysis, another new mechanism named resonant energy transfer (RET) was also proposed recently. The RET process is proposed as an alternative, nonradiative mechanism of LSPR-induced charge separation in semiconductors, which can directly excite electron–hole pairs in the semiconductor nonradiatively through the relaxation of the LSPR dipole. Different from the DET and LEMF mechanisms, which depend on the electronic band structure alignment and energy matching, RET is an effectively upward energy transfer process that can overcome the charge equilibration issues while still enhancing carrier creation at energies below the bandgap, which is more flexible for designing solar energy materials and optoelectronic devices.

4 Plasmonic Photocatalysts

Due to the strong LSPR of silver and gold nanoparticles, present plasmonic photocatalysts are mainly consisted of Au or Ag nanoparticles and semiconductor materials, such as Ag(or Au)/TiO_2, Au/CeO_2(Au/ZrO_2), Pt/Bi_2O_3, Au/SiO_2, and Ag@AgX (X=Cl, Br, I) [29–43]. And recently, LSPR has been also discovered in some heavily doped semiconductors and metal oxide nanocrystals such as WO_{3-x} and $Cu_{2-x}S$ nanocrystals [44, 45]. Due to the unique character of their outer-d valence electrons, a strong absorption in the visible and near-infrared region was observed. And their LSPR can be easily tuned by varying the compositions, which provides us the possibility of designing new types of plasmonic photocatalysts.

4.1 Noble Metal–Semiconductor Composites

The first plasmonic photocatalyst Ag@SiO_2–TiO_2 was explored by Awazu's group. They deposited TiO_2 films on a layer of Ag nanoparticles coated by SiO_2 shells to prevent their oxidation [28]. Ag nanoparticles' surface plasmonic resonance extended sample's absorption in the near-UV region, and the decomposition rate of methylene blue dye was enhanced by a factor of seven compared to pure TiO_2. They also demonstrated that protective SiO_2's thickness shells can affect the photocatalytic activity. As the thickness of the SiO_2 increased, the photocatalytic activity gradually decreased. Photocatalytic activity with 100 nm SiO_2 was observed to be close to that of pure TiO_2, which indicated that the enhanced photocatalytic activity can be attributed to the surface plasmonic resonance of Ag nanoparticles. Kumar et al. also investigated the role of the distance between TiO_2 and Ag nanoparticles on the influence of the photocatalytic activity and pointed out that the photocatalytic efficiency gradually increased with a decrease of SiO_2 shell thickness [46]. Linic's group also did some research in Ag/N–TiO_2 plasmonic photocatalyst. In their work, the evidence that the localized electric fields of the Ag NPs interact with TiO_2 to selectively form electron–hole (e^-–h^+) pairs in the near-surface region of TiO_2 was firstly presented, as shown in Fig. 1 [25]. The advantage of forming e^-–h^+ pairs near the semiconductor surface is that these charge carriers are readily separated from each other and easily migrate to the surface, where they can perform photocatalytic reactions.

In order to maximize charge separation of excited electrons from the nanoparticle surface, the nanoparticles can be coupled with an efficient electron acceptor to form the metal–semiconductor junction. The Schottky junction forms at the interface of the noble metal and the semiconductor, which can build up an internal electric field in the space-charge region. This internal electric field could force the electrons and holes to transfer to opposite directions and hence suppress the charge recombination. Such plasmonic nanocomposites could possess drastically enhanced visible-light absorption due to the SPR effect as well as the efficient charge separation.

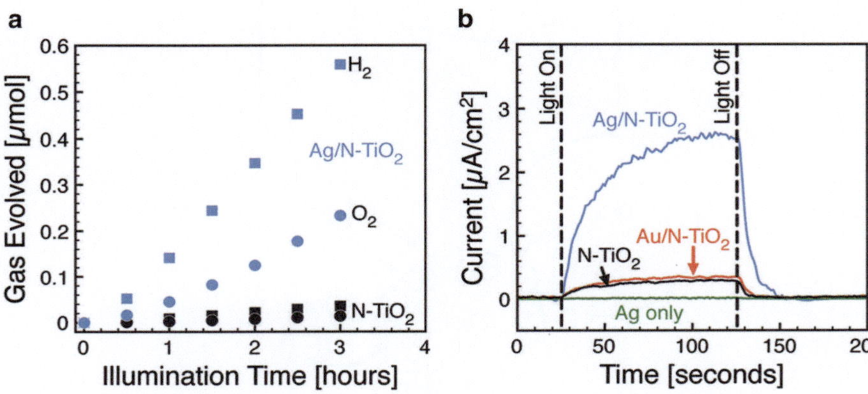

Fig. 1 (**a**) H_2 and O_2 (**b**) production upon visible illumination of N–TiO_2 (*black symbols*) and Ag/N–TiO_2 (*blue symbols*) photocatalysts, as measured by GC. (**b**) Photocurrent responses (per macroscopic electrode area) upon illumination with a broadband visible light source (400–900 nm) (published with permission of ACS) (26)

Extensive studies have shown that composite plasmonic metal–semiconductor photocatalysts achieved significantly higher rates in various photocatalytic reactions compared with pure semiconductor photocatalysts. Plasmon-induced photoelectrochemistry in the visible region was studied by Tatsuma and co-workers [47]. They found that the gold nanoparticles are photoexcited due to plasmon resonance, and charge separation is accomplished by the transfer of photoexcited electrons from the gold particle to the TiO_2 conduction band and the simultaneous transfer of compensative electrons from a donor in the solution to the gold particle (Fig. 2). Garcia and co-workers also showed that the rate of hydrogen-evolution half-reaction was higher on plasmonic Au/TiO_2 composites than on TiO_2 under broadband visible illumination (>400 nm) (Fig. 3) [48].

Furthermore, LSPR-enhanced photocatalytic reactions were also reported in other semiconductors besides TiO_2. An experimental study of the influence of gold nanoparticles on α-Fe_2O_3 photoanodes for photoelectrochemical water splitting is carried out by Scott C. Warren et al., and a relative enhancement in the water-splitting efficiency at photon frequencies corresponding to the plasmon resonance in gold was observed (Fig. 4) [49]. Notably, this relative enhancement was observed only for electrode geometries with metal particles that were localized at the semiconductor–electrolyte interface, consistent with the observation that minor

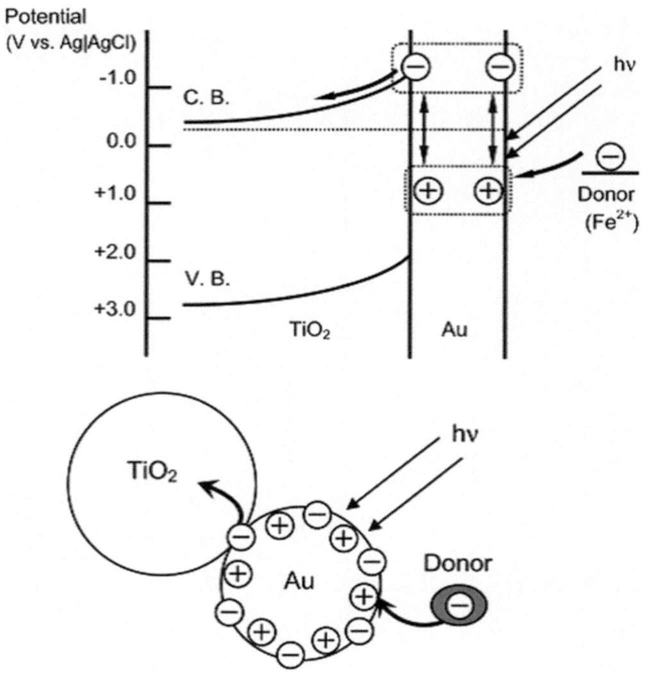

Fig. 2 Proposed mechanism for the photoelectrochemistry. Charges are separated at a visible-light-irradiated gold nanoparticle–TiO_2 system (published with permission of ACS [47])

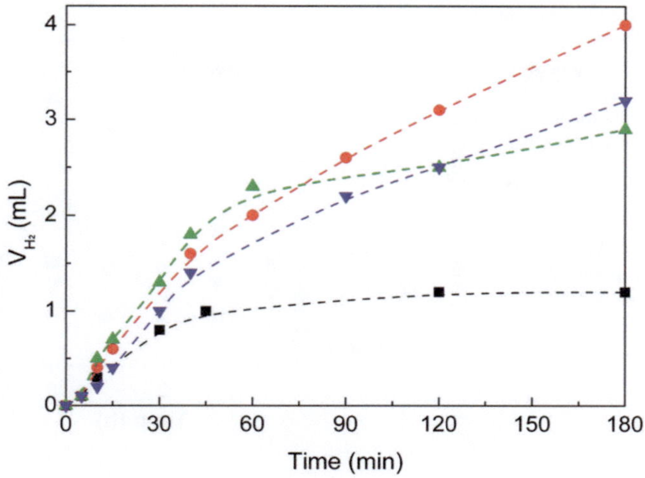

Fig. 3 The volume of evolved hydrogen on TiO_2 (*black squares*) and Au/TiO_2 composites with three different weight loadings of Au with respect to TiO_2 (*red circles*, 0.25 % Au; *green triangles*, 1.5 % Au; *blue triangles*, 2.2 % Au) on illumination with a broadband visible-light source (>400 nm) (published with permission of ACS)

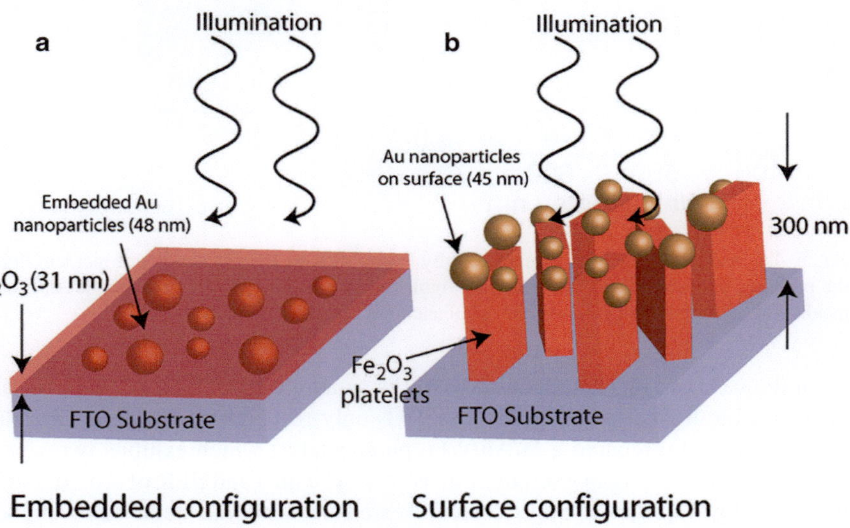

Fig. 4 Different electrode configurations. (**a**) Embedded and (**b**) surface. The electrolyte, counter electrode, and reference electrode are omitted for clarity (published with permission of ACS)

carrier transport to the electrolyte is the most significant impediment to achieving high efficiencies in this system. García and co-workers found that when irradiated with visible light ($\lambda > 400$ nm), 1 wt% gold-supported ceria nanoparticles generate oxygen from water (10.5 μmol h^{-1}) more efficiently than the standard WO_3 (1.7 μmol h^{-1}) even under UV irradiation (9.5 μmol h^{-1}) (Fig. 5) [38].

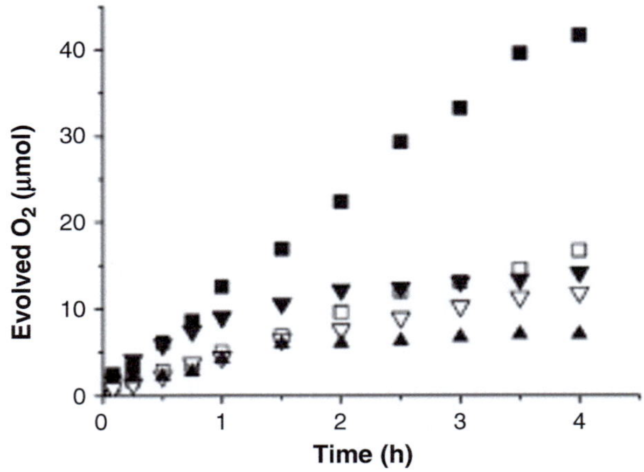

Fig. 5 Oxygen evolved upon visible-light ($\lambda > 400$ nm) illumination of an aqueous $AgNO_3$ suspension containing the photocatalyst. *Filled squares* Au(1.0 wt%)/CeO_2(A), *open squares* Au(1.0 wt %)/CeO_2(B), *inverted filled triangles* Au(3.0 wt%)/CeO_2, *filled triangles* WO_3 (published with permission of ACS)

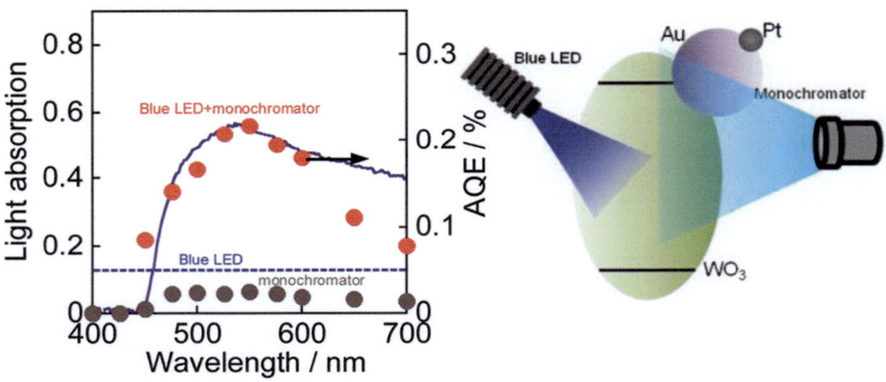

Fig. 6 Dependence of rates of H_2 evolution by Pt/Au/WO_3 on the wavelength of incident light in the presence and absence of irradiation by light from the blue LED. The *line* means AQE under irradiation by light from the blue LED (published with permission of ACS)

In the case of Au/TiO_2, TiO_2 only works as an electron transportation material under visible-light irradiation because TiO_2 only responds to UV light. Recently, Kominami et al. prepared a Pt/Au/WO_3 photocatalyst which exhibits two absorptions due to the bandgap excitation of WO_3 (<450 nm) and SPR of Au NPs (450–600 nm) (Fig. 6) [50]. Although former research on plasmonic metal–semiconductor photocatalysts also includes the visible-light-responding semiconductors, the working mechanism in the formation of H_2 and O_2 remains unclear. In this work, they investigated whether two types of photoabsorption contribute to the formation of H_2 and O_2 by using an action spectrum and found that both bandgap excitations of WO_3 and SPR of Au NPs simultaneously contribute to H_2 formation.

Among plasmonic composite photocatalysts, extensive efforts were focused on harvesting visible light around characteristic adsorption of coinage metal nanoparticles, while little report on harvesting longer-wavelength visible light which accounts for a larger proportion of solar energy. Recently, Ye et al. reported on a wide-range visible-light harvesting by TiO_2 via introducing Au nanorods (NRs) as antennas. They found that not only transversal plasma which is similar to sphere Au particles but also longitudinal plasma of Au NRs (with adsorption centered from 630 to 810 nm) could induce photocatalytic oxidation of 2-propanol (IPA) [51]. Almost at the same time, Li and co-workers demonstrated that the photoactivity of Au-decorated TiO_2 electrodes for photoelectrochemical water oxidation could be effectively enhanced in the entire UV–visible region from 300 to 800 nm by manipulating the shape of the Au nanostructures (Fig. 7) [52]. They proposed that the enhanced photoactivity in the UV region is due to the effective surface passivation as well as electric-field amplification effect, while the photoactivity enhancement in the visible region is mainly caused by the electric-field amplification effect and hot-electron generation upon SPR excitation.

There are several energy transfer mechanisms by which SPR can enhance the concentration of charge carriers and hence increase the photocatalytic activity [25]. The most discussed one is the plasmon-induced direct charge transfer from metal to semiconductor, which is similar to dye sensitization. The noble metal nanoparticles

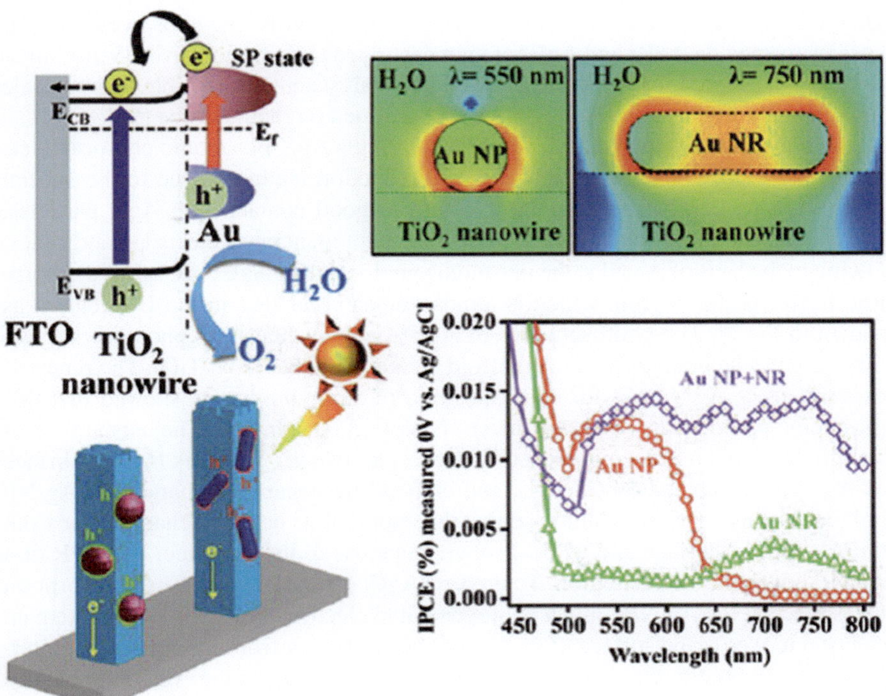

Fig. 7 Mechanism of SPR-induced charge transfer with approximate energy levels and IPCE plots collected at 0 V versus Ag/AgCl with the incident wavelength between 450 and 800 nm (published with permission of ACS)

act as a dye sensitizer, absorbing resonant photons and transferring the hot electrons, excited either by plasmon resonance or inter-/intraband transition, into the conductive band of the semiconductor, thus leading to electron–hole separation. The plasmonic nanocrystal then serves as an anode to conduct oxidizing reactions while the semiconductor functions as a cathode to induce reducing reactions.

4.2 *Silver Halide Plasmonic Photocatalysts*

Ag@AgX(X=Cl, Br, I) are fancy plasmonic photocatalysts that have been extensively studied. As we all know, silver halides are unstable under light irradiation. Upon absorbing a photon, a silver halide can get activated and generate an electron and a hole, and subsequently, the photogenerated electron reacts with a Ag^+ ion to form a Ag^0 atom. Ultimately, a cluster of silver atoms is formed within a silver halide particle upon durative light irradiation. Due to its instability under light irradiation, silver halide was seldom used as a photocatalyst. However, Calzaferri et al. succeeded in using AgCl to photocatalyze O_2 production from water in the presence of a small excess of silver ions in solution [53]. Kakuta's group also used AgBr to photocatalyze H_2 production from a CH_3OH–H_2O solution. They observed that Ag^0 species formed on AgBr and AgBr are not destroyed under successive UV illumination, which allowed for preparing a new photocatalyst active and stable under visible light, namely, AgX particles with silver NPs formed on their surface [54].

Recently, our group synthesized a series of Ag@AgX plasmonic photocatalysts by simple ion-exchange and light-induced reduction methods. Due to the special fabrication process, silver nanoparticles are in good contact with AgX particles, which leads to an efficient separation of the photogenerated electrons and holes. Owing to the SPR of Ag NPs, the plasmonic photocatalyst Ag@AgCl shows absorption in the visible region, which is almost the same as that in the UV region, as shown in Fig. 8. The photoactivities of Ag@AgCl, evaluated by photodecomposition of methylic orange dye, are eightfold greater than that of N–TiO_2. The repeated experiment of Ag@AgCl for decomposition of methylic orange showed that this plasmonic photocatalyst is stable under visible-light irradiation. The reason for this stability of AgCl has been proposed in our former papers. We think that the surface of AgCl is terminated by Cl^- ions, and is therefore negatively charged. A Ag NP polarizes its electron distribution so that the region of its negative charge is far from the Ag–AgCl interface and its positive charge close to the interface. The SPR of a Ag NP lies in the visible region. Thus, Ag@AgCl catalyst's visible-light absorption takes place at the Ag NPs, and a photogenerated electron–hole pair is separated; an electron moves to the surface of the Ag NP farthest away from the Ag@AgCl interface, while a hole to the surface of the AgCl particle bearing the Ag NP. The holes transferred to the AgCl surface oxidize Cl^- ions to Cl^0 atoms. The Cl^0 atoms will oxidize methylic orange and then become reduced to Cl^- ions again. During the photogenerated electron and hole separation process, the electrons remain in the Ag NPs rather than being transferred to the Ag^+ ions of AgCl, and the holes transferred to the AgCl surface oxidize Cl^- rather than the Ag NPs [30].

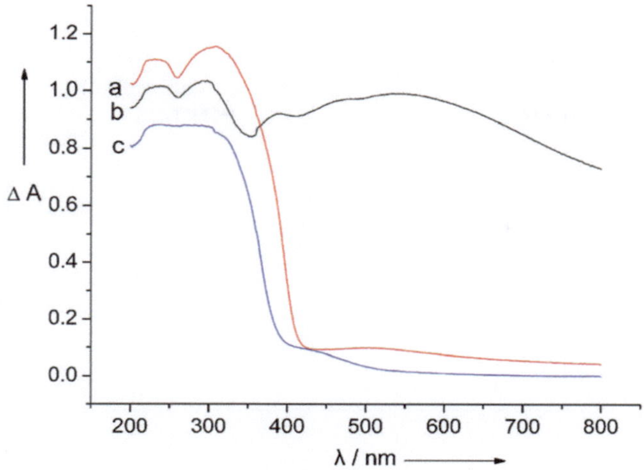

Fig. 8 UV/Vis diffuse reflectance spectra of (*a*) AgCl, (*b*) Ag@AgCl, and (*c*) N–TiO_2 (published with permission of Wiley)

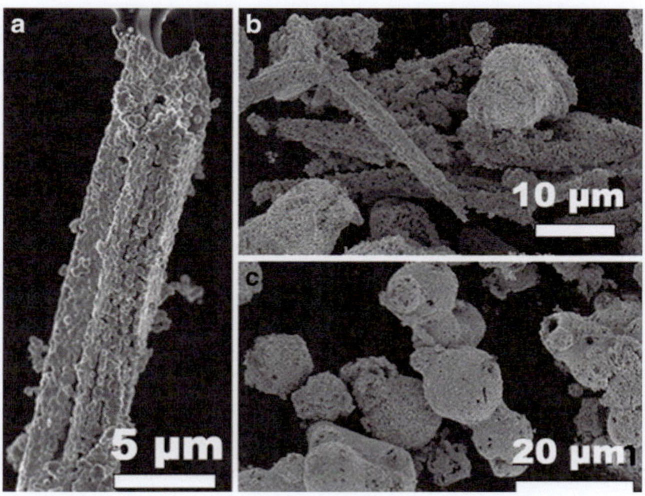

Fig. 9 SEM images of several as-prepared Ag@AgCl plasmonic photocatalysts with the morphology of (**a**) rods, (**b**) rods and irregular balls, and (**c**) hollow spheres (published with permission of Wiley)

Considering the morphology has a great effect on the performance of photocatalysts, Ag@AgCl with various different morphologies was fabricated just as shown in Fig. 9. The photocatalytic capabilities of these plasmonic photocatalysts were evaluated by monitoring the decomposition of methylic orange, showing that the Ag@AgCl samples with hollow sphere morphology are more efficient than other

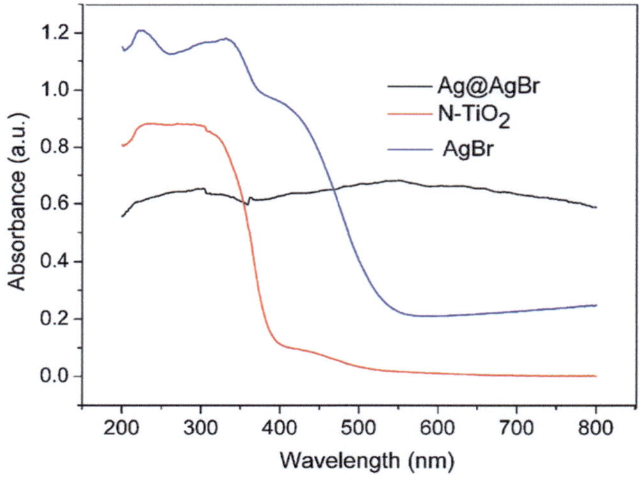

Fig. 10 UV/Vis diffuse reflectance spectra of AgBr, Ag@AgBr, and N-doped TiO_2 (published with permission of Wiley)

samples with other shapes [35]. Inspired by our work, various Ag@AgCl plasmonic photocatalysts were explored. For example, Sun et al. fabricated the AgCl:Ag (i.e., Ag–AgCl) hybrid NPs, in which a AgCl core is coated with Ag nanograins. These hybrid NPs have cubic shapes and uniform sizes with an average edge length of 130 nm. They exhibit high efficiency in photodecomposition of methylene blue under sunlight irradiation [55].

Recently, our group developed an ionic-liquid-assisted hydrothermal and photoreduction method to fabricate the Ag@AgCl and Ag@AgBr (i.e., Ag–AgBr) exposing different crystal facets. Photodecomposition of organic dye revealed that the plasmonic photocatalysts exposing higher-index facets exhibit higher photocatalytic activities [56, 57].

Taking into account of the effect of the surrounding medium on the SPR of Ag nanoparticles, we replaced AgCl with AgBr and fabricated the new plasmonic photocatalyst Ag@AgBr. Comparing to Ag@AgCl, Ag@AgBr has weaker visible-light absorption and the oxidation ability of Br^0 is weaker than that of Cl^0. However, Ag@AgBr was found to be more photocatalytically active than Ag@AgCl by a factor of 1.5. We think this can be attributed to the lower electron affinity of Br^0 than that of Cl^0 (i.e., 324 kJ mol^{-1} versus 349 kJ mol^{-1}). Furthermore, AgBr has smaller bandgap than does AgCl (2.25 eV versus 3.25 eV) and absorbs visible light just as shown in Fig. 10 (published with permission of Wiley [58]). The photocatalytic reducing ability of Ag@AgBr, checked by monitoring the reduction of $Cr^{(VI)}$ to $Cr^{(III)}$ ions under visible light, was lower. In order to improve it, iodine was doped in the AgBr lattice, i.e., Ag(Br,I). The photoreduction results reveal that Ag@Ag(Br,I) reduces $Cr^{(VI)}$ ions ninefold faster than Ag@AgBr does. The analogous plasmonic photocatalysts Ag–Ag(Cl,Br) and Ag–AgCl/AgI were also fabricated [39]. These

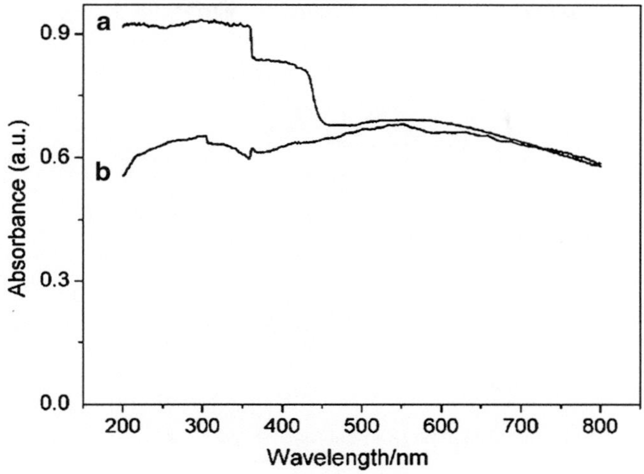

Fig. 11 UV/Vis diffuse reflectance spectra of (a) Ag@Ag(Br,I) and (b) Ag@AgBr (published with permission of Wiley [39])

plasmonic photocatalysts strongly absorb visible light (Fig. 11), oxidize methylic orange, and reduce $Cr^{(VI)}$ to $Cr^{(III)}$ under visible light with high efficiencies. Huang et al. [59] fabricated a series of photocatalysts by depositing Ag NPs on the surface of various Ag salts, Ag_nX ($X^{n-}=Cl^-$, Br^-, I^-, CrO_4^{2-}, PO_4^{3-}, $PW_{12}O_{40}^{3-}$, $SiW_{12}O_{40}^{4-}$). The photocatalytic activities of Ag–Ag*n*X can be tuned by altering the charge of the counteranions X_n^-. The photocatalytic abilities of these photocatalysts $Ag–Ag_nX$ were evaluated by measuring the decomposition of methylic orange. Among these samples, $Ag–Ag_4SiW_{12}O_{40}$ has the strongest photocatalytic ability among the Ag–Ag salt catalysts, and the decomposition of methylic orange is complete in less than 5 min, pointing out that the photocatalytic ability increases with an increase in the charge of the counteranion. This can be attributed to that a more negative anion of the Ag salt will lead to a stronger charge polarization in the Ag NP and hence to a more efficient separation of the photogenerated electrons and holes.

Composite photocatalysts is another group of important photocatalysts that have been extensively studied. Any part of the composite photocatalysts can absorb light energy separately, and the total photon energy absorbed by the composite photocatalysts can be enhanced. Furthermore, composite photocatalysts with suit bandgap positions can lead to separation of the photogenerated electrons and holes, improving the photocatalytic performance. In order to take advantage of composite photocatalyst and plasmonic photocatalysts, the composite photocatalyst $Ag–AgBr/WO_3{\cdot}H_2O$ was synthesized. The conductive band and valence band of AgBr are located at −3.7 and −5.95 eV [60], while the conductive band and valence band of $WO_3{\cdot}H_2O$ are located at −4.404 and −6.574 eV, respectively. Due to the suit bandgap position, the photogenerated electrons transfer to Ag NPs and are therefore separated

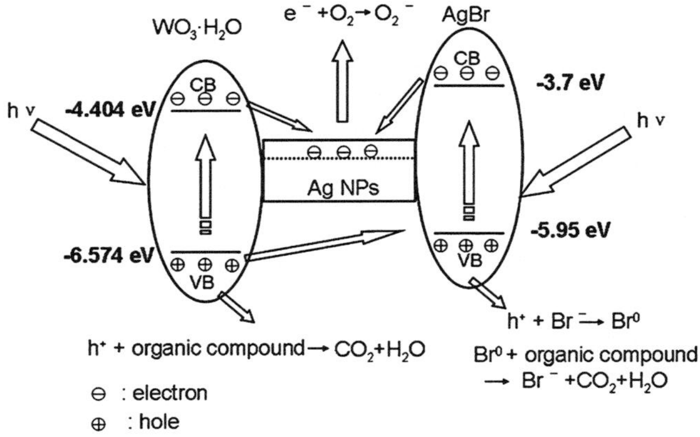

Fig. 12 Schematic diagram illustrating the photogenerated electrons–holes and their migration and reactions in the composite plasmonic semiconductor Ag/AgBr/$WO_3 \cdot H_2O$ (published with permission of ACS)

from the holes in the valence band of AgBr and $WO_3 \cdot H_2O$, as depicted in Fig. 12 (published with permission of ACS [61]). This photocatalyst Ag–AgBr/$WO_3 \cdot H_2O$ possesses the advantages of a plasmonic photocatalyst and a composite photocatalyst, strongly absorbs visible light, and exhibits a high photocatalytic activity. Compared with the plasmonic photocatalyst Ag@AgBr, it shows higher efficiency in photo-killing *E. coli* bacteria. Almost all of the *E. coli* is killed after 20 min of irradiation while more than 95 % of *E. coli* is still alive with Ag@AgBr. The research on Ag@AgBr/$WO_3 \cdot H_2O$ provides a useful guideline to design efficient composite photocatalysts. Inspired by this work, other plasmonic photocatalysts containing Ag NPs and composite semiconductors were explored, such as Ag–AgBr/Al_2O_3, TiO_2–Ag/AgBr, Ag/AgCl/TiO_2, and so on [62–64]. Recently, our group also fabricated plasmonic photocatalyst Ag/AgBr/BiOBr [65], displaying superior visible-light-driven photocatalytic activities in the sterilization of pathogenic organisms and the decomposition of organic dyes. Yu's group also fabricated composite photocatalyst Ag–AgCl/TiO_2 through depositing AgCl NPs onto TiO_2 nanotube arrays and then reducing some Ag^+ ions in the surface region of the AgCl particles to form Ag^0 species under lamp irradiation. The plasmonic photocatalyst exhibits high visible-light photocatalytic activity for the degradation of methylic orange [66]. Hu et al. employed the deposition–precipitation and photoreduction methods to synthesize the plasmonic photocatalyst Ag–AgI/Al_2O_3 [66, 67]. Due to the SPR of Ag nanoparticles, the as-prepared Ag–AgI/Al_2O_3 photocatalyst shows high absorption in the visible-light region as shown in Fig. 13. It has a high photocatalytic activity for the degradation and mineralization of toxic-persistent organic pollutants (2-chlorophenol, 2,4-dichlorophenol, and trichlorophenol) under visible-light or simulated solar light irradiation.

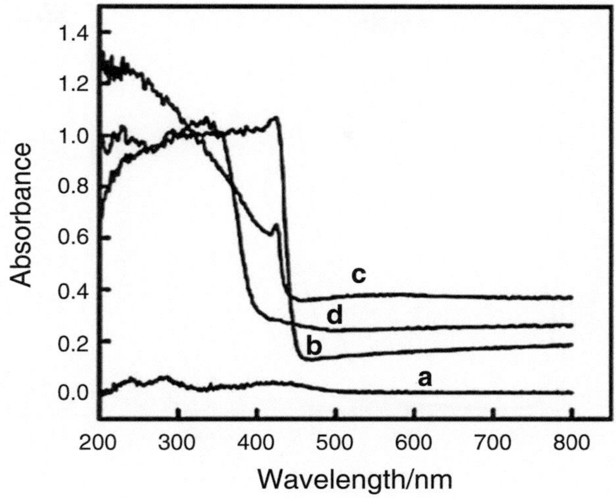

Fig. 13 UV/Vis diffuse reflectance spectra of (**a**) Al_2O_3, (**b**) AgI/Al_2O_3, (**c**) $Ag–AgI/Al_2O_3$, and (**d**) $TiO_{2\_}xNx$ (published with permission of ACS)

4.3 Nonmetallic Plasmonic Nanomaterials

Apart from the traditional noble metal nanostructures, LSPR has recently been discovered in some heavily doped semiconductor and metal oxide nanocrystals (NCs), where the high density of free charge carriers are introduced by doping. The density of free carriers in doped semiconductors is lower than that of noble metals, and as a result, their LSPR frequency is at longer wavelength [68]. Moreover, in the heavily doped semiconductors and metal oxides, the plasmonic resonance could be intriguingly tailored by changing the stoichiometric compositions, dopant concentrations, or phase transitions.

Generally, on the basis of the generation pathways for the free charge carriers, nonmetallic plasmonic nanomaterials can be classified into two predominant categories. One category is the extrinsic doping of the semiconductors and metal oxides by heterovalent atoms. In these extrinsically doped materials, the concentration of free charge carriers is usually of the same order to the dopant concentration introduced. Therefore, by increasing the dopant content, blue shift of the plasmonic frequency to higher energy will take place in the heterovalent-atom-doped semiconductors and metal oxides. The other alternative is nonstoichiometric semiconductors or metal oxides by intrinsic formation of deficient cation or anion vacancies, also called self-doping. To compensate for the sub-stoichiometry of the component, extra electrons in the conduction band or holes in the valence band are thus obtained to support the plasmonic resonance.

4.3.1 Extrinsic Doping of Semiconductors or Metal Oxides

Doping with heterovalent atoms into the semiconductor or metal oxides will generate extra free carriers, and considerable dopant concentration would give a birth to

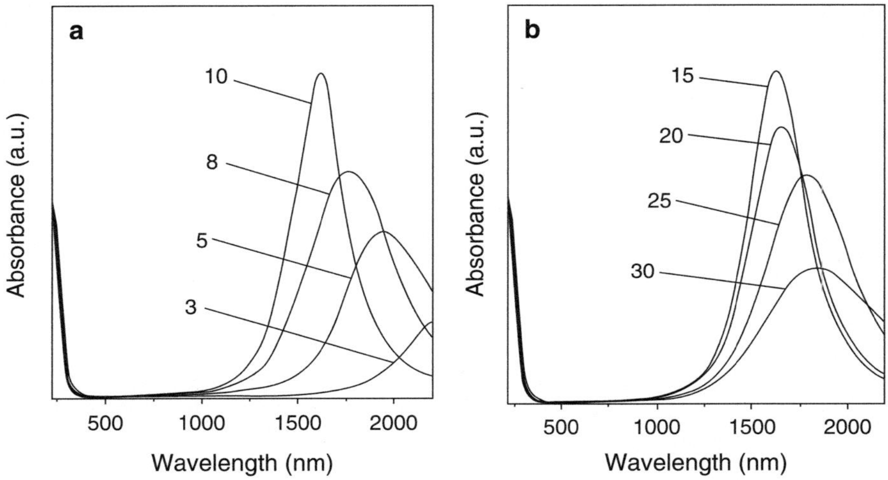

Fig. 14 UV–Vis–NIR spectra showing the dependence of LSPR wavelength of ITO NPs on the dopant (Sn%) concentration: (**a**) 3–10 % and (**b**) 15–30 % (published with permission of ACS)

apparent free charges to support LSPR. The extrinsic doping of semiconductor or metal oxides mainly involves Sn-doped In_2O_3 (ITO) [69, 70], Al-doped ZnO (AZO) [71], In-doped CdO (ICO) [72], Sb-doped SnO_2 (ATO) [73], and P-doped Si [74, 75].

Sn-doped In_2O_3 (ITO) is a transparent conductive oxide material, which is composed of a metal oxide semiconductor (In_2O_3) and doped heterovalent atom (Sn). The introduction of Sn (IV) atoms into the In_2O_3 lattice will lead to free charge carriers that are responsible for the conductivity, and considerable high concentration of free carriers could be expected to support LSPR in ITO NCs. Kanehara et al. synthesized colloidal ITO NCs with tunable In/Sn molar ratio by colloidal method [69]. As presented in Fig. 14, upon controlling the Sn concentration from 3 to 30 mol% in the precursor, LSPR in the ITO NCs could be tailored from 1,618 to >2,200 nm. Moreover, dynamic modulation of LSPR in ITO thin films could be controlled by electrochemical method [70]. By applying a bias voltage, the free charge carriers in ITO NCs can be manipulated and further affect the corresponding LSPR frequency.

Likewise, ZnO NCs doped with group III elements also have the potential to possess high density of free carriers. With respect to ITO, doped ZnO has the advantage of low cost with earth-abundant elements, which is important for displays. Buonsanti et al. reported the colloidal Al-doped ZnO (AZO) NCs with different sizes [71]. By changing the synthetic temperature and Al dopant precursor concentration, the AZO NCs prepared displayed tunable LSPR at the NIR region. Similarly, doping of CdO, SnO, and Si NCs with higher valent atoms including In, Sb, and P, respectively, also leads to sufficient free electrons to exhibit LSPR in the mid-IR region [72–75].

4.3.2 Self-Doped Nonstoichiometric Semiconductors or Metal Oxides

In stoichiometric semiconductors or metal oxides, LSPR is not observed due to the absence of free charge carriers. Intriguingly, in some semiconductors or metal oxides, the presence of deficiencies could lead to excessive free carriers (electrons or holes) to support the LSPR, which mainly includes the copper-deficient chalcogenides ($Cu_{2-x}E$, E = S, Se, Te) [44, 68, 76–86] and oxygen-deficient transition metal oxides (e.g., TiO_{2-x} [87], WO_{3-x} [45] and MoO_{3-x} [88, 89]).

The first discovery of NIR LSPR in $Cu_{2-x}S$ NCs was reported by Zhao et al. in 2009 [44]. With increasing the copper deficiencies, the LSPR frequency of the $Cu_{2-x}S$ NCs moved to shorter wavelength with higher energy (blue shift). The holes associated with the copper vacancies were considered to be the free charge carriers in Cu_{2-x} NCs. Later, Luther et al. adopted a colloidal method to synthesize monodisperse $Cu_{2-x}S$ NCs with sizes ranging from 2 to 6 nm (Fig. 15a) by reacting copper (II) acetylacetonate and ammonium diethyldithiocarbamate in the presence of oleic acid (OA) and dodecanethiol (DDT) [68]. The free carriers which arose from the excess holes in the valence band (VB) were estimated to be about 2×10^{21} cm^{-3}, corresponding to an LSPR frequency in the mid-IR to NIR region (Fig. 15b). $Cu_{2-x}S$ NCs with a composition of $Cu_{1.93}S$ displayed an LSPR absorption peak at 0.69 eV (corresponding to 1,800 nm). Besides composition, other factors involving media dielectric constant, shape, and oxidizing/reducing agents can also tailor the hole concentration in $Cu_{2-x}E$ (E = S, Se, Te) NCs to affect the LSPR frequencies [76–81, 83–86].

In addition to the copper-deficient chalcogenides, some oxygen-deficient metal oxide materials also exhibit LSPR that arises from the presence of oxygen vacancies. Different from the $Cu_{2-x}E$ (E = S, Se, Te) NCs, in which copper vacancies yield holes in the VB, in nonstoichiometric transition metal oxides, the free charge carriers are the electrons in the CB from the outer-d metal orbitals. TiO_2 as a wide bandgap (3.0–3.2 eV) semiconductor has been widely used in cosmetics, coating, and photocatalysis as well as in dye-sensitized solar cells. Murray et al. reported the colloidal synthesis of blue TiO_2 NCs with TiF_4 as the precursor, different from the traditional TiO_2 with white color [87]. The blue TiO_2 NCs showed an LSPR at mid-IR region ($\lambda_{max} \approx 3{,}400$ nm). However, no LSPR was found in the TiO_2 products prepared by using $TiCl_4$ or $TiBr_4$ precursors. The LSPR in the blue TiO_2 NCs was attributed to oxygen vacancies induced by fluorine doping and the presence of Ti^{3+}, which was confirmed by X-ray photoelectron spectroscopy (XPS) and electron paramagnetic resonance (EPR) measurements.

As the photo- and electrochromic materials, oxygen-deficient WO_3 and MoO_3 can also generate LSPR. Nonstoichiometric $WO_{2.83}$ nanorods (Fig. 15c) were prepared by decomposition of tungsten (V) ethoxide in oleic acid and trioctylamine (TOA) [45]. As shown in Fig. 15d, the as-prepared $WO_{2.83}$ nanorods exhibit a strong LSPR with an absorption peak around 900 nm. Huang et al. prepared MoO_3_x–oleylamine composite nanosheets with 3.3 nm thickness by hydrothermal decomposition of ammonium molybdate in an oil/water emulsion [88]. The composite nanosheets show LSPR ranging from 600 to 1,200 nm. Upon oxidation, the blue

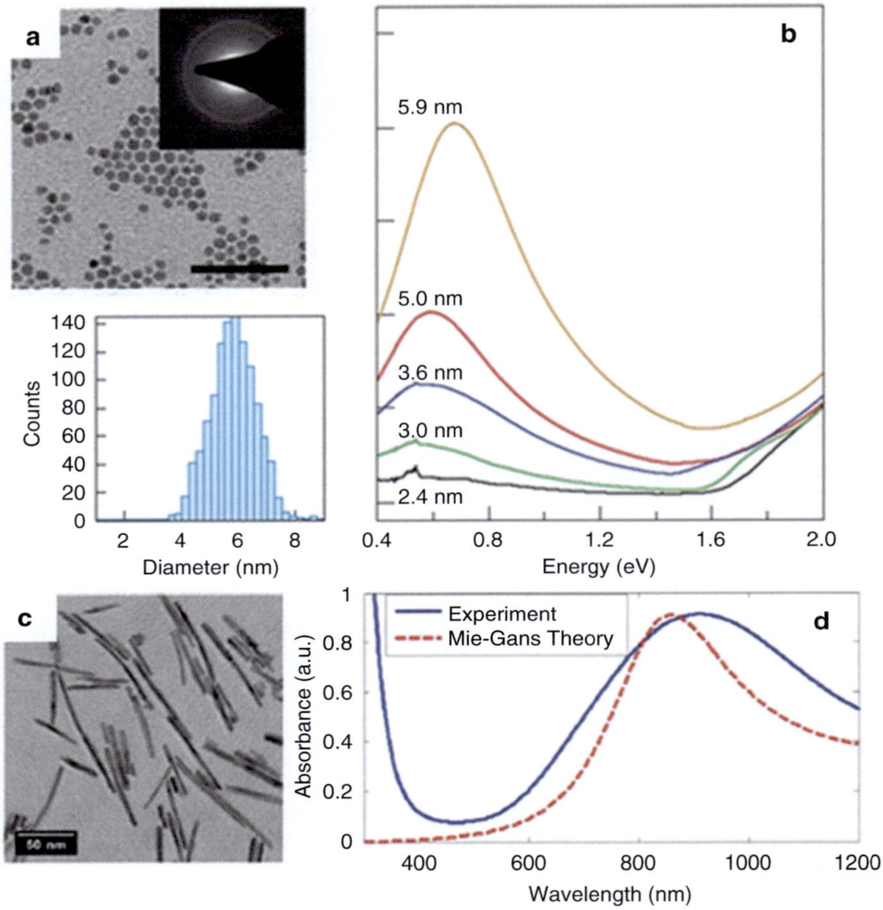

Fig. 15 TEM image (**a**) of the $Cu_{2-x}S$ NCs with average size of 6 nm (scale bar: 50 nm) and size dependence (**b**) of $Cu_{2-x}S$ NCs on their LSPR energy (published with permission of Nature publishing group) TEM image (**c**) and experimental measurement (**d**) of LSPR for $WO_{2.83}$ nanorods (published with permission of Wiley ACS [45])

sheets were converted into MoO_3 nanotubes with white color. In most cases, syntheses of plasmonic semiconductor or metal oxides NCs are resorted to the colloidal routes by using surfactants and capping agents. However, the use of surface-capping agents may lead to unpredictable cytotoxic effects and also lower their surface feasibility, restricting their large-scale applications. Cheng et al. recently reported the surfactant-free nonaqueous approach to the well-defined MoO_{3-x} nanosheets [89]. The MoO_{3-x} nanosheets were prepared by oxidizing metal molybdenum in H_2O_2 followed by solvothermal treatment in ethanol, which displayed strong LSPR at around 680 nm. As shown in Fig. 16, by changing the reacting temperatures and solvents, the LSPR frequency of MoO_{3-x} nanosheets could be tuned from visible (680 nm) to NIR (950 nm) region via tailoring the oxygen vacancies.

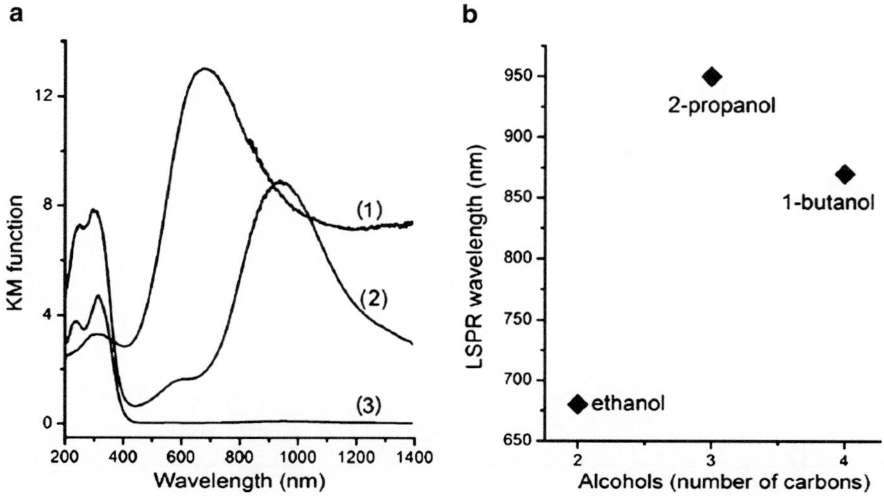

Fig. 16 (**a**) UV–Vis–NIR spectra of the MoO_{3-x} products prepared at 160 °C (*1*) and 140 °C (*2*) for 12 h, and commercial MoO_3 (*3*). (**b**) The dependence of LSPR wavelength of the products on the alcohols used in the solvothermal synthesis (published with permission of Wiley [89])

Due to the near-field enhancement, LSPR is capable of harvesting light at the nanoscale and is found in a wide range of applications as diverse as single-molecule spectroscopy, photothermal therapy, transparent conductive films, etc. [68]. For instance, the requirement of the plasmonic nanomaterials used for biomedical applications (e.g., bio-imaging and photothermal therapy) is LSPR in the NIR region, where tissue and biomolecules show relatively weak scattering and absorbance to allow deep penetration of light in vivo [90]. As expected, plasmonic semiconductors of $Cu_{2-x}E$ (E = S, Se, Te) [91–93] with LSPR in NIR have shown effective and promising application in photoacoustic imaging and cell photothermal therapy.

With respect to the photocatalytic applications, LSPR enables the doped semiconductors and metal oxides to absorb more Vis–NIR light. On the other hand, the plasmonic semiconductors and metal oxides should be efficient for charge separation, which further participate in the oxidation and reduction process. In this context, extrinsically doped semiconductors and metal oxides are not the ideal candidates, in which the impurity sites can easily become the combination centers for the photogenerated electrons and holes. Fortunately, for the self-doped plasmonic semiconductors and metal oxides including copper-deficient chalcogenides $Cu_{2-x}E$ (E = S, Se, Te) and oxygen-deficient transition metal oxides (TiO_2, WO_3, and MoO_3), excellent photocatalytic activities have been discovered in their stoichiometric forms [50, 94–98]. For example, derived from the calcination of the acid-treated $PbWO_4$ and $SrWO_4$ precursors, the hollow WO_3 spheres showed highly efficient visible-light photocatalytic activity toward Rhodamine B degradation and 2-propanol decomposition [94].

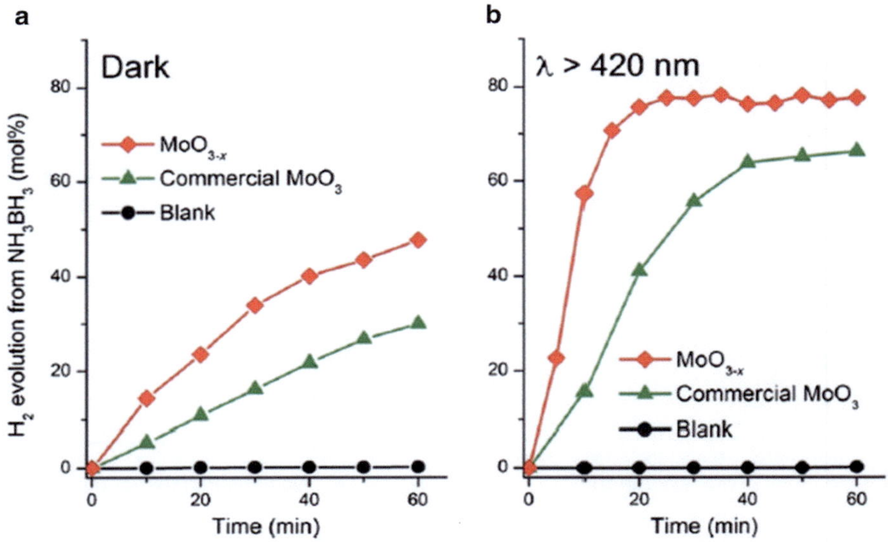

Fig. 17 Time course of H_2 evolution from NH_3BH_3 aqueous solution at room temperature for different samples: (**a**) in the dark and (**b**) under visible-light irradiation (published with permission of Wiley)

One typical example for highly efficient catalysis by plasmonic semiconductors is MoO_{3-x} nanosheets [89]. Ammonia borane (NH_3BH_3), which contains as high as 19.6 wt% of hydrogen and exceeds that of gasoline, is an attractive candidate for the next-generation hydrogen storage materials [80], and with a suitable catalyst, the hydrolysis of NH_3BH_3 would take place and H_2 is thus obtained following (1):

$$NH_3BH_3 + 2H_2O \xrightarrow{\text{catalyst}} NH_4^+ + BO_2^- + 3H_2 \tag{1}$$

Up to now, various noble metal nanostructures (e.g., Ru, Rh, Pt) have been found to show high activity for H_2 evolution from NH_3BH_3 solutions [99]. In terms of practical applications, development of earth-abundant catalysts with low cost and high efficiency is crucially important. Instead of noble metals, plasmonic MoO_{3-x} nanosheets were used as the catalyst for NH_3BH_3 dehydrogenation at room temperature. As shown in Fig. 17, NH_3BH_3 was quite stable without catalysts. In contrast to the dark condition, H_2 production was about fourfold enhanced under visible-light irradiation over the plasmonic MoO_{3-x} nanosheets. The initial H_2 yield rate over MoO_{3-x} nanosheets (5.74 mol % min^{-1}) was 3.7 times faster than that of commercial MoO_3 (1.55 mol % min^{-1}), even faster than the Ag/SBA-15 samples with the best activity (ca. 2.67 mol % min^{-1}) [100] by a factor of 2 in the same circumstances. In addition, the plasmonic MoO_{3-x} showed the structural and catalytic stability during the recycling tests, which demonstrates the potential application of the plasmonic semiconductor nanostructures as efficient and stable catalysts.

Though few researches have been done to date, in terms of the promising highly efficient heterogeneous photocatalysts, self-doped nonstoichiometric semiconductors and metal oxides are ideal candidates. On the one hand, the plasmonic absorption could enhance their light harvesting from the visible to NIR region, which could utilize the abundant solar energy efficiently. On the other hand, the plasmonic resonance can accelerate the deviation or separation of the free carriers on the surface of the semiconductors or metal oxides, thus facilitating the transfer of the charge carriers to participate in the following photocatalytic oxidation/reduction process. Moreover, with the aid of voltage bias, it offers new opportunities to enhance their catalytic properties by the photoelectrocatalysis. In the past two decades, the advent of nanoscience and nanotechnology has provided protocols and guidance to design highly active photocatalysts. With the exponential in-depth studies, it is expected that the non-noble metal plasmonic nanostructures will find more and more green, robust, and efficient photocatalytic applications in energy exploitation as well as environmental protection.

5 Conclusions

Owing to the unique LSPR of noble metal nanoparticles, plasmonic photocatalysts have recently attracted increasing interests due to their intense light absorption and high photocatalytic activities. Over the past few years, considerable efforts have been made on the exploration of highly efficient plasmonic photocatalysts, such as metal–semiconductor hybrids, silver halides, and recently developed nonmetallic plasmonic photocatalysts. Although great efforts have been made, the efficiencies of plasmonic photocatalysts are still too low to meet the requirement for practical applications. Therefore, it is still necessary to further improve the photocatalytic efficiencies for plasmonic photocatalysts. Present investigations are mainly focused on the expansion of the light absorption spectra by adjusting the size, shape, or surrounding medium of nanoparticles. However, the relatively low charge separation efficiency could be the main issue that limits the photocatalytic efficiency for plasmonic photocatalysts. Therefore, to further improve the charge separation efficiencies could be the hot spot in plasmonic photocatalysis. Furthermore, due to the strong LSPR of Au and Ag nanoparticles, present plasmonic photocatalysts have mainly focused on noble metal nanoparticles, which are expensive and in short supply. Nonmetallic plasmonic photocatalysts, such as extrinsic doped or self-doped semiconductors or metal oxides, would be more favorable for practical applications. Due to the unique character of the outer-d valence electrons, the LSPR properties could be easily modulated by varying the compositions, which provides us the possibility of designing new types of plasmonic photocatalysts.

Acknowledgments Parts of work described in Sect. 4 were financially supported by the National Basic Research Program of China (the 973 Program: No. 2013CB632401) and the National Natural Science Foundation of China (no. 21333006, 11374190, and 51021062). Peng Wang thanks the Center for Nanoscale Materials, a US Department of Energy, Office of Science, Office of Basic Energy Sciences user facility under contract no. DE-AC02-06CH11357.

References

1. Sherry LJ, Jin RC, Mirkin CA, Schatz GC, and Van Duyne RP (2006) Localized surface plasmon resonance spectroscopy of single silver triangular nanoprisms. *Nano Lett* **6**, 2060-5.
2. Amendola V, Bakr OM, and Stellacci F (2010) A Study of the Surface Plasmon Resonance of Silver Nanoparticles by the Discrete Dipole Approximation Method: Effect of Shape, Size, Structure, and Assembly. *Plasmonics* **5**, 85-97.
3. Beyene HT, Chakravadhanula VSK, Hanisch C, Elbahri M, Strunskus T, Zaporojtchenko V et al. (2010) Preparation and plasmonic properties of polymer-based composites containing Ag-Au alloy nanoparticles produced by vapor phase co-deposition. *Journal of Materials Science* **45**, 5865-71.
4. Pastoriza-Santos I, Alvarez-Puebla RA, and Liz-Marzan LM (2010) Synthetic Routes and Plasmonic Properties of Noble Metal Nanoplates. *European Journal of Inorganic Chemistry* 4288-97.
5. Gupta MK, Konig T, Near R, Nepal D, Drummy LF, Biswas S et al. (2013) Surface Assembly and Plasmonic Properties in Strongly Coupled Segmented Gold Nanorods. *Small* **9**, 2979-90.
6. Amendola V, Scaramuzza S, Agnoli S, Polizzi S, and Meneghetti M (2014) Strong dependence of surface plasmon resonance and surface enhanced Raman scattering on the composition of Au-Fe nanoalloys. *Nanoscale* **6**, 1423-33.
7. Chan GH, Zhao J, Hicks EM, Schatz GC, and Van Duyne RP (2007) Plasmonic properties of copper nanoparticles fabricated by nanosphere lithography. *Nano Lett* **7**, 1947-52.
8. Jain PK and El-Sayed MA (2008) Noble Metal Nanoparticle Pairs: Effect of Medium for Enhanced Nanosensing. *Nano Lett* **8**, 4347-52.
9. Huang XH, Neretina S, and El-Sayed MA (2009) Gold Nanorods: From Synthesis and Properties to Biological and Biomedical Applications. *Advanced Materials* **21**, 4880-910.
10. Jain PK and El-Sayed MA (2010) Plasmonic coupling in noble metal nanostructures. *Chemical Physics Letters* **487**, 153-64.
11. Tian LM, Chen E, Gandra N, Abbas A, and Singamaneni S (2012) Gold Nanorods as Plasmonic Nanotransducers: Distance-Dependent Refractive Index Sensitivity. *Langmuir* **28**, 17435-42.
12. Verbruggen SW, Keulemans M, Martens JA, and Lenaerts S (2013) Predicting the Surface Plasmon Resonance Wavelength of Gold-Silver Alloy Nanoparticles. *Journal of Physical Chemistry C* **117**, 19142-5.
13. Wang Z, Liu J, and Chen W (2012b) Plasmonic Ag/AgBr nanohybrid: synergistic effect of SPR with photographic sensitivity for enhanced photocatalytic activity and stability. *Dalton Transactions* **41**, 4866-70.
14. Zhang X, Chen YL, Liu R-S, and Tsai DP (2013) Plasmonic photocatalysis. *Reports on Progress in Physics* **76**, 046401.
15. Gordon R, Sinton D, Kavanagh KL, and Brolo AG (2008) A new generation of sensors based on extraordinary optical transmission. *Accounts of chemical research* **41**, 1049-57.
16. Wang P, Huang BB, Dai Y, and Whangbo MH (2012a) Plasmonic photocatalysts: harvesting visible light with noble metal nanoparticles. *Physical Chemistry Chemical Physics* **14**, 9813-25.
17. Atwater HA and Polman A (2010) Plasmonics for improved photovoltaic devices. *Nat Mater* **9**, 205-13.
18. Barnes WL, Dereux A, and Ebbesen TW (2003) Surface plasmon subwavelength optics. *Nature* **424**, 824-30.
19. Jain P, Huang X, El-Sayed I, and El-Sayed M (2007) Review of Some Interesting Surface Plasmon Resonance-enhanced Properties of Noble Metal Nanoparticles and Their Applications to Biosystems. *Plasmonics* **2**, 107-18.
20. Tian Y and Tatsuma T (2005a) Mechanisms and applications of plasmon-induced charge separation at TiO2 films loaded with gold nanoparticles. *J Am Chem Soc* **127**, 7632-7.

21. Yu K, Tian Y, and Tatsuma T (2006) Size effects of gold nanoparticles on plasmon-induced photocurrents of gold-TiO2 nanocomposites. *Physical Chemistry Chemical Physics* **8**, 5417-20.
22. Furube A, Du L, Hara K, Katoh R, and Tachiya M (2007) Ultrafast plasmon-induced electron transfer from gold nanodots into TiO2 nanoparticles. *J Am Chem Soc* **129**, 14852-3.
23. Kowalska E, Remita H, Colbeau-Justin C, Hupka J, and Belloni J (2008) Modification of Titanium Dioxide with Platinum Ions and Clusters: Application in Photocatalysis. *The Journal of Physical Chemistry C* **112**, 1124-31.
24. Wu XF, Song HY, Yoon JM, Yu YT, and Chen YF (2009) Synthesis of core-shell Au@TiO2 nanoparticles with truncated wedge-shaped morphology and their photocatalytic properties. *Langmuir* **25**, 6438-47.
25. Linic S, Christopher P, and Ingram DB (2011) Plasmonic-metal nanostructures for efficient conversion of solar to chemical energy. *Nat Mater* **10**, 911-21.
26. Ingram DB and Linic S (2011) Water splitting on composite plasmonic-metal/semiconductor photoelectrodes: evidence for selective plasmon-induced formation of charge carriers near the semiconductor surface. *Journal of the American Chemical Society* **133**, 5202-5.
27. Cushing SK, Li J, Meng F, Senty TR, Suri S, Zhi M et al. (2012) Photocatalytic activity enhanced by plasmonic resonant energy transfer from metal to semiconductor. *Journal of the American Chemical Society* **134**, 15033-41.
28. Awazu K, Fujimaki M, Rockstuhl C, Tominaga J, Murakami H, Ohki Y et al. (2008) A plasmonic photocatalyst consisting of silver nanoparticles embedded in titanium dioxide. *J Am Chem Soc* **130**, 1676-80.
29. Chen X, Zhu HY, Zhao JC, Zheng ZF, and Gao XP (2008) Visible-Light-Driven Oxidation of Organic Contaminants in Air with Gold Nanoparticle Catalysts on Oxide Supports. *Angewandte Chemie* **120**, 5433-6.
30. Wang P, Huang BB, Qin XY, Zhang XY, Dai Y, Wei JY et al. (2008) Ag@AgCl: A highly efficient and stable photocatalyst active under visible light. *Angewandte Chemie-International Edition* **47**, 7931-3.
31. Sun S, Wang W, Zhang L, Shang M, and Wang L (2009) Ag@ C core/shell nanocomposite as a highly efficient plasmonic photocatalyst. *Catalysis Communications* **11**, 290-3.
32. Chen J-J, Wu JC, Wu PC, and Tsai DP (2010) Plasmonic photocatalyst for H2 evolution in photocatalytic water splitting. *The Journal of Physical Chemistry C* **115**, 210-6.
33. Li R, Chen W, Kobayashi H, and Ma C (2010) Platinum-nanoparticle-loaded bismuth oxide: an efficient plasmonic photocatalyst active under visible light. *Green Chemistry* **12**, 212-5.
34. Naya S-i, Inoue A, and Tada H (2010) Self-assembled heterosupramolecular visible light photocatalyst consisting of gold nanoparticle-loaded titanium (IV) dioxide and surfactant. *Journal of the American Chemical Society* **132**, 6292-3.
35. Wang P, Huang BB, Lou ZZ, Zhang XY, Qin XY, Dai Y et al. (2010) Synthesis of Highly Efficient Ag@AgCl Plasmonic Photocatalysts with Various Structures. *Chemistry-a European Journal* **16**, 538-44.
36. Hou W, Hung WH, Pavaskar P, Goeppert A, Aykol M, and Cronin SB (2011a) Photocatalytic conversion of CO2 to hydrocarbon fuels via plasmon-enhanced absorption and metallic interband transitions. *ACS Catalysis* **1**, 929-36.
37. Ingram DB, Christopher P, Bauer JL, and Linic S (2011) Predictive model for the design of plasmonic metal/semiconductor composite photocatalysts. *ACS Catalysis* **1**, 1441-7.
38. Primo A, Marino T, Corma A, Molinari R, and Garcia H (2011) Efficient visible-light photocatalytic water splitting by minute amounts of gold supported on nanoparticulate CeO2 obtained by a biopolymer templating method. *Journal of the American Chemical Society* **133**, 6930-3.
39. Wang P, Huang B, Zhang X, Qin X, Dai Y, Wang Z et al. (2011) Highly Efficient Visible Light Plasmonic Photocatalysts Ag@Ag(Cl,Br) and Ag@AgCl-AgI. *ChemCatChem* **3**, 360-4.
40. Zhai W, Xue S, Zhu A, Luo Y, and Tian Y (2011) Plasmon-Driven Selective Oxidation of Aromatic Alcohols to Aldehydes in Water with Recyclable Pt/TiO2 Nanocomposites. *ChemCatChem* **3**, 127-30.

41. Zhang P, Shao C, Zhang Z, Zhang M, Mu J, Guo Z et al. (2011a) Core/shell nanofibers of TiO2@ carbon embedded by Ag nanoparticles with enhanced visible photocatalytic activity. *Journal of Materials Chemistry* **21**, 17746-53.
42. Zhang Q, Lima DQ, Lee I, Zaera F, Chi M, and Yin Y (2011b) A Highly Active Titanium Dioxide Based Visible-Light Photocatalyst with Nonmetal Doping and Plasmonic Metal Decoration. *Angewandte Chemie* **123**, 7226-30.
43. Zheng Z, Huang B, Qin X, Zhang X, Dai Y, and Whangbo M-H (2011) Facile in situ synthesis of visible-light plasmonic photocatalysts M@ TiO2 (M=Au, Pt, Ag) and evaluation of their photocatalytic oxidation of benzene to phenol. *Journal of Materials Chemistry* **21**, 9079-87.
44. Zhao Y, Pan H, Lou Y, Qiu X, Zhu J, and Burda C (2009) Plasmonic Cu2− x S Nanocrystals: Optical and Structural Properties of Copper-Deficient Copper (I) Sulfides. *Journal of the American Chemical Society* **131**, 4253-61.
45. Manthiram K and Alivisatos AP (2012) Tunable localized surface plasmon resonances in tungsten oxide nanocrystals. *Journal of the American Chemical Society* **134**, 3995-8.
46. Kumar MK, Krishnamoorthy S, Tan LK, Chiam SY, Tripathy S, and Gao H (2011) Field effects in plasmonic photocatalyst by precise SiO2 thickness control using atomic layer deposition. *ACS Catalysis* **1**, 300-8.
47. Tian Y and Tatsuma T (2005b) Mechanisms and applications of plasmon-induced charge separation at TiO2 films loaded with gold nanoparticles. *Journal of the American Chemical Society* **127**, 7632-7.
48. Gomes Silva Cu, Juárez R, Marino T, Molinari R, and García H (2010) Influence of excitation wavelength (UV or visible light) on the photocatalytic activity of titania containing gold nanoparticles for the generation of hydrogen or oxygen from water. *Journal of the American Chemical Society* **133**, 595-602.
49. Thimsen E, Le Formal F, Grätzel M, and Warren SC (2010) Influence of plasmonic Au nanoparticles on the photoactivity of Fe2O3 electrodes for water splitting. *Nano letters* **11**, 35-43.
50. Tanaka A, Hashimoto K, and Kominami H (2014) Visible light-induced hydrogen and oxygen formation over Pt/Au/WO3 photocatalyst utilizing two types of photoabsorption due to surface plasmon resonance and band-gap excitation. *Journal of the American Chemical Society* **136**, 586-9.
51. Liu L, Ouyang S, and Ye J (2013a) Gold-Nanorod-Photosensitized Titanium Dioxide with Wide-Range Visible-Light Harvesting Based on Localized Surface Plasmon Resonance. *Angewandte Chemie* **125**, 6821-5.
52. Pu Y-C, Wang G, Chang K-D, Ling Y, Lin Y-K, Fitzmorris BC et al. (2013) Au Nanostructure-Decorated TiO2 Nanowires Exhibiting Photoactivity Across Entire UV-visible Region for Photoelectrochemical Water Splitting. *Nano letters* **13**, 3817-23.
53. Schürch D, Currao A, Sarkar S, Hodes G, and Calzaferri G (2002) The Silver Chloride Photoanode in Photoelectrochemical Water Splitting. *The Journal of Physical Chemistry B* **106**, 12764-75.
54. Kakuta N, Goto N, Ohkita H, and Mizushima T (1999) Silver Bromide as a Photocatalyst for Hydrogen Generation from CH3OH/H2O Solution. *The Journal of Physical Chemistry B* **103**, 5917-9.
55. An C, Peng S, and Sun Y (2010) Facile Synthesis of Sunlight-Driven AgCl:Ag Plasmonic Nanophotocatalyst. *Advanced Materials* **22**, 2570-4.
56. Lou Z, Huang B, Qin X, Zhang X, Wang Z, Zheng Z et al. (2011a) One-step synthesis of AgBr microcrystals with different morphologies by ILs-assisted hydrothermal method. *CrystEngComm* **13**, 1789-93.
57. Lou Z, Huang B, Wang P, Wang Z, Qin X, Zhang X et al. (2011b) The synthesis of the near-spherical AgCl crystal for visible light photocatalytic applications. *Dalton Transactions* **40**, 4104-10.
58. Wang P, Huang B, Zhang X, Qin X, Jin H, Dai Y et al. (2009b) Highly Efficient Visible-Light Plasmonic Photocatalyst Ag@ AgBr. *Chemistry-A European Journal* **15**, 1821-4.

59. Huang H, Li X, Kang Z, Liu Y, Li H, He X et al. (2010) Tuning metal@metal salt photocatalytic abilities by different charged anions. *Dalton Transactions* **39**, 10593-7.
60. Seki K, Yanagi H, Kobayashi Y, Ohta T, and Tani T (1994) UV photoemission study of dye/AgBr interfaces in relation to spectral sensitization. *Phys Rev B Condens Matter* **49**, 2760-7.
61. Wang P, Huang B, Qin X, Zhang X, Dai Y, and Whangbo M-H (2009a) Ag/AgBr/WO3·H2O: Visible-Light Photocatalyst for Bacteria Destruction. *Inorganic Chemistry* **48**, 10697-702.
62. Yu J, Dai G, and Huang B (2009) Fabrication and Characterization of Visible-Light-Driven Plasmonic Photocatalyst Ag/AgCl/TiO2 Nanotube Arrays. *The Journal of Physical Chemistry C* **113**, 16394-401.
63. Hou Y, Li X, Zhao Q, Quan X, and Chen G (2011b) TiO2 nanotube/Ag-AgBr three-component nanojunction for efficient photoconversion. *Journal of Materials Chemistry* **21**, 18067-76.
64. Song L, Yang Y, Zhang Q, Tian H, and Zhu W (2011) Synthesis and photochromism of naphthopyrans bearing naphthalimide chromophore: predominant thermal reversibility in color-fading and fluorescence switch. *J Phys Chem B* **115**, 14648-58.
65. Cheng H, Huang B, Wang P, Wang Z, Lou Z, Wang J et al. (2011) In situ ion exchange synthesis of the novel Ag/AgBr/BiOBr hybrid with highly efficient decontamination of pollutants. *Chemical Communications* **47**, 7054-6.
66. Hu C, Peng T, Hu X, Nie Y, Zhou X, Qu J et al. (2009) Plasmon-Induced Photodegradation of Toxic Pollutants with Ag–AgI/Al2O3 under Visible-Light Irradiation. *Journal of the American Chemical Society* **132**, 857-62.
67. Hu X, Hu C, Peng T, Zhou X, and Qu J (2010) Plasmon-Induced Inactivation of Enteric Pathogenic Microorganisms with Ag–AgI/Al2O3 under Visible-Light Irradiation. *Environmental Science & Technology* **44**, 7058-62.
68. Luther JM, Jain PK, Ewers T, and Alivisatos AP (2011) Localized surface plasmon resonances arising from free carriers in doped quantum dots. *Nature materials* **10**, 361-6.
69. Kanehara M, Koike H, Yoshinaga T, and Teranishi T (2009) Indium tin oxide nanoparticles with compositionally tunable surface plasmon resonance frequencies in the near-IR region. *Journal of the American Chemical Society* **131**, 17736-7.
70. Garcia G, Buonsanti R, Runnerstrom EL, Mendelsberg RJ, Llordes A, Anders A et al. (2011) Dynamically modulating the surface plasmon resonance of doped semiconductor nanocrystals. *Nano letters* **11**, 4415-20.
71. Buonsanti R, Llordes A, Aloni S, Helms BA, and Milliron DJ (2011) Tunable infrared absorption and visible transparency of colloidal aluminum-doped zinc oxide nanocrystals. *Nano letters* **11**, 4706-10.
72. Gordon TR, Paik T, Klein DR, Naik GV, Caglayan H, Boltasseva A et al. (2013) Shape-Dependent Plasmonic Response and Directed Self-Assembly in a New Semiconductor Building Block, Indium-Doped Cadmium Oxide (ICO). *Nano letters* **13**, 2857-63.
73. Xu J, Li L, Wang S, Ding H, Zhang Y, and Li G (2013) Influence of Sb doping on the structural and optical properties of tin oxide nanocrystals. *CrystEngComm* **15**, 3296-300.
74. Chou L-W, Shin N, Sivaram SV, and Filler MA (2012) Tunable mid-infrared localized surface plasmon resonances in silicon nanowires. *Journal of the American Chemical Society* **134**, 16155-8.
75. Chou LW and Filler MA (2013) Engineering Multimodal Localized Surface Plasmon Resonances in Silicon Nanowires. *Angewandte Chemie International Edition* **52**, 8079-83.
76. Dorfs D, Härtling T, Miszta K, Bigall NC, Kim MR, Genovese A et al. (2011) Reversible Tunability of the Near-Infrared Valence Band Plasmon Resonance in Cu2–x Se Nanocrystals. *Journal of the American Chemical Society* **133**, 11175-80.
77. Hsu S-W, On K, and Tao AR (2011) Localized surface plasmon resonances of anisotropic semiconductor nanocrystals. *Journal of the American Chemical Society* **133**, 19072-5.
78. Hsu S-W, Bryks W, and Tao AR (2012) Effects of Carrier Density and Shape on the Localized Surface Plasmon Resonances of Cu2–x S Nanodisks. *Chemistry of Materials* **24**, 3765-71.
79. Kriegel I, Jiang C, Rodríguez-Fernández J, Schaller RD, Talapin DV, Da Como E et al. (2012) Tuning the excitonic and plasmonic properties of copper chalcogenide nanocrystals. *Journal of the American Chemical Society* **134**, 1583-90.

80. Zhao Y and Burda C (2012) Development of plasmonic semiconductor nanomaterials with copper chalcogenides for a future with sustainable energy materials. *Energy & Environmental Science* **5**, 5564-76.
81. Jain PK, Manthiram K, Engel JH, White SL, Faucheaux JA, and Alivisatos AP (2013) Doped Nanocrystals as Plasmonic Probes of Redox Chemistry. *Angewandte Chemie International Edition* **52**, 13671-5.
82. Li W, Zamani R, Rivera Gil P, Pelaz B, Ibáñez M, Cadavid D et al. (2013) CuTe Nanocrystals: Shape and Size Control, Plasmonic Properties, and Use as SERS Probes and Photothermal Agents. *Journal of the American Chemical Society* **135**, 7098-101.
83. Liu X, Lee C, Law W-C, Zhu D, Liu M, Jeon M et al. (2013c) Au–Cu2–x Se Heterodimer Nanoparticles with Broad Localized Surface Plasmon Resonance as Contrast Agents for Deep Tissue Imaging. *Nano letters* **13**, 4333-9.
84. Liu X, Wang X, Zhou B, Law WC, Cartwright AN, and Swihart MT (2013d) Size-Controlled Synthesis of Cu2-xE (E=S, Se) Nanocrystals with Strong Tunable Near-Infrared Localized Surface Plasmon Resonance and High Conductivity in Thin Films. *Advanced Functional Materials* **23**, 1256-64.
85. Xie Y, Carbone L, Nobile C, Grillo V, D'Agostino S, Della Sala F et al. (2013a) Metallic-like Stoichiometric Copper Sulfide Nanocrystals: Phase-and Shape-Selective Synthesis, Near-Infrared Surface Plasmon Resonance Properties, and Their Modeling. *ACS nano* **7**, 7352-69.
86. Xie Y, Riedinger A, Prato M, Casu A, Genovese A, Guardia P et al. (2013b) Copper Sulfide Nanocrystals with Tunable Composition by Reduction of Covellite Nanocrystals with Cu+Ions. *Journal of the American Chemical Society* **135**, 17630-7.
87. Gordon TR, Cargnello M, Paik T, Mangolini F, Weber RT, Fornasiero P et al. (2012) Nonaqueous synthesis of TiO2 nanocrystals using TiF4 to engineer morphology, oxygen vacancy concentration, and photocatalytic activity. *Journal of the American Chemical Society* **134**, 6751-61.
88. Huang Q, Hu S, Zhuang J, and Wang X (2012) MoO3–x-Based Hybrids with Tunable Localized Surface Plasmon Resonances: Chemical Oxidation Driving Transformation from Ultrathin Nanosheets to Nanotubes. *Chemistry-A European Journal* **18**, 15283-7.
89. Cheng H, Kamegawa T, Mori K, and Yamashita H (2014) Surfactant-Free Nonaqueous Synthesis of Plasmonic Molybdenum Oxide Nanosheets with Enhanced Catalytic Activity for Hydrogen Generation from Ammonia Borane under Visible Light. *Angewandte Chemie International Edition*
90. Liu X and Swihart MT (2014) Heavily-doped colloidal semiconductor and metal oxide nanocrystals: an emerging new class of plasmonic nanomaterials. *Chemical Society Reviews*
91. Tian Q, Jiang F, Zou R, Liu Q, Chen Z, Zhu M et al. (2011) Hydrophilic Cu9S5 nanocrystals: A photothermal agent with a 25.7% heat conversion efficiency for photothermal ablation of cancer cells in vivo. *ACS nano* **5**, 9761-71.
92. Ku G, Zhou M, Song S, Huang Q, Hazle J, and Li C (2012) Copper sulfide nanoparticles as a new class of photoacoustic contrast agent for deep tissue imaging at 1064 nm. *Acs Nano* **6**, 7489-96.
93. Liu X, Law WC, Jeon M, Wang X, Liu M, Kim C et al. (2013b) Cu2–xSe Nanocrystals with Localized Surface Plasmon Resonance as Sensitive Contrast Agents for In Vivo Photoacoustic Imaging: Demonstration of Sentinel Lymph Node Mapping. *Advanced healthcare materials* **2**, 952-7.
94. Chen D and Ye J (2008) Hierarchical WO3 hollow shells: dendrite, sphere, dumbbell, and their photocatalytic properties. *Advanced Functional Materials* **18**, 1922-8.
95. Yang HG, Sun CH, Qiao SZ, Zou J, Liu G, Smith SC et al. (2008) Anatase TiO2 single crystals with a large percentage of reactive facets. *Nature* **453**, 638-41.
96. Kim Y, Park KY, Jang DM, Song YM, Kim HS, Cho YJ et al. (2010) Synthesis of Au–Cu2S Core–Shell Nanocrystals and Their Photocatalytic and Electrocatalytic Activity. *The Journal of Physical Chemistry C* **114**, 22141-6.
97. Chen X, Liu L, Peter YY, and Mao SS (2011) Increasing solar absorption for photocatalysis with black hydrogenated titanium dioxide nanocrystals. *Science* **331**, 746-50.

98. Liu Y, Deng Y, Sun Z, Wei J, Zheng G, Asiri AM et al. (2013e) Hierarchical Cu2S Microsponges Constructed from Nanosheets for Efficient Photocatalysis. *Small* **9**, 2702-8.
99. Yadav M and Xu Q (2012) Liquid-phase chemical hydrogen storage materials. *Energy & Environmental Science* **5**, 9698-725.
100. Fuku K, Hayashi R, Takakura S, Kamegawa T, Mori K, and Yamashita H (2013) The Synthesis of Size-and Color-Controlled Silver Nanoparticles by Using Microwave Heating and their Enhanced Catalytic Activity by Localized Surface Plasmon Resonance. *Angewandte Chemie International Edition* **52**, 7446-50.

Soft X-Ray Spectroscopy and Electronic Structure of 3d Transition Metal Compounds in Artificial Photosynthesis Materials

Mukes Kapilashrami, Yi-Sheng Liu, Per-Anders Glans, and Jinghua Guo

Abstract The surge to address the most pressing energy-related concerns has brought together researchers from different disciplines to common platforms of joint forces, which has led to significant advancements in the field materials science and in the development of new and improved analytical techniques for in situ studies of matter at atomic level with high precision. This chapter presents an overview on the role of transition metal elements in the field of energy research, and their contribution towards realization of efficient and sustainable photoelectric materials. In addition, this section also outlines the importance (as well as recent development) of synchrotron based X-ray spectroscopy technique (in the soft X-ray range) in the evaluation of transition metal complexes at nanoscale by reflecting over recent studies on the local *chemical environment*, *charge symmetry*, and *charge transfer* therein as a function of e.g. *particle size* (in nanoparticles) and/or *impurity level* (in doped systems).

Keywords Artificial photosynthesis • Conduction band minimum • Valence band maximum • Transition metals • Synchrotron radiation • Soft X-ray absorption spectroscopy • Soft X-ray emission spectroscopy • In situ

1 Introduction

Fossil fuels account for ca. 80 % of the world's energy supply, but given the fact that their reserves are not infinite, and considering the harmful by-products produced from combustion of the same (that contributes towards global warming and other air pollutants that are injurious to the natural environment and human health), they are not viable in the long run. Instead, we need to find sustainable and cost-effective technologies that address our energy-related concerns and the associated global climate challenges [1–4].

M. Kapilashrami
Center For Engineering Concepts Development, Department of Mechanical Engineering, University of Maryland, College Park, MD 20742, USA

Advanced Light Source, Lawrence Berkeley National Laboratory, Berkeley, CA 94720, USA

Y.-S. Liu • P.-A. Glans • J. Guo (✉)
Advanced Light Source, Lawrence Berkeley National Laboratory, Berkeley, CA 94720, USA
e-mail: jguo@lbl.gov

E.A. Rozhkova, K. Ariga (eds.), *From Molecules to Materials*,
DOI 10.1007/978-3-319-13800-8_10

Solar power, which is an abundant source of energy, provides us more energy every hour than what we consume p.a., and, moreover, it does not produce any non-environmental-friendly pollutants.

To add additional clarity, the rate at which we consume energy is today ca. 4×10^{20} J p.a., equivalent to ca. 13 TW p.a., while the energy we receive from the sun exceeds 110,000 TW p.a. (i.e., $> 4 \times 10^{20}$ J/h) [1]. Seemingly, harvesting of solar energy and its translation to any other means of applicable energy that is *eco-friendly*, *cost efficient*, *sustainable*, and *contributing towards decomposition of air- and waterborne pollutants* is desirable and is expected to hold great promise even if it cannot be used in its full capacity.

The solar radiation spectrum range from ca. 10^2 to 10^6 nm, i.e., from ultraviolet - (UV) to infrared (IR) wavelength range, wherein maximum radiation is emitted in the visible wavelength range (ca. 400–700 nm), as illustrated in Fig. 1 [5]. Clearly, for far most efficient utilization of solar power, it is highly desirable to operate in the visible wavelength range (corresponding to ca. 1.8–3.0 eV).

Photocatalysis is a technique that offers a green and cost-efficient path for, e.g., *production of solar-hydrogen* and/or *decomposition of air- and waterborne pollutants* by decomposition of a reactant (e.g., H_2O and/or CO_2) in the presence of a photosensitive catalyst fueled by solar energy [6–13].

For instance, hydrogen gas is the key ingredient in the so-called zero-emission fuel cells and is typically ensued from electrocatalytic reactions of hydrocarbons [14–19]. The ability to produce solar-hydrogen opens up new avenues for effective utilization of hydrogen-based fuel cells since it enables the production of hydrogen gas at larger scale, higher yield, and low production cost. Moreover, it excludes the production of carbon emission to the atmosphere, as otherwise in the case of electrocatalytic reactions of hydrocarbons.

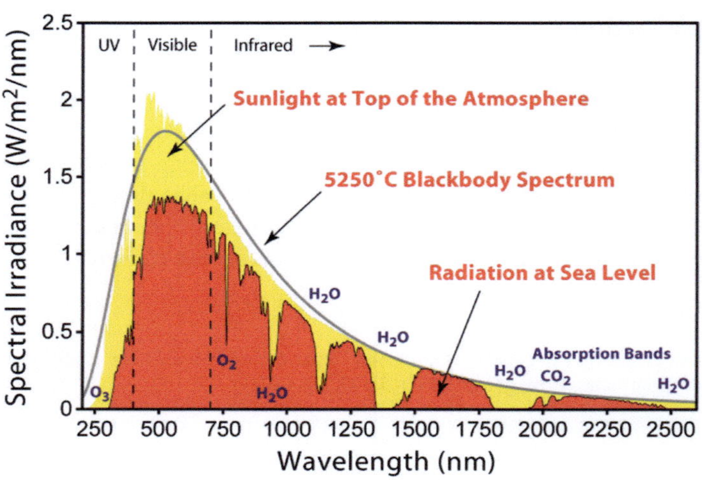

Fig. 1 Solar radiation spectrum featuring the variation in the intensity in the solar irradiation at different wavelengths [5]

1.1 Principles of Photocatalysis

Photocatalysis, which is the light-induced acceleration of a reaction in the presence of a photosensitive catalyst (a photosensitive semiconductor), is often compared to the way plants use chlorophyll to convert water and carbon dioxide into oxygen and glucose fueled by sunlight (thereby often referred as to *artificial photosynthesis*) [6]. The typical photocatalytic reaction is recognized by the following steps: (1) absorption of light (photons) with higher energy than the band gap of the catalyst, upon which (2) electrons in the occupied electronic states (valence band, VB) are excited into the unoccupied electronic states (conduction band, CB) forming electron–hole pairs, and (3) decomposition of a reactant by its reduction and oxidation close to the energy band edges of the catalyst via charge transfer between the two, Fig. 2 [6, 7].

Since photocatalysis is a surface property that relies on the effective charge transfer between a catalyst and a reactant, it requires careful alignment of the relative positions of the energy band edges in the catalyst with respect to the redox potential of the reactant. For instance, for production of water hydrogen, the position of the conduction band minimum (CBM) of the catalyst needs to be above the redox potential for H_2/H_2O, while its valence band maximum (VBM) needs to be below the redox potential for O_2/H_2O. Given the normal hydrogen electrode (NHE), the band gap of the catalyst needs to be close to/above 1.23 eV (Figs. 2 and 3) [27–29]. A big challenge often experienced in any photoelectric application is the event of recombination of the photoexcited electrons (in the conduction band) with their associated

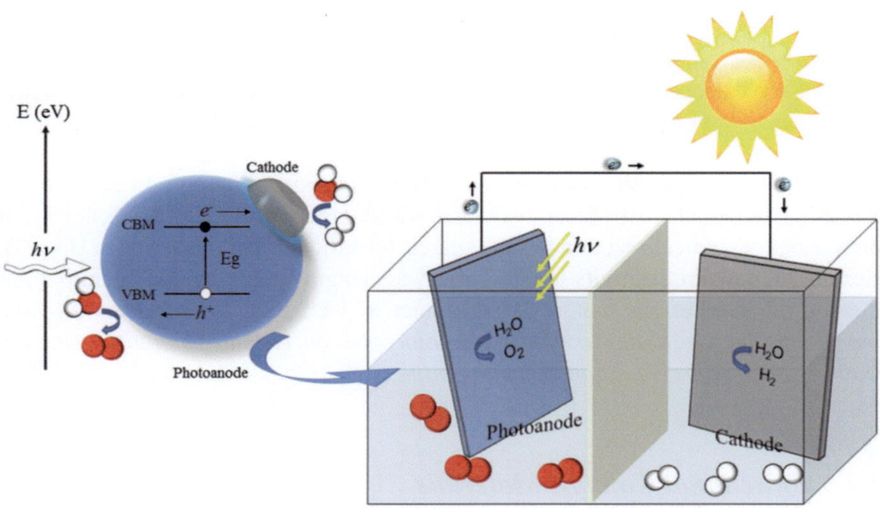

Fig. 2 The typical *photoelectrochemical cell* (PEC) comprised of a photosensitive anode (e.g., Fe_2O_3, TiO_2, ZnO, or GaN) and a counter electrode (typically Pt or Rh), for the production of water hydrogen by decomposition of water molecules [20–26]

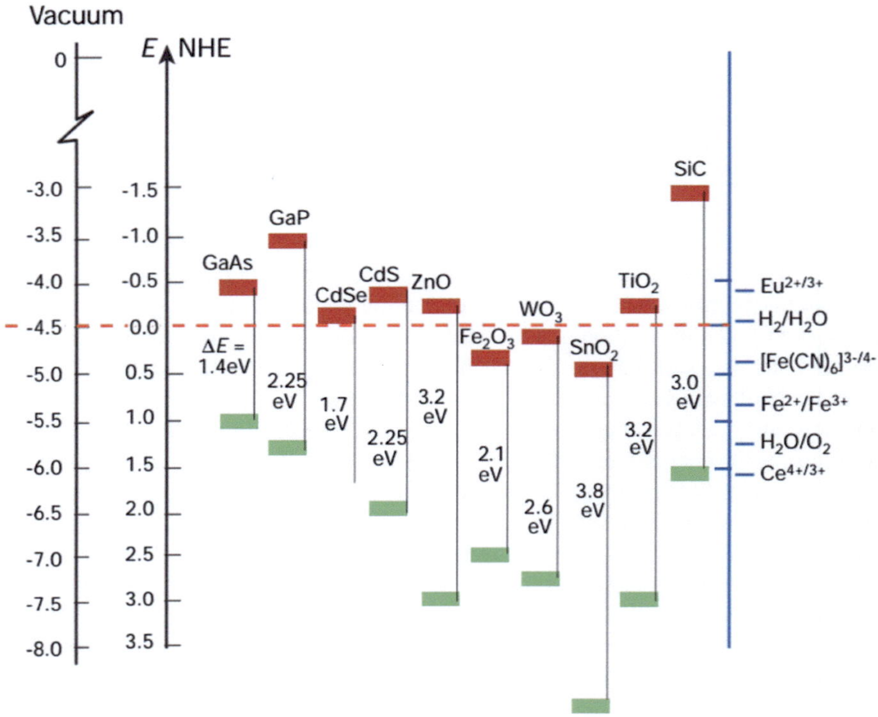

Fig. 3 Variation in the energy band positions of few selected semiconductors in contact with aqueous electrolyte at pH 1 with respect to normal hydrogen electrode [27]

holes (in the valence band) that lowers the overall photoelectric effect. Details on the different pathways developed to improve the electron-hole pair longevity is discussed in Sec. 1.2.

A wide range of semiconductor materials (with band gap ranging from ca. 1.4 to 3.8 eV) have extensively been evaluated and compared for their applicability in photoelectrochemical reactions in the search for efficient photoelectrodes (Fig. 3). Besides high photosensitivity in the visible wavelength range (which is an important factor), other criterions that need to be met as well include: *low electron–hole recombination rate*, *ability to withstand photocorrosion*, *chemical stability, nontoxic, stability in aqueous conditions*, and *tunability of electronic properties* [7, 28–34].

1.2 Functional Catalyst Materials

Reflecting over the myriad of available elements in the periodic table (Fig. 4), the transition metals (Co, Ni, Pd, Ti, Pt, Fe, and Mn) and in particular the intrinsic and transition metal-doped complexes of TiO_2 and Fe_2O_3 are to date among the

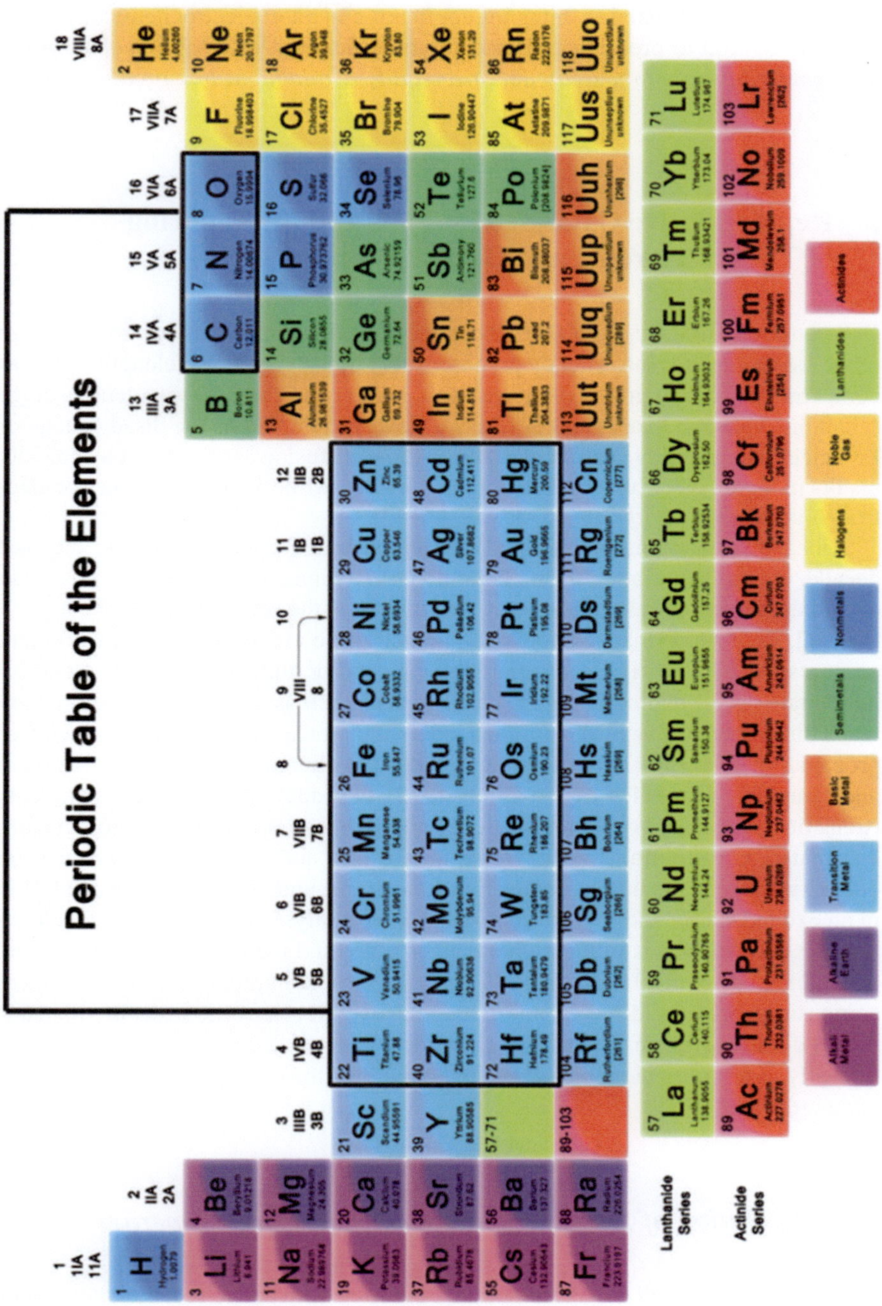

Fig. 4 The highlighted elements in the periodic table represent the most commonly occurring, and/or most promising materials applicable in light – matter (e.g. PVs and LEDs) related applications [35]

most widely investigated systems for application in photoelectrochemical cells as they, in addition to promising photocatalytic efficiency, offer *chemical stability, bio-compatibility, zero toxicity and are readily available at low production cost* [28–32, 36–44].

2 Representative Catalyst Materials and the Tailoring of Their Electronic and Optical Properties

Different from TiO_2, Fe_2O_3 has the advantage of having a relatively narrow band gap (ca. 2.1 eV), i.e., it is intrinsically photosensitive in the visible wavelength range, as compared to TiO_2 which is photosensitive maninly in the UV wavelength range (given its wide band gap, ca. 3.0-3.2 eV). However, Fe_2O_3 has the disadvantage of low electrical mobility and short electron–hole diffusion length (which typically results in higher electron–hole recombination rate), while TiO_2 offers superior electron–hole pair longevity [28, 36, 37, 39, 45–47].

Albeit this emerging technology has demonstrated great progress in the past decade, its transformation from laboratory to commercial platforms has been decelerated since current efficiencies are still below the set benchmark of ca. 10 % [48–51]. Poor performance/photocatalytic activity commonly root in insufficient photoresponse in the visible wavelength range and/or high electron–hole recombination rate (as the photoelectrons recombine with their associated holes before they reach the surface). Typical approach often undertaken to alter/improve charge propagation efficiency and direction includes (1) alteration of the morphological and structural properties of the catalyst; (2) introduction of impurity elements (transition metals and/or nonmetals, e.g., nitrogen, carbon, fluorine) into the catalyst matrix (substituting for either the host cation and-/or anion to enhance the disparity between the number of available electrons, and holes when the catalyst is in an excited state); and (3) introduction of a co-catalyst (as indicated in Fig. 1) to trap the photoelectrons in the higher energy state (and thereby extend the lifetime of the excited state) [31, 36, 52–59]. Nevertheless, manipulation of the chemical composition provides direct means to alter the overall electronic structure (depending on doping level) of the catalyst since local charge *-distribution* and *-symmetry* typically varies with the ionic *-state*, and *-radius* of the constituent elements in the lattice [60].

2.1 Alteration of Structural and Morphological Properties

Structural disorder and defects are often associated with active recombination sites (located at the grain boundaries) that obstruct the propagation of the photoelectrons to the surface. Controlling the structural properties, such as crystallinity and/or porosity, reduces the overall concentration of grain boundaries and active charge trapping sites therein [61]. Another important feature to consider is the *surface-to-volume ratio* that relates to the relative distance photoelectrons need to travel before they reach the surface. Along this line, it is important to remember that at reduced dimensions (bulk → few clusters of atoms), the energy levels in matter become more discrete and well defined; hence, by controlling, e.g., the particle/cluster size (in the case of nanoparticles), one can directly alter the electronic/band properties and the various possible pathways a photoelectron may propagate for efficient charge transfer. Besides, with increasing *surface-to-volume ratio*, the impurity elements reside closer to the surface instead of being embedded in the bulk and thereby participate actively in the photoelectric reactions [62–66].

2.2 Alteration of the Electronic Properties by Introduction of Transition Metal Impurity Elements

Transition metal doping of wide band gap semiconductors, substituting for the host cations, introduces localized energy states within the band gap that position either closely above the VBM or below the CBM of the host matrix [67–71]. In the case of TiO_2, wherein the CBM and VBM are mainly comprised of Ti 3*d* and O 2*p*, respectively, these impurity states contribute towards the alteration of the optical properties of the oxide by extending either the *valence band* or the *conduction band region* closer to the Fermi level via (1) cation (impurity)–cation (host) $d \rightarrow d$ transitions (close to the CBM) or (2) anion (host)–cation (impurity) $p \rightarrow d$ transitions (close to the VBM) [67, 72, 73]. To further understand the correlation between transition metal doping and alteration of the optical properties, let us briefly remind ourselves of the principles of crystal field theory.

The crystal field theory details the splitting of the 3*d* orbitals ($3d \rightarrow t_{2g}$ and e_g) in transition metals that occurs in the presence of an electrostatic field arising from repulsive interaction between a transition metal ion and surrounding negative ions/molecules (ligands). Compared to the t_{2g} orbitals, the e_g orbitals experience stronger repulsive forces towards the ligands as the lobes of the e_g orbitals point towards the ligands (while the t_{2g} orbitals point between the ligands). This leads to an elevation in the energy of the e_g with respect to the t_{2g} by Δ_0 (commonly referred to as the crystal field splitting energy) [74, 75]. For instance, in the case of 3*d* transition metal doped TiO_2, the magnitude of Δ_0 depends on the overall charge population in the t_{2g} and e_g that further diverge the *transition metal–ligand* interaction. In spectroscopy studies, these features of the 3*d* orbitals emerge as two separated peaks at the absorption edge (right below the CBM) where the t_{2g} (representing π bonding) is situated below the e_g (representing σ bonding). The relative intensity of the two peaks and the splitting energy relates to the oxidation state of the transition metal and the overall crystal structure of the host matrix (latter defines the charge distribution surrounding the transition metal and further the transition *metal–ligand* interaction) [74, 76].

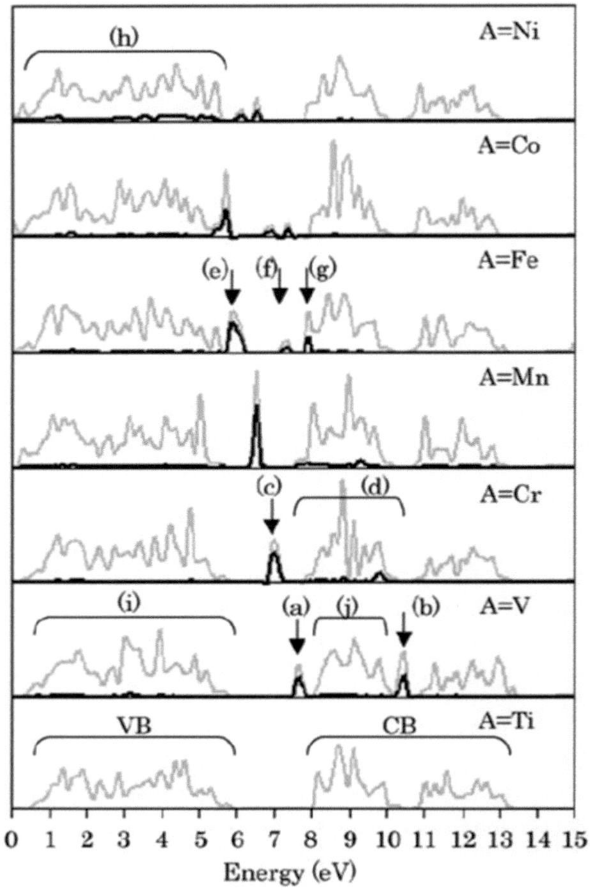

Fig. 5 Predicted variation in the density of states of TiO_2 upon introduction of transition metal impurity ions (substituting for the host cation) with different electron occupancy in their 3*d* orbitals, ranging from $3d^3$ for vanadium to $3d^8$ for nickel [67]

From ab initio calculations, based on the density functional theory (DFT), the variation in the density of states of TiO_2 upon the introduction of transition metal impurity elements has previously been predicted and indicates a sequential shift in the energy position of the impurity-induced localized energy states towards lower energies up on doping with impurities carrying higher charge density in their partially occupied 3*d* states (i.e., from $3d^3$ for vanadium to $3d^8$ for nickel); see Fig. 5 [67].

If we look at the two boundary conditions in Fig. 5 (V and Ni doping), Fig. 6 summarizes the correlation between electron occupancy in the 3*d* orbitals of the

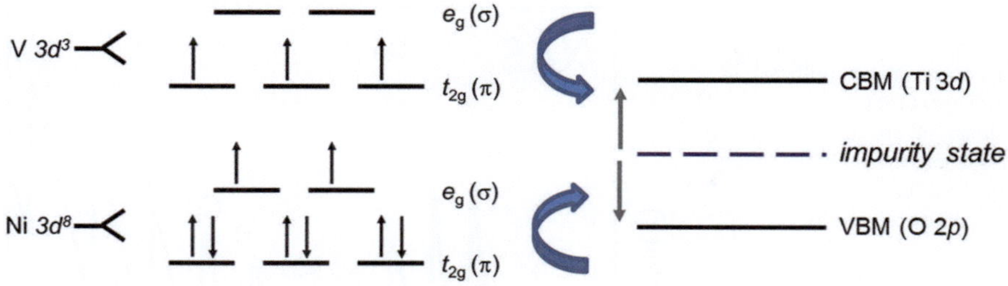

Fig. 6 Illustration presenting the correlation between charge occupancy in the 3*d* orbital of the transition metal impurity element, and the relative energy position of the impurity-induced localized energy state with respect to the energy band edges in TiO_2

transition metal ion and the relative energy position of the localized impurity state (with respect to the CBM and VBM) in TiO_2 (given the presented DOS calculations in Fig. 5).

Given the fact that the atomic configuration and electronic structure of matter eventually define their fundamental and functional properties, it is very important to establish a sound understanding for these properties in order to further advance in areas where the intrinsic properties of materials are tailored, and new exotic properties are induced for enhanced performance that involves charge generation and transportation between different phases. Different from conventional experimental techniques commonly employed to study the optical and electronic properties (e.g., *photoluminescence*, *ellipsometry–reflectance*, *quantum efficiency*, *and/or optical transmittance*), synchrotron based X-ray spectroscopy technique offers the unique capability to project the *electronic structure*, *charge symmetry* and *local chemical environment* in matter at high precision (<1 eV resolution) [76–79]. Since this technique involves excitation of core electrons into the unoccupied density of states, element selectivity is achieved by the virtue of differences in the energy separation at the core levels which is element specific. Spectroscopy studies in the soft X-ray energy range (<1 keV), in particular, has established certain degree of prominence in the field of materials science and energy research wherein the transition metal elements serve as an active ingredient. Sine the L-edge of transition metals is represented within the energy range 0.3–1.1 keV (Fig. 7), operating in the soft X-ray range does not only enable the projection of $2p \rightarrow 3d$ transitions but also discern the variation in the $3d \rightarrow t_{2g}$ and e_g splitting energy as the surrounding chemical environment is altered.

Traditionally, synchrotron studies are restricted to UHV operational conditions. However, following recent progress in the field we are today able to perform in-situ studies of e.g., complete photoelectrochemical cells in-operando that are nearly identical to real operational conditions (at few tens of mBar pressure) [80].

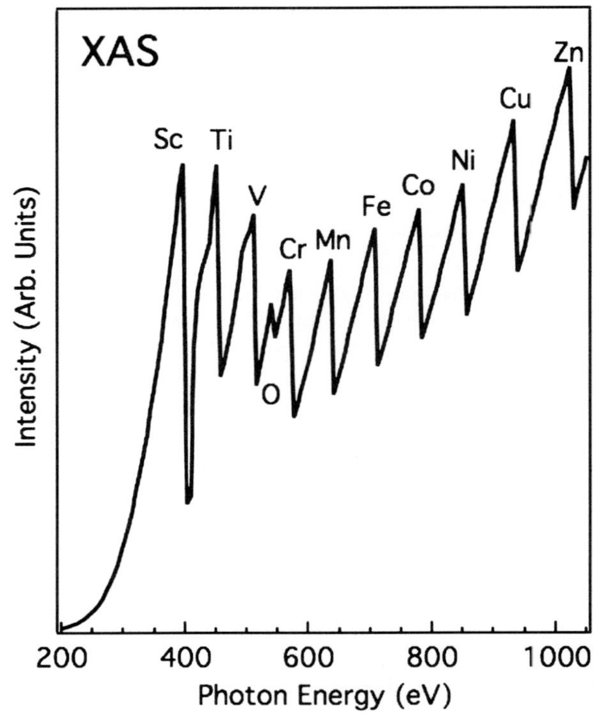

Fig. 7 Simulated wide energy X-ray absorption spectrum of a hypothetic 3*d* transition metal oxide. Observed spectral features represent the typical L-edge spectral feature of the respective element

3 Synchrotron-Based X-Ray Spectroscopy

3.1 Principles

Synchrotron radiation refers to polarized light/radiation of high brilliance, ranging from *soft X-ray* to *hard X-ray* (ca. 10–120 keV), produced from particles traveling at relativistic speed along a curved trajectory [77, 81]. Since the light produced from the synchrotron arises from a narrow and well-confined beam of electrons (confined by means of a magnetic field) it offers high spatial and spectral resolution, and can be treated as a point source [77]. Although the development of synchrotron technique can be dated back to mid-1940s, its prominence has evolved as the dimensionality of matter has been reduced from 3D → 0D (i.e., from bulk down to quantum dots) owing to the miniaturization of semiconductor electronic devices (wherein more functionalities are added on to the same unit in order to, e.g., achieve higher data transfer rate and lower energy consumption) and the surge for an experimental capability that couples chemical selectivity and high resolution at lattice distances of few tens of angstroms [82]. X-ray absorption spectroscopy (XAS) and X-ray emission spectroscopy (XES) represent two principle spectroscopic capabilities commonly employed for detailed projection of the *density of unoccupied* and *occupied states*, respectively. Resonantly, inelastic X-ray spectroscopy is an additional emission spectroscopic technique for identification of local energy levels that reflect the chemical and physical properties, and phase transformations in semiconductors [77].

3.2 X-Ray Absorption and Emission Spectroscopy

Absorption spectra emanate from the excitation of core electrons into empty states in the conduction band (Fig. 8). The transition probability of a core-level electron into an unoccupied state is governed by the dipole selection rule. Further, the transition probability for single crystal samples varies under polarization dependent condition: (i) higher if the direction of incident beam is polarized such that it aligns parallel to the axis of the available orbitals to be occupied in the conduction band and (ii) lower as the incident angle of the beam propagates towards perpendicular alignment with respect to the axis of the orbitals in the conduction band [82, 83].

The absorption spectra are typically recorded either in the surface-sensitive (typically <10 nm) total electron yield (TEY) detection mode by measuring the drain current on the surface, or in the bulk-sensitive (typically 100–200 nm) total fluorescence yield (TFY) detection mode by detecting the photoemission using a photodetector. The combination of TEY and TFY detection mode offers crucial information that eases our understanding for the difference in the surface and bulk electronic properties [84]. This also carries information on how to improve the charge transfer from bulk to surface in photoelectric materials to prevent the event of electron–hole recombination and/or how to improve charge transfer through interfaces in multilayer charge tunneling devices (e.g., magnetic tunneling junctions and light-emitting diodes).

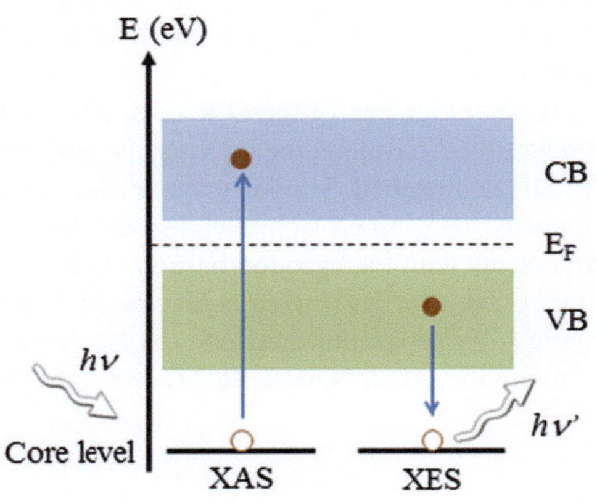

Fig. 8 Schematic presentation of the principle X-ray absorption and emission processes to project the total density of *unoccupied* and *occupied* states, respectively

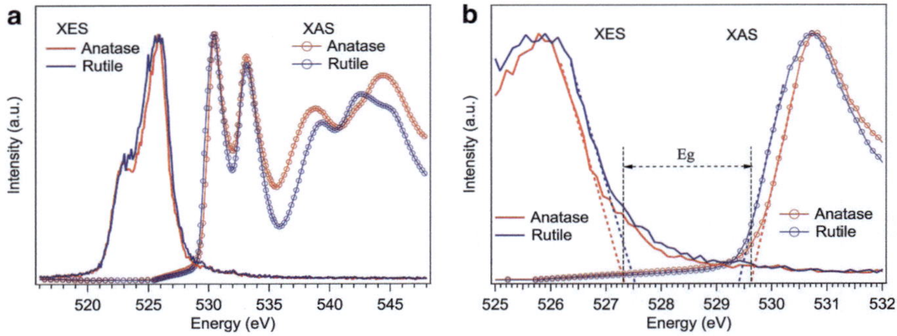

Fig. 9 (**a**) Typical O K-edge XAS and XES of rutile- and anatase TiO2, (**b**) the band gap of the two oxide complexes has been extracted by combining their respective XAS and XES and estimating the relative energy separation between the CBM and VBM

Emission spectroscopy is a second-order process evolving from electronic transition between the valence states and localized core states, acquired by monitoring the emission (as core holes are re-populated by electrons in the valence orbitals) probability as a function of photon-out (Fig. 8) [77]. XAS and XES are commonly treated as two independent techniques, wherein XAS introduces an excited state and XES arises as a result of a decay of the same. However, by combining the two one can estimate the band gap by directly comparing the relative energy separation between the lowest total density of unoccupied states (CBM) and the highest total density of occupied states (VBM) [78].

Figure 9a presents the typical O K-edge XAS and XES of rutile and anatase TiO_2 arising from excitation of O 1*s* into the unoccupied O 2*p* states and the repopulation of the core levels from the valence state, respectively. The energy region ca. 530–536 eV mainly reflects the O 2*p*–Ti 3*d* hybridization, wherein spectral features at ca. 531 and 534 eV represent the t_{2g} (σ-antibonding states) and e_g σ-antibonding states (arising from Ti $3d \rightarrow t_{2g}$ and e_g splitting), respectively. Features >536 eV prevail from O 2*p* and Ti 4*sp* hybridization. The O K-edge emission spectra arise from O 2*p*–Ti 3*d* hybridizations (the low energy band at ca. 523 eV represents O 2*p* states) [85].

Upon extension of the O K-edge absorption edge to where it intersects with the background/baseline, the relative energy position of the conduction band minimum can be determined. Similarly, the relative energy position of the valence band maximum is determined by extending the linear section (O K-edge XES) to where it intersects with background/baseline (Fig. 9b). An alternative method to extract the energy band edges is to calculate the first derivative of the two spectra and measure the energy separation between the two upon alignment (for details, see studies by Kronawitter et al. Nano Lett. 11, 3855 (2011)).

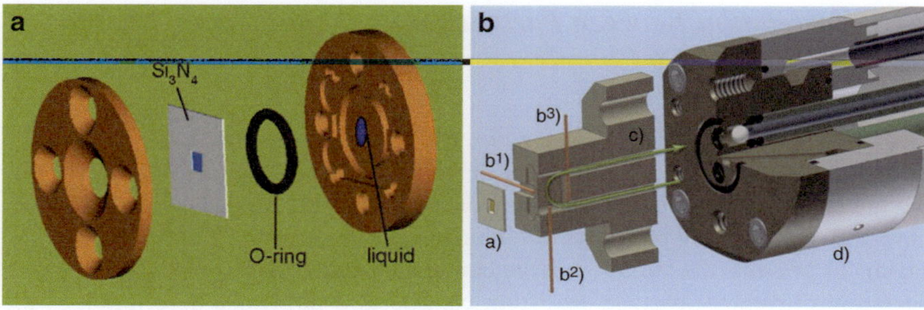

Fig. 10 3D presentation of the typical (**a**) static liquid cell and (**b**) flow cell developed at Lawrence Berkeley National Laboratory for in situ spectroscopy studies nearly identical to operational conditions [80]

3.3 In Situ Experiments

Ex situ studies hold the drawback of introducing artifacts arising from, e.g., sample handling/transfer between sample preparation/synthesis and different characterization tools and difficulties with connecting materials' properties to device performance due to the unavoidable time delay between sample *synthesis*, *characterization* and *evaluation of the performance of a device* [86].

Understanding the material properties under ambient conditions and/or close to operational conditions is highly desirable and very important to enable further optimization of the same by foreseeing plausible structural degradation in matter that may occur as a function of time, and variation in surrounding conditions.

Figure 10 presents two novel experimental capabilities, developed at Lawrence Berkeley National Laboratory, that have opened up new horizons for in situ spectroscopy studies of electrochemical cells [86–90]. While the static liquid cell (Fig. 10a) is used to investigate sample properties in non-operational mode under ambient conditions, the flow liquid cell (Fig. 10b) is used for in operando studies as it allows for a continuous flow of a fluid at the rate of 50–200 nl/s. In addition, the flow cell encompasses a work electrode (b_1), a reference electrode (b_2) and a counter electrode (b_3) which allows repetitive cycling of an electrochemical cell (if a voltmeter is connected through an external circuitry) to monitor plausible events of evolution in the structural/electronic properties whilst under operational mode [86, 87].

4 Soft X-Ray Spectroscopy Studies on Nanostructured Metal Oxide Complexes

This section presents few representative synchrotron-based X-ray spectroscopy studies on key materials' properties e.g., *electronic, structural, and optical properties* of TiO_2 (intrinsic, and Zr doped), Fe_2O_3, ZnO (intrinsic and Al doped), and Co complexes at reduced scale, all that strongly correlate to their photoelectric properties.

4.1 Correlation Between Particle Size and Electronic Structure in TiO_2 and Co Complexes

With increasing *surface-to-volume ratio*, the constituent atoms in matter are situated closer to the surface, and the overall band structure is altered such that the different band levels therein become more discrete and well defined which introduces new and more direct pathways for charge to migrate from "bulk" to surface. Figure 11 presents the variation in the conduction band of TiO_2 nanoparticles with decreasing (average) particle size from ca. 200 to 2 nm (for a description of the spectral feature, see Sect. 2.2). As evident from the O K-edge XAS in Fig. 11, the O 2*p*–Ti 3*d* hybridization and the $3d \rightarrow t_{2g}$ and e_g splitting become more obscured as the particle size approaches the exciton radius (ca. 1 nm), whereas the O 2*p*–Ti 4*s* hybridization becomes more pronounced. The observed variation in the O K-edge XAS with decreasing particle size has been attributed to a decrease in the octahedral symmetry of TiO_2 as the constituent atoms reside closer to the surface, and the contracted (core-like) nature of the 3*d* orbital which exhibits reduced radial distribution as compared to both thes and porbitals [91, 92].

Cobalt has previously shown some very interesting properties that relates its size to its catalytic property [93–98]. Along this line, systematic studies on the correlation between particle size effect and electronic structure of cobalt nanocrystals suspended in 1,2-dichlorobenzene (to prevent agglomeration of the nanoparticles) have

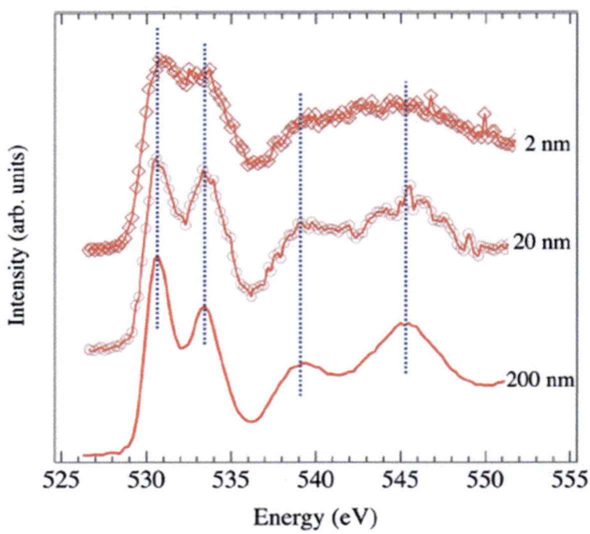

Fig. 11 Variation in the O K-edge XAS of anatase TiO_2 nanoparticles with decreasing particle from 200 → 2 nm. As evident, the $3d \rightarrow t_{2g}$ and e_g splitting and the overall spectral features become less resolved as the nanoparticles reach an average particle size close to the exciton radius [91]

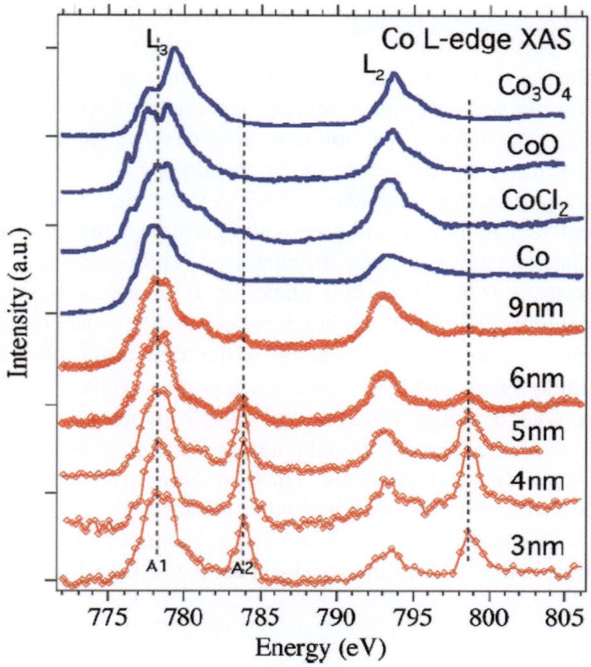

Fig. 12 Variation in Co 2*p* XAS of cobalt nanoparticles (suspended in 1,2-dichlorobenzene) as a function of decreasing particles size (from bulk → few nm). Feature A2 reflects metal-to-ligand charge transfer to unoccupied ligand orbitals, and becomes more pronounced with decreasing particle size. This spectral feature is not evident in the cobalt reference samples [93]

previously been presented by Liu et al. in [93]. Figure 12 presents the Co L-edge XAS that projects the $2p \rightarrow L_{3,2}$ excitation and the variation in the same as Co is reduced in size from bulk to few nm (as for reference, oxide complexes as well as chloride complex of Co have been included). The two main spectral features ($L_{3,2}$) arise from core-level spin splitting of the $2p_{3/2,1/2}$ orbitals. A third feature (A2) evolves as the particle size is reduced to <9 nm, which becomes more pronounced as the particle size if further reduced. This additional spectral feature arises as a result of metal-to-ligand charge transfer (to unoccupied ligand orbitals) and becomes more resolved at reduced scale as a result of increased proportion of cobalt surfactant molecular interaction, compared to bulk cobalt or when bound to a ligand (Co_3O_4, CoO, $CoCl_2$) [93].

4.2 *Orbital Anisotropy and Polarization-Dependent XAS*

Present-day solid-state electronic devices are typically composed of multilayers of nanomaterials, each of highly oriented crystallographic characteristic and (independently) carefully optimized for improved performance and durability. Since surface properties vary with surface structure/-geometrics (e.g. atomic arrangement and orbital symmetry) and thereby with the orientation of crystal facets therein [99–104], there is clearly room for further optimization and functionalization of the surface properties of nanostructure complexes than just manipulating dimensionality. Now, this raises the question(s): *how does one evaluate and optimize the surface properties of nanomaterials to identify e.g. the most favorable crystallographic characteristics/orientation for optimum photoelectric activity, especially if the size of the specimen is nearly close to the exciton radius*?

Polarization dependent XAS reveals the difference in the surface structure between nanomaterials grown along different preferential directions. In other words, one can carefully probe the surface crystallographic properties (and allowed charge transitions along certain plane/facet) in nanostructure materials by varying the angular direction of the incident beam with respect to the surface plane of the specimen and recording any change in the transition probability as a function of incident angle. From sole materials science point of view, it is of great interest to understand to what extent (surface) orbital anisotropy may leverage the efficiency of a catalyst as it presents an additional parameter that may need to be considered in the material's design and synthesis. Thus functionalized nanomaterials can be designed for higher chemical reactivity and/or selectivity.

One example of such study is presented in Fig. 13 which compares the variation in the XAS of microcrystal ZnO as a function of incident angle, from gracing

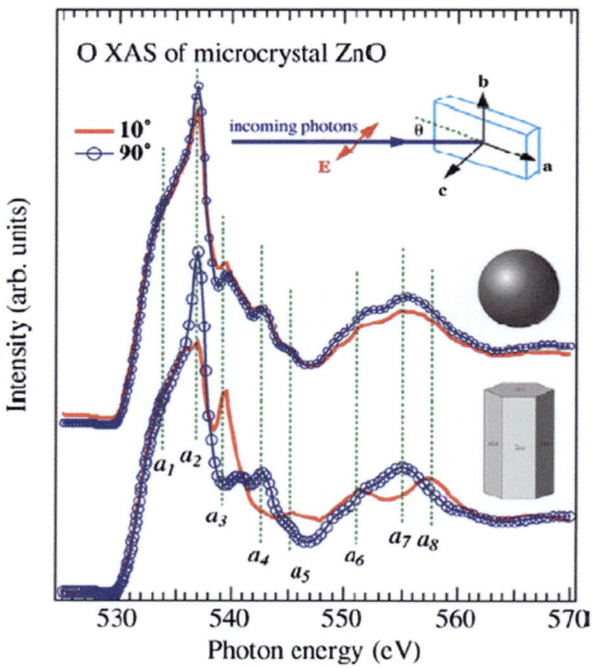

Fig. 13 A comparison of polarization-dependent XAS of ZnO particulate thin films with isotropic (spherical) and anisotropic (rod) morphological characteristics. At 90° incident angle (θ), the incoming photons irradiate the sample perpendicular to the *c*-axis (direction of crystal growth), while at 10° incident angle, beam is close to parallel to the *c*-axis [105]

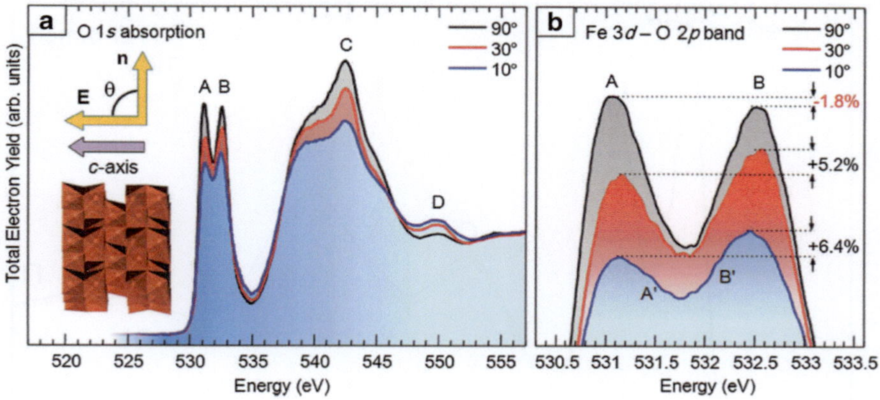

Fig. 14 (**a**) A comparison of polarization-dependent XAS of α-Fe_2O_3 at photon incident angles 90°, 30°, and 10° with respect to the sample surface. Stronger Fe 3*d*–O 2*p* hybridization along the perpendicular direction relates to the out-of-plane [110] directional growth of the hematite nanorods. (**b**) Detailed presentation of the $3d \rightarrow t_{2g}$ and e_g splitting intensity at different beam incident angle [107]

incident (10°) to out-of-plane (90°) with respect to the sample surface [105]. As seen in Fig. 13, while the particulate ZnO thin film sample does not manifest any notable anisotropic effect the multi-facet ZnO microrods show an angular dependency such that: (1) featues a_3, a_5, and a_8 are stronger at gracing incident angle, and (2) features a_2, a_4, and a_7 are stronger normal incident angle.

Similarly, studies on the orbital anisotropy in [110] oriented hematite (α-Fe_2O_3) nanorod arrays have indicted on enhanced hybridization along the direction of sample growth [107]. Spectral features A and B in Fig. 14a arise from the splitting of Fe $3d \rightarrow t_{2g}$ (A) and e_g (B) by virtue of Fe 3*d*–O 2*p* hybridization, while features C and D arise from Fe 4*sp*–O 2*p* hybridizations. Different from spectral features A–C, feature D is found to become more pronounced with decreasing photon incident angle. As indicated in Fig. 14b, the relative difference in the intensity of t_{2g} and e_g orbitals is reduced at higher incident angles (although the overall crystal field splitting energy remains the same). The enhanced intensity of the t_{2g} peak with increasing incident angle implies that the direction of the lobe of the t_{2g} orbital is closely aligned along the direction of the polarization vector at 90° incident angle (i.e., out of plane) [107].

4.3 Evolution in the Electronic Structure of TiO_2 and ZnO upon Introduction of Transition Metal Impurity Elements

Moderate Zr doping (1–5 at.%) of TiO_2 has demonstrated to improve the overall photoefficiency of TiO_2-based dye-sensitized solar cells (DSSCs) [108, 109]. The underlying reason behind such observations has been attributed to an increase in the

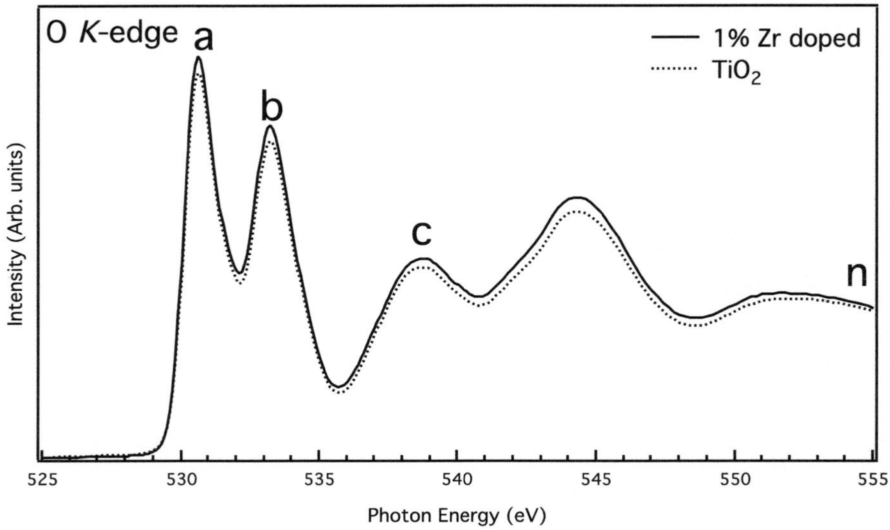

Fig. 15 Comparison between the O K-edge XAS of intrinsic and 1 at.% Zr-doped anatase TiO_2. The nearly identical spectral features imply that the local charge symmetry remains the same at moderate Zr doping, and the overall structural properties of the oxide is also improved by the same (as evident from the increase in the overall spectral intensity)

active surface area of the oxide by the same (typically evaluated using the Brunauer–Emmett–Teller theory). Besides, Zr has also been found to stabilize the structural properties of TiO_2; this latter benefit of Zr doping is an important factor to consider since (as previously discussed) structural defect introduces active *electron–hole* recombination sites that prevent the migration of photoelectrons to the surface that reduces the overall photoelectric effect. Figure 15 presents a comparison between the O K-edge XAS of intrinsic and 1 at.% Zr-doped anatase TiO_2 thin films wherein spectral features a (t_{2g}) and b (e_g), which reflect the O 2p and Ti 3d hybridization states are found to become more pronounced upon Zr doping. In the higher energy range, >538 eV, the spectral features predominantly reflect O 2p – Ti 4sp hybridization. Similarity in spectral features suggests that the local charge symmetry (octahedral structure) remains the same since Zr substitutes for Ti and not O.

The overall increase in the absorption intensity upon Zr doping indicates that introduction of Zr into the oxide matrix, substituting for host cations, (i) improves the structural properties and (ii) enhances the hybridization strength between Ti and O.

The O K-edge normal emission (n) and RIXS (a–c) spectra in Fig. 16 (reflecting O 2p–1s transitions) do not indicate on any obvious change in the O 2p valence state upon Zr doping. This suggest that improvements in the solar cell efficiency upon Zr doping may most likely be due to a change in the local charge density in the Ti 3d band structure rather than in the O 2p band. To further understand how/if Zr doping affects the local charge distribution surrounding the cation in the oxide matrix, Figs. 17 and 18 further present the Ti L-edge XAS and RIXS of the intrinsic and Zr doped TiO_2.

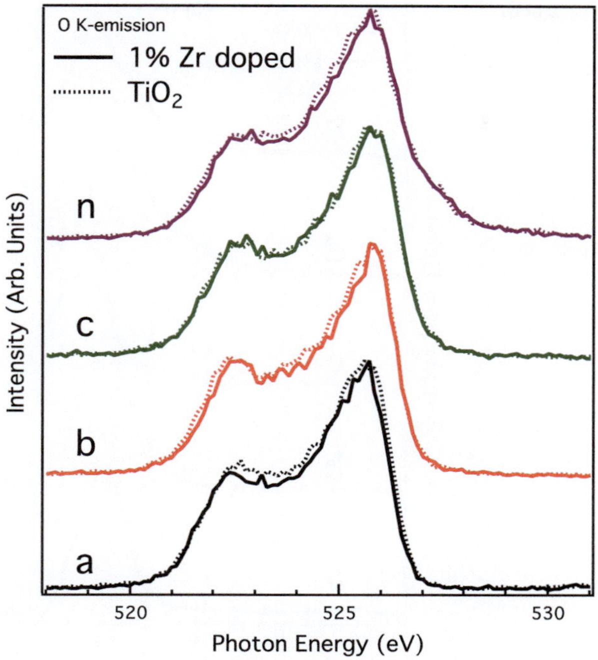

Fig. 16 O 1*s* XES of intrinsic and 1 at.% Zr doped anatase TiO_2 at different excitation energies (as indicated in Fig. 15). The nearly identical spectral features indicate that Zr doping does not disturb the local charge symmetry in the valence band of the oxide

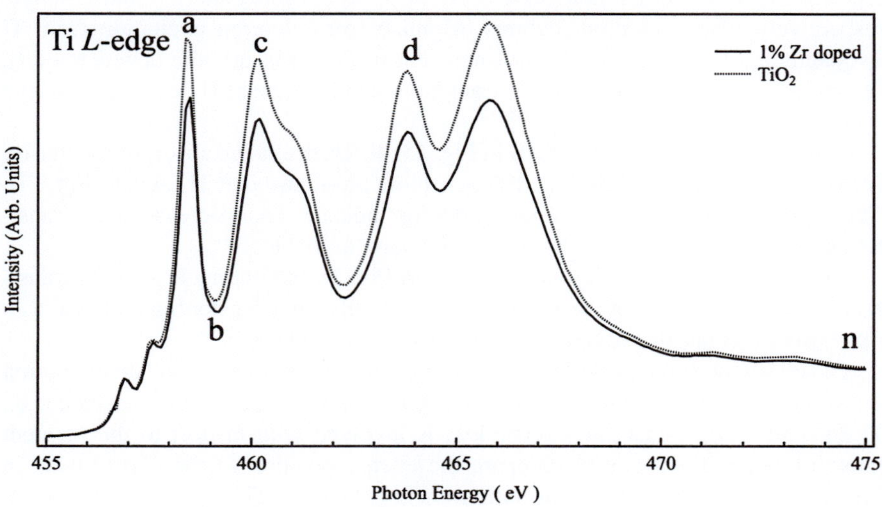

Fig. 17 Comparison between the Ti L-edge XAS of intrinsic and 1 at.% Zr doped anatase TiO_2 which shows a reduction in the Ti $2p - 3d$ transitions as Ti ions are substituted by Zr impurity ions

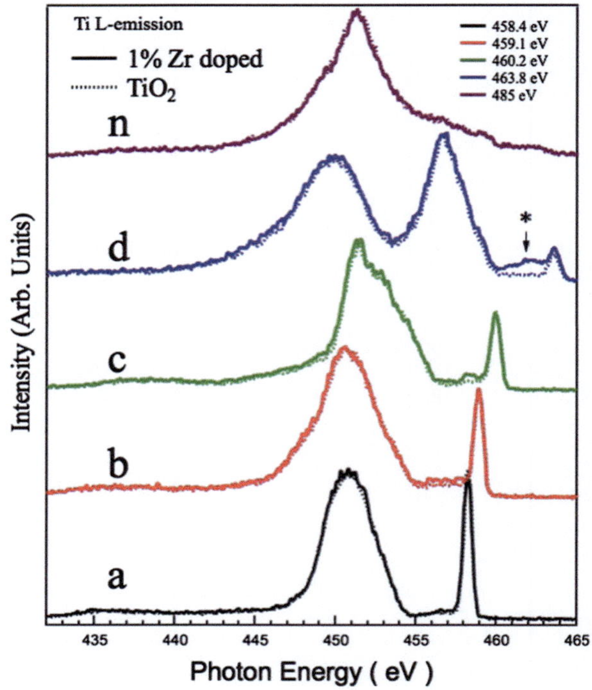

Fig. 18 A comparison between Ti 2*p* XES of intrinsic and 1 at.% Zr-doped anatase TiO_2 at excitation energies as indicated in Fig. 17. The additional feature (indicated "*") at excitation energy 463.8 eV arises from enhanced *d*–*d* transition in Ti^{4+} plausibly as the Ti 3*d*–O 2*p* hybridization is modified due to additional charge transfer between Zr and O

The Ti L-edge XAS presented in Fig. 17 can be divided into two main energy regions: (1) 458–462 eV and (2) 463–467 eV, representing the L_3- and L_2-edge, respectively. These spectral features originate from electron excitation from Ti $2p_{3/2}$ and $2p_{1/2}$ to unoccupied Ti 3*d* states that further split into the degenerated t_{2g} (a) and e_g (c) states by the virtue of transition metal - ligand (Ti 3*d*→O 2*p*) charge transfer.

Compared to the O K-edge XAS in Fig. 15, difference in the absorption intensity between the intrinsic and Zr doped TiO_2 is more pronounced at Ti L-edge (Fig. 17). The reduced spectral intensity upon Zr doping indicates on a decrease in the transition probability (plausibly) following a reduction of Ti^{4+} to Ti^{3+}.

Further, the Ti L-edge emission spectra in Fig. 18 (reflecting Ti 2*p* – 3*d* transitions) show a variation in the valence state not only upon Zr doping, but also as a function of increasing excitation energy (from 458.4–485 eV).

The elastic peak (at ca. 458 eV in spectrum "a") corresponds to the de-excitation threshold where the electrons in the valence band repopulate the ground state (i.e., Ti $2p \rightarrow 3d \rightarrow 2p$); hence, no energy loss is involved with respect to the incident photon energy (for instance, compare the energy position of the elastic peak in Fig. 18a with respect to the excitation energy therein). The main broad peak in Fig. 18a represents the energy-loss features arising from inelastic transitions, e.g., charge transfer or relaxation to pre-ground state energy levels [110]. Zr doping contributes towards an alteration of the electron occupancy in the Ti 3*d* orbitals which affects the repopulation of the ground state as it introduces an intermediate energy state above the ground state in the valence band. The relative energy position of the additional energy state (with respect to the energy band edges of the host oxide) relates to the electron population in the t_{2g}/e_g states (for detailed description, see [110]). This energy-loss feature has been indicated with an asterix in Fig. 18, and confirms the instance of charge transfer between Zr and Ti via Zr–O hybridization by which Ti^{4+} $3d^0$ is promoted to $3d^1$.

Transparent conducting oxides (or TCOs) play a pivotal role in *light-matter* related applications, as they, different from conventional metal electrodes, offer light weight, optical transparency (both in the visible and UV wavelength), and tunable transport properties [106, 111, 112]. Although both $In_{1-x}Sn_xO$ (ITO) and F_xSnO_{1-x} (FTO) are two well established TCO complexes, there is a continuous search for new and improved TCOs that offer higher out-put at lower production cost [113].

Among these, ZnO is a prime candidate TCO material since it is intrinsically an *n*-type semiconductor which offers high solubility of impurity elements, and crystallizes into the hexagonal wurtzite structure at temperatures closely above room temperature [61, 114, 115]. Huang et al. [116] have recently demonstrated on a broadening in the band gap of RF-sputtered ZnO thin films upon Al doping (doping concentration was controlled by varying the Ne sputter flux pressure), from ca. 3.3 eV (at 5 mTorr Ne) to 4.3 (at 100 mTorr Ne) eV (estimated from combined XAS– XES studies); see Fig. 19a. In addition, the authors reported that with increasing Ne flux pressure the oxide underwent a phase transition from a *wurtzite crystalline phase* to an *amorphous phase* that contributed towards improved transmittance of the thin films.

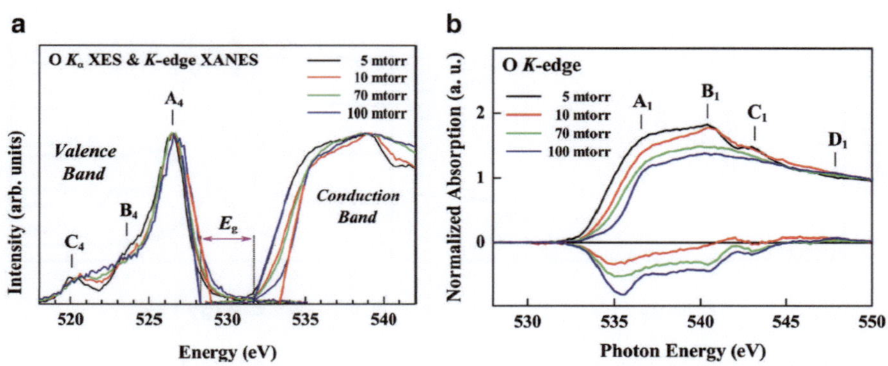

Fig. 19 (**a**) Variation in the O K-edge XAS and XES in ZnO thin films upon Al doping under different Ne sputter flux pressure conditions. Combined XAS-XES studies reveal an increase in the band gap from 3.3→4.3 eV follwing an increase in the Al/Zn ratio from 0.02→0.05. (**b**) Detailed presentation of the variation in the XAS and the Zn–O orbital hybridizations with increasing Al/Zn ratio [116]

The deacrease in the relative intensity of the absorption spectra in Fig. 19b has been attributed to a decrease in the number of availabe O 2*p* and Zn 3*d* unoccupied states as Al (interstitially positioned) is introduced into the oxide matrix that also enhances the ionic character of the oxide complex [116].

4.4 Reduction in the Band Gap of TiO_2 by Hydrogenation of the Oxide Surface

The recent discovery of black TiO_2 nanocrystals has opened many new and interesting opportunities for TiO_2 in *light–matter*-related applications since it demonstrates the great ability to reduce the band gap of intrinsic anatase TiO_2, from ca. 3.2 to 1.54 eV, exclusively in the absence of any cation impurity element [72, 73]. Black TiO_2 is comprised of a crystalline core and an amorphous surface; the latter is achieved upon surface hydrogenation that gives rise to surface disorder. There is a general consensus that change in color and in the band gap of TiO_2 root in oxygen-related lattice defects and/or in the existence of Ti^{3+} impurities in the oxide lattice [73, 117–119]. However, fingerprints on the local electronic structure of white and black TiO_2, attained from combined XAS and RIXS studies disclose the absence of Ti^{3+} species in black TiO_2, as the two manifest nearly identical XAS spectral features at Ti L-edge (Fig. 20). However, the disparity in the RIXS features between the two oxides at excitation energy 458.8 eV has been attributed to differences in the local chemical environment surrounding the cation. In white TiO_2 the Ti ion is

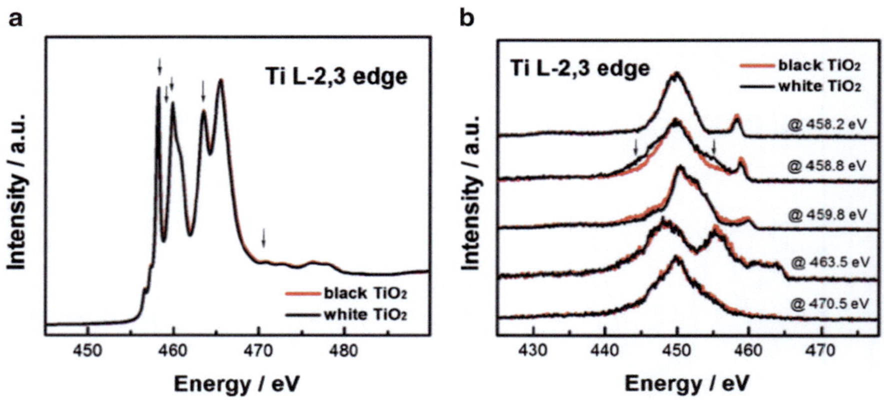

Fig. 20 (**a**) Typical Ti L-edge XAS of black and white TiO_2 nanoparticles. The nearly identical spectral features of the two oxides argues for the absence of any Ti^{3+} that may be the underlying reason behind the improved optical properties of TiO_2 upon surface hydrogenation. (**b**) Ti 2*p* XES with excitation energies as indicated in (**a**). The modest difference in the emission spectra observed at excitation energy 458.8 eV has been attributed to the difference in the local charge symmetry surrounding the Ti ion upon hydrogenation that changes the local matrix from O–Ti–O to H–Ti–O [73]

exclusively surrounded by oxygen, but shares bonds with hydrogen in black TiO_2 [73]. Given the absence of Ti^{3+} species in black TiO_2, improved optical properties of black TiO_2 have been reported to relate to surface disorder which gives rise to mid-gap states residing closely above the VBM. These mid-gap states reduce the energy separation between the unoccupied and occupied density of states by extending the valence band region closer to the conduction band region [73].

Acknowledgment The work at the ALS is supported by the US Department of Energy under Contract No. DE-AC02-05CH11231. The authors would like to thank Kaiqi Nie for her timely help editing this chapter.

References

1. U.S. Department of Energy (2013) International Energy Outlook 2013. U.S. Energy Information Administration: Washington, DC
2. AAAS Climate Science Panel (2014) What we know, the reality, risks, and response to climate change. American Association for the Advancement of Science
3. International Bank for Reconstruction and Development. (2012) Turn down the Heat: Why a 4 °C Warmer World Must Be Avoided. The World Bank
4. Office of science. (2005) Basic research needs for solar energy utilization. DOE Available via http://www.sc.doe.gov/bes/reports/files/SEU_rpt.pdf
5. Pagliaro M, Ciriminna R, Palmisano G (2008) Flexible solar cells. Chem. Sus. Chem. 1(11):880-891
6. Tachibana Y, Vayssieres L, Durrant J R (2012) Artificial photosynthesis for solar water-splitting. Nat. Photonics 6(8): 511-518
7. Ni M, Leung M K, Leung D Y, Sumathy K (2007) A review and recent developments in photocatalytic water-splitting using TiO2 for hydrogen production. Renew. Sust. Energ. Rev. 11(3): 401-425
8. Warren S C, Voïtchovsky K, Dotan H, Leroy C M, Cornuz M, Stellacci F, Grätzel M (2013) Identifying champion nanostructures for solar water-splitting. Nature mater. 12(9): 842-849
9. Kudo A, Miseki Y (2009) Heterogeneous photocatalyst materials for water splitting. Chem. Soc. Rev. 38(1): 253-278
10. Litter M I (1999) Heterogeneous photocatalysis: transition metal ions in photocatalytic systems. Appl. Catal. B: Environ. 23(2): 89-114
11. Schultz D M, Yoon T P (2014) Solar Synthesis: Prospects in Visible Light Photocatalysis. Science 343(6174): 1239176
12. Linsebigler A L, Lu G, Yates Jr J T (1995) Photocatalysis on TiO_2 surfaces: principles, mechanisms, and selected results. Chem. Rev. 95(3): 735-758
13. Hoffmann M R, Martin S T, Choi W, Bahnemann D W (1995) Environmental applications of semiconductor photocatalysis. Chem. Rev. 95(1): 69-96
14. Mao J, Li K, Peng T (2013) Recent advances in the photocatalytic CO_2 reduction over semiconductors. Catal. Sci. Technol. 3(10): 2481-2498
15. Hong J, Zhang W, Ren J, Xu R (2013) Photocatalytic reduction of CO_2: a brief review on product analysis and systematic methods. Anal. Methods 5(5): 1086-1097
16. Mori K, Yamashita H, Anpo M (2012) Photocatalytic reduction of CO_2 with H_2O on various titanium oxide photocatalysts. RSC Advances 2(8): 3165-3172
17. Habisreutinger S N, Schmidt-Mende L, Stolarczyk J K (2013) Photocatalytic Reduction of CO_2 on TiO_2 and Other Semiconductors. Angew. Chem. Int. Ed. 52(29): 7372-7408
18. Tahir M, Amin N S (2013) Advances in visible light responsive titanium oxide-based photocatalysts for CO_2 conversion to hydrocarbon fuels. Energ. Convers. Manage. 76: 194-214

19. Zhu W, Michalsky R, Metin O, Lv H, Guo S, Wright C J, Sun S (2013) Monodisperse Au Nanoparticles for Selective Electrocatalytic Reduction of CO_2 to CO. J. Am. Chem. Soc. 135(45): 16833-16836
20. Melo M D O, Silva L A (2011) Photocatalytic production of hydrogen: an innovative use for biomass derivatives. J. Braz. Chem. Soc. 22(8): 1399-1406
21. Karunakaran C, Senthilvelan S (2006) Fe_2 O_3-photocatalysis with sunlight and UV light: Oxidation of aniline. Electrochem. Commun. 8(1): 95-101
22. Hisatomi T, Kubota J, Domen K (2014) Recent advances in semiconductors for photocatalytic and photoelectrochemical water splitting., Chem. Soc. Rev. DOI: 10.1039/c3cs60378d
23. Qu Y, Duan X (2013) Progress, challenge and perspective of heterogeneous photocatalysts., Chem. Soc. Rev. 42(7): 2568-2580
24. Wang J, Pan S, Chen M, Dixon D A (2013) Gold Nanorod-Enhanced Light Absorption and Photoelectrochemical Performance of α-Fe_2O_3 Thin-Film Electrode for Solar Water Splitting. J. Phys. Chem. C 117(42): 22060-22068
25. Yang X, Wolcott A, Wang G, Sobo A, Fitzmorris R C, Qian F, Li Y (2009) Nitrogen-doped ZnO nanowire arrays for photoelectrochemical water splitting. Nano Lett. 9(6): 2331-2336
26. Fujishima A (1972) Electrochemical photolysis of water at a semiconductor electrode. nature 238: 37-38
27. Grätzel M (2001) Photoelectrochemical cells. Nature 414(6861): 338-344
28. Kim J Y, Magesh G, Youn D H, Jang J W, Kubota J, Domen K, Lee J S (2013) Single-crystalline, wormlike hematite photoanodes for efficient solar water splitting. Sci. Rep. 3. DOI: 10.1038/srep02681
29. Carp O, Huisman C L, Reller A, (2004) Photoinduced reactivity of titanium dioxide. Prog. Solid State Chem. 32(1): 33-177
30. Yu J, Yu X, Huang B, Zhang X, Dai Y (2009) Hydrothermal synthesis and visible-light photocatalytic activity of novel cage-like ferric oxide hollow spheres. Cryst. Growth Des. 9(3): 1474-1480
31. Wang Y, Wang Q, Zhan X, Wang F, Safdar M, He J (2013) Visible light driven type II heterostructures and their enhanced photocatalysis properties: a review. Nanoscale 5(18): 8326-8339
32. Ni M, Leung M K, Leung D Y, Sumathy K (2007) A review and recent developments in photocatalytic water-splitting using TiO_2 for hydrogen production. Renew. Sust. Energ. Rev. 11(3): 401-425
33. Lan Y, Lu Y, Ren Z (2013) Mini review on photocatalysis of titanium dioxide nanoparticles and their solar applications. Nano Energy 2(5): 1031-1045
34. Fu H, Xu T, Zhu S, Zhu Y (2008) Photocorrosion inhibition and enhancement of photocatalytic activity for ZnO via hybridization with C_{60}. Environ. Sci. Technol. 42(21): 8064-8069
35. Periodic table (2014) http://www.tempstreet.com/periodic-table/
36. Bohn C D, Agrawal A K, Walter E C, Vaudin M D, Herzing A A, Haney P M, Szalai V A (2012) Effect of Tin Doping on α-Fe_2O_3 Photoanodes for Water Splitting. J. Phys. Chem. C 116(29): 15290-15296
37. Wei Q, Zhang Z, Li Z, Zhou Q, Zhu Y (2008) Enhanced photocatalytic activity of porous α-Fe_2O_3 films prepared by rapid thermal oxidation. Journal of Physics D: Appl. Phys. 41(20): 202002
38. Rashad M M, Elsayed E M, Al-Kotb M S, Shalan A E (2013) The structural, optical, magnetic and photocatalytic properties of transition metal ions doped TiO_2 nanoparticles. J. Alloys Compd. 581: 71-78
39. Townsend T K, Sabio E M, Browning N D, Osterloh F E (2011) Photocatalytic water oxidation with suspended alpha-Fe_2O_3 particles-effects of nanoscaling Energy Environ. Sci. 4(10): 4270-4275
40. Kakuta S, Abe T (2009) Photocatalysis for water oxidation by Fe_2O_3 nanoparticles embedded in clay compound: correlation between its polymorphs and their photocatalytic activities. J Mater. Sci. 44(11): 2890-2898

41. Kim H S, Kang S H (2013) Effect of Hydrogen Treatment on Anatase TiO_2 Nanotube Arrays for Photoelectrochemical Water Splitting. Bull. Korean Chem. Soc 34(7): 2067
42. Boucle J, Ackermann J (2012) Solid state dye sensitized and bulk heterojunction solar cells using TiO_2 and ZnO nanostructures: recent progress and new concepts at the borderline. Polym. Int.61(3): 355-373
43. Barnes R. J Molina R, Xu J, Dobson P J, Thompson I P (2013) Comparison of TiO_2 and ZnO nanoparticles for photocatalytic degradation of methylene blue and the correlated inactivation of gram-positive and gram-negative bacteria. J. Nanopart. Res.15(2): 1-11
44. Gupta S M, Tripathi M (2011) A review of TiO_2 nanoparticles. Chin. Sci. Bull. 56(16): 1639-1657
45. Braun A, Sivula K, Bora D K, Zhu J, Zhang L, Grätzel M, Constable E C (2012) Direct Observation of Two Electron Holes in a Hematite Photoanode during Photoelectrochemical Water Splitting. J. Phys. Chem. C 116(32): 16870-16875
46. Scanlon D O, Dunnill C W, Buckeridge J, Shevlin S A, Logsdail A J, Woodley S M, Sokol A A (2013) Band alignment of rutile and anatase TiO_2. Nat. Mater. 12(9): 798-801
47. Reyes-Coronado D, Rodriguez-Gattorno G, Espinosa-Pesqueira M E, Cab C, De Coss R, Oskam G (2008) Phase-pure TiO_2 nanoparticles: anatase, brookite and rutile. Nanotechnology 19(14): 145605
48. Hu Y H (2012) A Highly Efficient Photocatalyst-Hydrogenated Black TiO_2 for the Photocatalytic Splitting of Water. Angew. Chem. Int. Ed. 51(50): 12410-12412
49. Bard A J, Fox M A (1995) Artificial photosynthesis: solar splitting of water to hydrogen and oxygen. Acc. Chem. Res. 28(3): 141-145
50. Teoh W Y, Scott J A, Amal R (2012) Progress in heterogeneous photocatalysis: from classical radical chemistry to engineering nanomaterials and solar reactors. J. Phys. Chem. Lett. 3(5): 629-639
51. Abe R (2010) Recent progress on photocatalytic and photoelectrochemical water splitting under visible light irradiation. Journal of Photochemistry and Photobiology C: Photochemistry Reviews 11(4): 179-209
52. Chen X, Burda C (2008) The electronic origin of the visible-light absorption properties of C-, N-and S-doped TiO_2 nanomaterials. J. Am. Chem. Soc. 130(15): 5018-5019
53. In S, Orlov A, Berg R, García F, Pedrosa-Jimenez S, Tikhov M S, Lambert R M (2007) Effective visible light-activated B-doped and B, N-codoped TiO_2 photocatalysts. J. Am. Chem. Soc. 129(45): 13790-13791
54. Wu G, Nishikawa T, Ohtani B, Chen A (2007) Synthesis and characterization of carbon-doped TiO_2 nanostructures with enhanced visible light response. Chem. Mater. 19(18): 4530-4537
55. Liu G, Zhao Y, Sun C, Li F, Lu G Q, Cheng H M (2008) Synergistic Effects of B/N Doping on the Visible Light Photocatalytic Activity of Mesoporous TiO_2. Angew. Chem. Int. Ed. 47(24): 4516-4520
56. Kundu S, Ciston J, Senanayake S D, Arena D A, Fujita E, Stacchiola D, Rodriguez J A (2012) Exploring the Structural and Electronic Properties of Pt/Ceria-Modified TiO_2 and Its Photocatalytic Activity for Water Splitting under Visible Light. J. Phys. Chem. C 116(26): 14062-14070
57. Iwaszuk A, Nolan M (2011) Electronic Structure and Reactivity of Ce-and Zr-Doped TiO_2: Assessing the Reliability of Density Functional Theory Approaches. J. Phys. Chem. C 115(26): 12995-13007
58. Pereira A L, Gracia L, Beltrán A, Lisboa-Filho P N, Da Silva J H, Andres J (2012) Structural and electronic effects of incorporating Mn in TiO_2 films grown by sputtering: anatase versus rutile. J. Phys. Chem. C 116(15): 8753-8762
59. Breault T M, Bartlett B M (2012) Lowering the band gap of anatase-structured TiO_2 by coalloying with Nb and N: electronic structure and photocatalytic degradation of methylene blue dye. J. Phys. Chem. C 116(10): 5986-5994
60. Kudo A (2007) Photocatalysis and solar hydrogen production. Pure Appl. Chem. 79(11): 1917-1927

61. Kapilashrami M, Xu J, Biswas A, Tamaki T, Sharma P, Rao K V, Belova L (2010) Coexistence of ultraviolet photo-response and room-temperature ferromagnetism in polycrystalline ZnO thin films. Mater. Lett. 64(11): 1291-1294
62. Leutwyler W K, Bürgi S L, Burgl H B (1996) Semiconductor clusters, nanocrystals, and quantum dots. Science 271(5251): 933-937
63. Alivisatos A P (1996) Perspectives on the physical chemistry of semiconductor nanocrystals. J. Phys. Chem. 100(31): 13226-13239
64. Zhai H J, Wang L S (2007) Probing the Electronic Structure and Band Gap Evolution of Titanium Oxide Clusters (TiO_2) n-(n=1-10) Using Photoelectron Spectroscopy. J. Am. Chem. Soc. 129(10): 3022-3026
65. Sato H, Ono K, Sasaki T, Yamagishi A (2003) First-principles study of two-dimensional titanium dioxides. J. Phys. Chem. B 107(36): 9824-9828
66. Shevlin S A, Woodley S M (2010) Electronic and optical properties of doped and undoped (TiO_2) n nanoparticles. J. Phys. Chem. C 114(41): 17333-17343
67. Umebayashi T, Yamaki T, Itoh H, Asai K (2002) Analysis of electronic structures of 3d transition metal-doped TiO_2 based on band calculations. J. Phys. Chem. Solids 63(10): 1909-1920
68. Thompson T L, Yates J T (2006) Surface science studies of the photoactivation of TiO2 new photochemical processes. Chem. Rev. 106(10): 4428-4453
69. Meng Q, Wang T, Liu E, Ma X, Ge Q, Gong J (2013) Understanding electronic and optical properties of anatase TiO_2 photocatalysts co-doped with nitrogen and transition metals. Phys. Chem. Chem. Phys. 15(24): 9549-9561
70. Nikolay T, Larina L, Shevaleevskiy O, Ahn B T (2011) Electronic structure study of lightly Nb-doped TiO_2 electrode for dye-sensitized solar cells. Energy Environ. Sci. 4(4): 1480-1486
71. Yu J C, Lin J, Kwok R W (1997) Enhanced photocatalytic activity of $Ti_{1-x}V_xO_2$ solid solution on the degradation of acetone. J. Photochem. Photobiol.; A: Chem. 111(1): 199-203
72. Chen X, Liu L, Peter Y Y, Mao S S (2011) Increasing solar absorption for photocatalysis with black hydrogenated titanium dioxide nanocrystals. Science 331(6018): 746-750
73. Chen X, Liu L, Liu Z, Marcus M A, Wang W C, Oyler N A, Mao S S (2013) Properties of disorder-engineered black titanium dioxide nanoparticles through hydrogenation. Sci. rep. 3. DOI: 10.1038/srep01510
74. De Groot F M F, Grioni M, Fuggle J C, Ghijsen J, Sawatzky G A, Petersen H (1989) Oxygen 1 s x-ray-absorption edges of transition-metal oxides. Phys. Rev. B 40(8): 5715
75. R.G. Burns (1993) Mineralogical Applications of Crystal Field Theory, 2nd ed. Cambridge University Press, Cambridge
76. Kapilashrami M, Zhang H, Fang M, Li X, Sun X, Rao K V, Guo J (2012) Electronic structure of room-temperature ferromagnetic $Mg_{1-x}Fe_xO_y$ thin films. Appl. Phys. Lett. 101(8): 082411-082411
77. Guo J-h, Glans P A, Liu Y-S, Chang C, L. Vayssieres (2009) Electronic Structure Study of Nanostructured Transition Metal Oxides Using Soft X-Ray Spectroscopy, in On Solar Hydrogen & Nanotechnology. John Wiley & Sons (Asia) Pte Ltd.
78. Kotani A, Shin S (2001) Resonant inelastic x-ray scattering spectra for electrons in solids. REVIEWS OF MODERN PHYSICS, VOLUME 73, JANUARY
79. Guo, J (2009) Interface science in nanoparticles: An electronic structure view of photon-in/photon-out soft-X-ray spectroscopy. International Journal of Quantum Chemistry, 109(12): 2714-2721 Wiley InterScience 2009. DOI:10.1002/qua.22174
80. Guo, J.-H. (2013) The development of in situ photon-in/photon-out soft X-ray spectroscopy on beamline 7.0.1 at the ALS. J. of Electron Spec. and Rel. Phenom., 188: 71
81. David Attwood (1999) Soft X-rays and Extreme Ultraviolet Radiation Principles and Applications. Cambridge University Press, ISBN-13 978-0-521-65214-8
82. Mino L, Agostini G, Borfecchia E, Gianolio D, Piovano A, Gallo E, Lamberti C (2013) Low-dimensional systems investigated by x-ray absorption spectroscopy: a selection of 2D, 1D and 0D cases. J. Phys. D: Appl. Phys. 46(42): 423001

83. Himpsel F J, Cook P L, de la Torre G, Garcia-Lastra J M, Gonzalez-Moreno R, Guo J H, Hamersg R J, Kronawitterh C X, Johnsona P S, Ortegad J E, Pickupd D, Ragoussic M E, Rogerod C, Rubiod A, Rutherg R E, Vayssieresi L, Yangf W, Zegkinoglou I (2013) Design of solar cell materials via soft X-ray spectroscopy. J. Electron. Spectrosc. Relat. Phenom. 190: 2-11
84. Singh J P, Gautam S, Singh B B, Raju M, Chaudhary S, Kabiraj D, Chae K H (2013) Magnetic and Electronic Structure study of Fe/MgO/Fe/Co Multilayer Stack Deposited by E-Beam Evaporation. arXiv preprint arXiv:1305.2717
85. Guo J-h, Glans P A, Liu Y-S, Chang C, L. Vayssieres (2009) Electronic Structure Study of Nanostructured Transition Metal Oxides Using Soft X-Ray Spectroscopy, in On Solar Hydrogen & Nanotechnology. John Wiley & Sons (Asia) Pte Ltd.
86. Jiang P, Chen J L, Borondics F, Glans P A, West M W, Chang C L, Guo J (2010) In situ soft X-ray absorption spectroscopy investigation of electrochemical corrosion of copper in aqueous $NaHCO_3$ solution. Electrochem. Commun. 12(6): 820-822
87. Guo J H, Augustsson A, Kashtanov S, Spångberg D, Nordgren J, Hermansson K, Augustsson A (2005) The interaction of cations and liquid water studied by resonant soft-X-ray absorption and emission spectroscopy. J. of Electron Spec. and Rel. Phenom.144, 287-290
88. Herranz T, Deng X, Cabot A, Guo J, Salmeron M (2009) Influence of the cobalt particle size in the CO hydrogenation reaction studied by in situ X-ray absorption spectroscopy. J. Phys. Chem. B 113(31): 10721-10727
89. Liu H, Guo J, Yin Y, Augustsson A, Dong C, Nordgren J, Salmeron M (2007) Electronic structure of cobalt nanocrystals suspended in liquid. Nano Lett. 7(7): 1919-1922
90. Kvashnina K O, Butorin S M, Modin A, Soroka I, Marcellini M, Guo J H, Nordgren J (2007) Changes in electronic structure of copper films in aqueous solutions. J. Phys.: Condens. Matter 19(22): 226002
91. Vayssieres L, Persson C, Guo J H (2011) Size effect on the conduction band orbital character of anatase TiO_2 nanocrystals. Appl. Phys. Lett. 99(18): 183101
92. Woicik J C, Nelson E J, Kronik L, Jain M, Chelikowsky J R, Heskett D, Herman G S (2002) Hybridization and Bond-Orbital Components in Site-Specific X-Ray Photoelectron Spectra of Rutile TiO_2. Phys. Rev. Lett. 89(7): 077401
93. Liu H, Guo J, Yin Y, Augustsson A, Dong C, Nordgren J, Salmeron M (2007) Electronic structure of cobalt nanocrystals suspended in liquid. Nano Lett. 7(7): 1919-1922
94. Yang J, Tveten E Z, Chen D, Holmen A (2010) Understanding the Effect of Cobalt Particle Size on Fischer-Tropsch Synthesis: Surface Species and Mechanistic Studies by SSITKA and Kinetic Isotope Effect†. Langmuir 26(21): 16558-16567
95. Tyo E C, Yin C, Di Vece M, Qian Q, Kwon G, Lee S, Vajda S (2012) Oxidative Dehydrogenation of Cyclohexane on Cobalt Oxide (Co_3O_4) Nanoparticles: The Effect of Particle Size on Activity and Selectivity. ACS Catal. 2(11): 2409-2423
96. Melaet G, Lindeman A E, Somorjai G A. Cobalt Particle Size Effects in the Fischer–Tropsch Synthesis and in the Hydrogenation of CO_2 Studied with Nanoparticle Model Catalysts on Silica. Top Catal 1-8
97. Michalek F, Lagunas A, Jimeno C, Pericàs M A (2008) Synthesis of functional cobalt nanoparticles for catalytic applications. Use in asymmetric transfer hydrogenation of ketones J. Mater. Chem. 18(39): 4692-4697
98. Wender H, Gonçalves R V, Dias C S B, Zapata M J, Zagonel L F, Mendonça E C, Garcia F (2013) Photocatalytic hydrogen production of $Co(OH)_2$ nanoparticle-coated α-Fe_2O_3 nanorings. Nanoscale 5(19): 9310-9316
99. Liu S, Yu J, Jaroniec M (2011) Anatase TiO_2 with dominant high-energy {001} facets: synthesis, properties, and applications. Chem. Mater. 23(18): 4085-4093
100. Ji Y, Wang B, Luo Y (2012) Location of Trapped Hole on Rutile-TiO_2 (110) Surface and Its Role in Water Oxidation. J. Phys. Chem. C 116(14): 7863-7866
101. Tian F, Zhang Y, Zhang J, Pan C (2012) Raman spectroscopy: a new approach to measure the percentage of anatase TiO_2 exposed (001) facets. J. Phys. Chem. C 116(13): 7515-7519

102. Liu S, Yu J, Jaroniec M (2010) Tunable photocatalytic selectivity of hollow TiO_2 microspheres composed of anatase polyhedra with exposed {001} facets. J. Am. Chem. Soc. 132(34): 11914-11916
103. Nguyen C K, Cha H G, Kang Y S (2011) Axis-oriented, anatase TiO_2 single crystals with dominant {001} and {100} facets. Cryst. Growth Des. 11(9): 3947-3953
104. Xiang Q, Yu J, Wang W, Jaroniec M (2011) Nitrogen self-doped nanosized TiO_2 sheets with exposed {001} facets for enhanced visible-light photocatalytic activity. Chem. Commun 47(24): 6906-6908
105. Guo J H, Vayssieres L, Persson C, Ahuja R, Johansson B, Nordgren J (2002) Polarization-dependent soft-x-ray absorption of highly oriented ZnO microrod arrays., J. Phys.: Condens. Matter 14(28): 6969
106. Pern F J, Glick S H, Li X, DeHart C, Gennett T, Contreras M, Gessert T (2009) Stability of TCO window layers for thin-film CIGS solar cells upon damp heat exposures: part III.In SPIE Solar Energy + Technology (pp. 74120 K-74120K). International Society for Optics and Photonics. Doi:10.1117/12.826559
107. Kronawitter C X, Zegkinoglou I, Shen S, Guo J.-H, Himpsel F J, Mao S S, Vayssieres L (2013) On the orbital anisotropy in hematite nanorod-based photoanodes. Phys. Chem. Chem. Phys. 15: 13483–13488
108. Österlund L, Mattsson A, Leideborg M, Westin G (2007) Photodecomposition of acetone on ZrOx-TiO2 thin films in O2 excess and deficient conditions, Proc. 31:st Int. Conf. Adv. Ceram. Comp., Nanostructured Materials and Nanotechnology, S7: 175-186
109. Mattson A, Leideborg M, Larsson K, Westin G, Österlund L (2006) Solar light decomposition of acetone on TiO2 and Nb-doped TiO2 thin films. J. Phys. Chem. B. 110: 1210-1220
110. Groot F D, Kotani A (2008) Core Level Spectroscopy of Solids. CRC Press
111. Kapilashrami M, Kronawitter C X, Törndahl T, Lindahl J, Hultqvist A, Wang W C, Guo J (2012) Soft X-ray characterization of $Zn_{1-x}Sn_xO_y$ electronic structure for thin film photovoltaics. Phys. Chem. Chem. Phys. 14(29): 10154-10159
112. Ding L, Boccard M, Bugnon G, Benkhaira M, Nicolay S, Despeisse M, Ballif C (2012) Highly transparent ZnO bilayers by LP-MOCVD as front electrodes for thin-film micromorph silicon solar cells. Sol. Energ. Mat. Sol. C. 98: 331-336
113. Aouaja M Ait, Diazb R, Belayachia A, Ruedab F, Abd-Lefdil M (2009) Comparative study of ITO and FTO thin films grown by spray pyrolysis. Mater. Res. Bull. 44: 1458-1461
114. Hongsingthong A, Krajangsang T, Yunaz I A, Miyajima S, Konagai M (2010) ZnO films with very high haze value for use as front transparent conductive oxide films in thin-film silicon solar cells. Appl. Phys. express 3(5): 051102
115. Calestani D, Pattini F, Bissoli F, Gilioli E, Villani M, Zappettini A (2012) Solution-free and catalyst-free synthesis of ZnO-based nanostructured TCOs by PED and vapor phase growth techniques. Nanotechnology 23(19): 194008
116. Huang W H, Sun S J, Chiou J W, Chou H, Chan T S, Lin H-J, Kumar K, Guo J-H, (2011) Electronic structure of Al-doped ZnO transparent conductive thin films studied by x-ray absorption and emission spectroscopies, J. Appl. Phys. Lett. 110:103705
117. Chen X, Mao S S, (2007) Titanium dioxide nanomaterials: Synthesis, properties, modifications, and applications. Chem. Rev. 107:2891-2959
118. Diebold U, (2003) The surface science of titanium dioxide. Surf. Sci. Rep. 48:53-229
119. Anpo M, Takeuchi M, (2003) The design and development of highly reactive titanium oxide photocatalysts operating under visible light irradiation. J. Catal. 216:505-516

Assessment of the Electronic Structure of Photo-electrodes with X-Ray and Electron Spectroscopy

Debajeet K. Bora and Artur Braun

Abstract The photoelectrodes are used in solar ware splitting reaction in order to generate future green fuel in the form of hydrogen. However the detailed assessment of the electronic structure of these materials is necessary in order to increase the performance of the electrode by tuning at the molecular level. The assessment of the photoelectrodes electronic structure with synchrotron spectroscopy is described within this chapter. The scope of the chapter is to give the reader a concise overview of the X- Ray and electron based spectrosocpic characterization techniques to shed light upon the electronic structure of solar water splitting photoelectrodes.

Keywords Hematite • Electronic structure • NEXAFS • PES • In situ spectroscopy

1 Introduction

The functionality of electrochemical energy conversion and storage devices (batteries, fuel cells, photoelectrochemical cells, and the like) depends at large on their transport properties. These include the electric transport by electrons, holes, and polarons, the ionic electric transport, and also the mass transport such as liquids or fluids and the heat transport. The photoelectrodes are the inorganic backbone in photoelectrochemical cells (PEC). At the molecular level, it is the electronic structure which determines the transport properties. The most important function of a photoelectrode is its light absorption properties owing to its optical bandgap energy. The optical properties are a direct reflection of the electronic properties. For solar applications, the bandgap should be in an energy range which spans the wavelengths of the solar radiation. The optical properties are typically studied with optical spectroscopy. The advantage of soft X-ray spectroscopy over optical spectroscopy in this

D.K. Bora • A. Braun (✉)
Laboratory for High Performance Ceramics, Empa. Swiss Federal Laboratories for Materials Science and Technology, CH – 8600 Dübendorf, Switzerland
e-mail: artur.braun@alumni.ethz.ch

E.A. Rozhkova, K. Ariga (eds.), *From Molecules to Materials*,
DOI 10.1007/978-3-319-13800-8_11

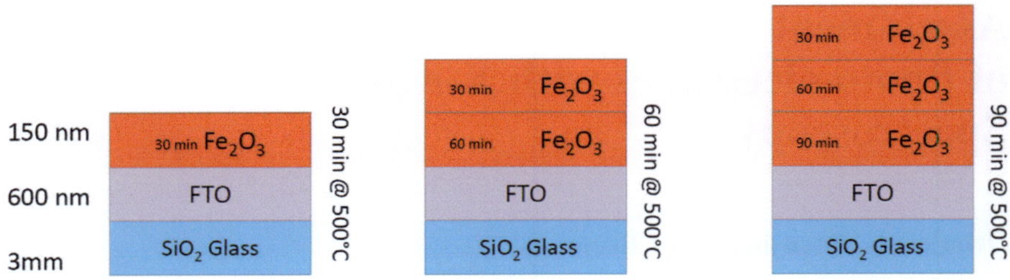

Fig. 1 Scheme showing the configuration electrode - electrolyte interface under operando configuration

respect is that the core levels can be assessed directly, which makes X-ray spectroscopy element specific. In addition to the optical properties, the photoelectrode should have a good electrical conductivity so as to allow a photocurrent density to be built up. In PEC, photoelectrodes are in contact with an electrolyte. The electrode–electrolyte interface must allow for charge transfer for electrons and holes (Fig. 1).

Hence, there is little exaggeration in saying that photoelectrochemical cells for water splitting or carbon dioxide reduction are not easily beaten by other systems in terms of experimental complexity. The bulk and the volume of the electrode constitute the light absorber, and unless particular photonic effects are employed, the absorbed light in the electrode scales with the volume of the absorber. The absorption properties and the optical properties of the electrode are primarily a problem of conventional solid-state physics or condensed matter physics.

The role of the absorber is to capture the sunlight and convert as much of it as possible into electron–hole pairs and also make sure that the electrons and holes remain separated as long as possible and in the ideal case move to the current collector and to the electrolyte. This too is a matter of solid-state physics.

Since the absorber is typically deposited on an electric current collector, often fluorine-doped tin oxide (FTO) deposited on a rigid glass support, the electronic bands of the absorber and current collector should somehow match. Since PEC systems often are built on metal oxides, their oxygen nonstoichiometry and readiness to interact with ambient humidity can pose electronic incompatibilities which amount in ohmic losses and drop of performance of the PEC system at the solid-state side.

It is therefore necessary to accurately determine the work function and *Austrittsarbeit* of the metal oxide layers and also control their stoichiometry during electrode synthesis and processing. A very good example for the empirical and quantitative assessment for this situation is exercised in [1, 2].

At the electrode side where the absorber is interfacing the (liquid) electrolyte, further problems will arise. For the photoanode, the holes produced in the absorber are expected to move toward the electrolyte and oxidize there the water according to the reaction $4\ h^+ + 2H_2O <=> 4H^+ + O_2$. Even if the hole has made it to the absorber/electrode surface, it is not for sure that there it will be able to oxidize a water molecule. Note that when the absorber or the electrode is inserted into the electrolyte, the Fermi level of the electrode will adjust so as to match the redox potential of the electrolyte. This is done by bending the energy band of the electrode toward the redox potential, which forms a charge carrier depletion layer, the spatial width of which scales with the applied bias potential by a parabolic law.

For iron oxide, for example, it is necessary to apply such bias in order to raise its conduction band high enough to be able to oxidize the water, regardless of the fact that the absorbed light energy quanta are energetically high enough to oxidize water.

Practically, however, the necessary bias voltage can be significantly higher than this aforementioned bias voltage, because the well-known overpotentials exist at the electrode–electrolyte interface, including the kinetic barriers. It is the task of electrocatalysis to lower these overpotentials. From the aforementioned outline, it is clear that the complete experimental assessment of the electronic structure of an actual realistic photoelectrode is a very difficult task. Below, we will step by step walk through the difficulties that can be faced.

2 Ex Situ Studies of the Electrode Bulk

A first assessment of the electrode bulk as it constitutes the absorber should be made with conventional (powder) X-ray diffraction. This allows the determination of the crystallographic phase for the major part of the electrode. It will be difficult to discriminate the top and bottom portions of the absorber from the bulk using XRD, but it is possible. One issue is that PEC electrodes have to be porous. The highly open porous structure then allows forming a large solid–liquid interface. For this, the absorber materials should be arranged in a self-similar branched structure, where every branch has a bulk and a surface layer. This will make it difficult to use, for example, low-angle X-ray diffraction to differentiate the surface structure from the bulk structure. In such a case, it can be necessary to use model systems of nonporous electrodes like physical vapor deposited films and in the extreme case even single crystals.

The optimum absorber layer thickness relates directly with the bandgap energy and increases with increasing bandgap and can range from several 100 nm to several microns. A hard X-ray absorption or emission spectrum of the metal K-edge would probe the entire sample and allow for the determination of the averaged metal ion oxidation state, potentially along with charge transfer satellite information when RIXS maps are recorded [3]. The same would hold for the metal L-edges, which come from an information depth of around 1 μm [4].

An example for a bulk study is explained here [5]. Let us stay with iron oxide as PEC electrode material. It can be synthesized in a wet, soft chemical method by dissolving iron nitrate in oleic acid, heat it to 125 °C to cause combustion, and dissolve the reaction product in tetrahydrofuran where nanoparticulates precipitate. Heating these nanoparticulates to around 550 °C will convert them to the hematite phase (α-Fe_2O_3). We monitor the structural evolution of the material during high-temperature

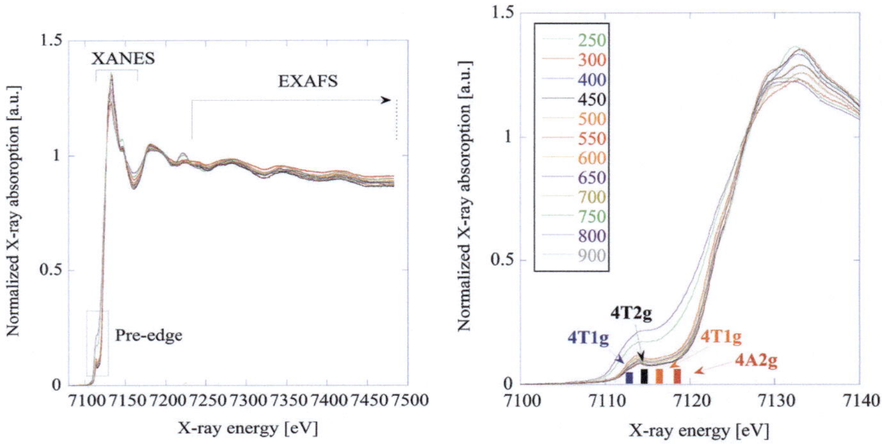

Fig. 2 Fe XAS spectra of iron oxide particles heated in air from 250 to 900 °C (*right*). The magnified region of the XANES and pre-edge show distinct spectral changes depending on heat treatment temperature. The spectra were recorded at the Swiss Norwegian Beamline of the ESRF in Grenoble, France

treatment from 250 to 900 °C ex situ with X-ray absorption near-edge spectroscopy (XANES) at the iron K-shell absorption edge at around 7100 eV. The attenuation length of X-rays in this energy range (6,000–8,000 eV) ranges from 80 to 10 μm, with the absorption edge just above 7,110 eV [6]. Hence, a photoelectrode with 0.5 μm thick iron oxide absorber layer can be completely probed with Fe K-shell XANES.

We do not consider in this Chapter the extended X-ray absorption fine structure (EXAFS) oscillations, which occur for iron at 7,200 eV and extend to far beyond 7,500 eV and can amount to a time-consuming synchrotron experiment. These oscillations account for the near-range periodicity and order of the oxygen and next–next neighbor iron atoms of the iron in its coordination shell, which can be extracted by Fourier transformation of that EXAFS part of the spectra [7]. Since the EXAFS spectroscopy is an element-specific method by definition, it can insofar outperform conventional XRD, particularly when the material lacks long-range order. Figure 2 shows a set of Fe XANES spectra of the particulates heated from 250 to 900 °C.

What we learn from the pre-edge peaks is that the as-prepared nanoparticles contain a significant amount of Fe^{2+}, as can be drawn from the gradual decrease of the t_{2g}/e_g peak height ratio of the two leading pre-edge resonances. It requires an additional heat treatment temperature so as to convert the nanoparticles in the necessary hematite phase, confirming oxidation of the Fe from Fe^{2+} toward Fe^{3+}. This result is paralleled by the XRD investigations [5].

The physicochemical properties of iron oxide and virtually all inorganic materials, particularly its electronic transport properties, originate not only from its elemental constituents, the iron and oxygen, respectively. Rather, the interplay of the oxygen and iron orbitals in the crystal lattice provides the electronic landscape in

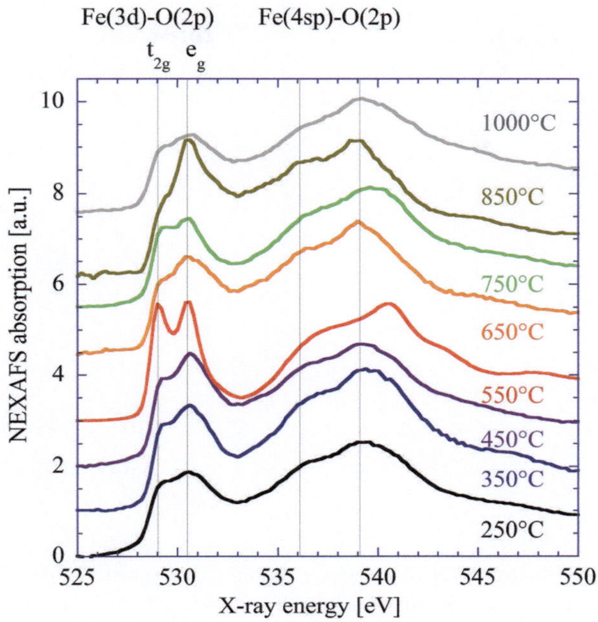

Fig. 3 X-ray spectra of the oxygen K-shell absorption edge of iron oxide nanoparticulates, depending on heat treatment temperature (HTT). Reproduced from Phys. Chem. Chem. Phys. 2013, 15, 1443–1451 with permission from the PCCP Owner Societies

which electrons and holes hop and diffuse so as to provide net charge transport. These (exchange) interactions where the O2p orbitals of the oxygen actually mediate the electron transport between the cations by virtue of the electron spins are summarized in the Goodenough–Kanamori rules [8].

It has therefore turned out by many soft X-ray studies that the ligand ion absorption edges, particularly the oxygen K-edges, carry valuable and visible information about the valence band of compounds, more so than the metal edges. This holds particular for electron holes, which sometimes show up spectacular in the oxygen pre-edge [9, 10].

Figure 3 shows the near-edge X-ray absorption fine structure (NEXAFS) spectra of the iron oxide nanoparticulates, recorded at the oxygen K-edge (oxygen NEXAFS spectra). The necessary heat treatment temperature (HTT) for iron oxide to be converted to hematite is around 550 °C in air. The corresponding spectrum in Fig. 3 shows a double peak at around 530 eV, indicative to hematite. The oxygen ligand spectrum in this energy range probes virtually the valence band of the compound and shows the hybridized states from Fe3d and O2p orbitals. In the range around 535–540 eV, the corresponding hybridized Fe(4sp)-O(2p) states show up. The structural rearrangement in the crystal lattice during HTT manifests in the spectral changes in the oxygen NEXAFS spectra.

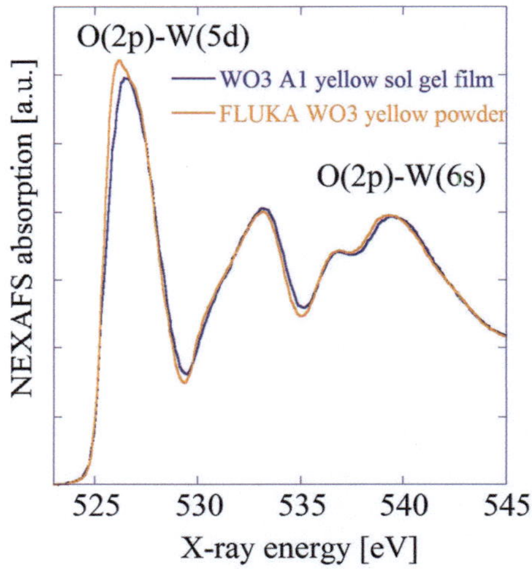

Fig. 4 NEXAFS spectrum of nominal WO_3 from commercial batch of FLUKA and from sol–gel-synthesized film (Samples courtesy of Dr. Renata Solarska, Empa. Spectra recorded at ALS Berkeley, BL 9.3.2)

The X-ray absorption edge of oxygen is located at around 525 eV. This is the soft X-ray energy range. The attenuation length of X-rays from 500 to 600 eV ranges from around 0.4 to 1.1 μm. Hence, an iron oxide photoelectrode with an optical absorption length of 0.5 μm can still be fully bulk probed in terms of its oxygen NEXAFS spectra.

When we want to investigate a tungsten oxide ($WO_{3-}\delta$) photoanode, we have to bear in mind that those electrodes must be thicker because of their larger optical absorption length, which can be around 3 μm. An X-ray beam of 525 eV will experience an extreme attenuation over such length. The electrode will be too thick for a bulk oxygen NEXAFS spectrum. The spectra below in Fig. 4 were recorded from a sol–gel-derived WO_3 film and from a layer of commercial WO_3 powder (FLUKA). Although the materials were synthesized by different routes, their normalized oxygen spectra match very well. Noticeable differences are seen in the valence band from O(2p)–W(5d) hybridized states. Particularly, a somewhat sharp resonance peaks off at around 526 eV for the commercial powder sample.

For TiO_2-based photoelectrodes, which have larger bandgap energy than WO_3 and Fe_2O_3, the optical absorption length is even larger, and thus, the electrodes tend to be thicker. Hence, a bulk oxygen NEXAFS is virtually impossible [4]. However, this does not rule out meaningful oxygen NEXAFS studies. One just has to be aware of the shortcomings just mentioned. Below we see the oxygen NEXAFs spectrum of nitrogen-doped TiO_2 photocatalyst nanoparticles [11]. Comparison with pure TiO_2 and various doping concentrations has shown that the nitrogen induces an additional, third spectral signature in the Ti(3d)–O(2p) doublet. This becomes obvious when we compare the different spectra in a graphical overlay, as is indicated in the insert in Fig. 5. Subsequent peak deconvolution then points to an additional peak, which can be quantified by the least square fit of the deconvoluted spectra.

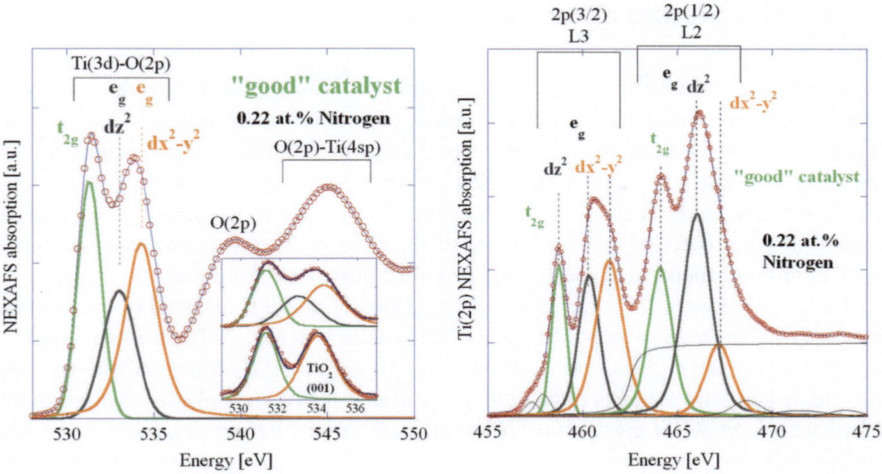

Fig. 5 Oxygen NEXAFS spectrum of nitrogen-doped titania. Ti L-edge spectrum. Spectra recorded by Dr. Ashley Harvey, ETH Zürich, with Peter Purton, PhD, at Daresbury, UK. Adapted with permission from J. Phys. Chem. C 2010, 114, 516–519. Copyright 2010 American Chemical Society

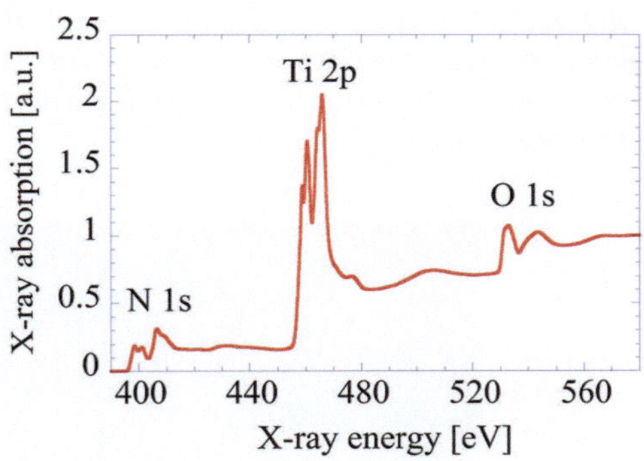

Fig. 6 Complete NEXAFS scan from 400 to 800 eV for a nitrogen-doped TiO2 powder samples. Spectra recorded by Dr. Ashley Harvey, ETH Zürich, at Daresbury, UK. Adapted with permission from Journal of Electron Spectroscopy and Related Phenomena 190 (2013) 93–105. Copyright 2012 Elsevier B.V

Before the X-ray absorption energy range for oxygen, the titanium metal L-edges have their orbital multiplet spectrum which originates from the spin orbit coupling, producing the $Ti2p_{3/2}$ and $Ti2p_{1/2}$ branches. The crystallographic crystal field causes an additional lift of the degeneracy of energy levels and produces a t_{2g} and e_g peak for every branch, making Ti compounds a gratifying system for L-edge spectroscopy. The right part in Fig. 5 shows the L-edge spectrum of the 0.22 % N-doped TiO_2.

Below the Ti L-edge, nitrogen has the K-shell absorption edge extending from 390 to 340 eV. Figure 6 shows thus an example where in one long scan, all three elements in N-doped TiO_2 can be probed. An extensive introduction and treatment of the metal L-edges in solar cell materials is given by Jinghua Guo in this book in Chapter 8.

3 Ex Situ Studies of the Electrode Surface

One essential difference between photovoltaic electrodes and electrodes for solar water splitting is that the latter are typically inserted in an aqueous electrolyte. There are other alternatives for solar water splitting, such as the combination of a separate PV unit with a separate water electrolyzer or the reverse operation of a PEM fuel cell. In the former, the PV electrode is exposed to the atmosphere, maybe coated with a protective layer. In the latter, the water-splitting electrode is in contact with a polymer membrane electrolyte in a very humid environment.

Here, we only consider the case that the photoelectrode is inserted in an aqueous electrolyte. An actual water-splitting device is shown in Fig. 7. The photoanode is a 10 cm × 10 cm large glass panel coated with a 10 nm thin FTO layer, on which a 500 μm thick hematite layer is deposited [12]. It is inserted in a plastic compartment filled with 1 M KOH electrolyte. A simple mechanical barrier separates this from the cathode compartment, where a platinum wire is the cathode inserted in the same electrolyte. Three silicon solar cells of 0.5 V each are connected with the electrodes and provide the necessary 1.5 V DC bias. Upon illumination by the sunlight, oxygen gas evolves at the iron oxide electrode and is collected in a vial. The hydrogen bubbles evolving at the platinum wire are collected in a separate vial.

In this natural operational environment, electrodes can be measured in situ or, *operando*, with conventional electrochemical methods, such as cyclic voltammetry and impedance spectroscopy [13], which is a routine work. However, it is not straightforward to conduct X-ray and electron spectroscopy measurements on PEC electrodes during operation.

Fig. 7 PEC device with 10 cm × 10 cm large iron oxide anode. The DC bias of 1.5 V is provided by three small silicon PV units with 0.5 V each and gas collection containers. The *right side* shows the hydrogen-evolving compartment with H_2 gas bubble evolution on the Pt counter electrode wire

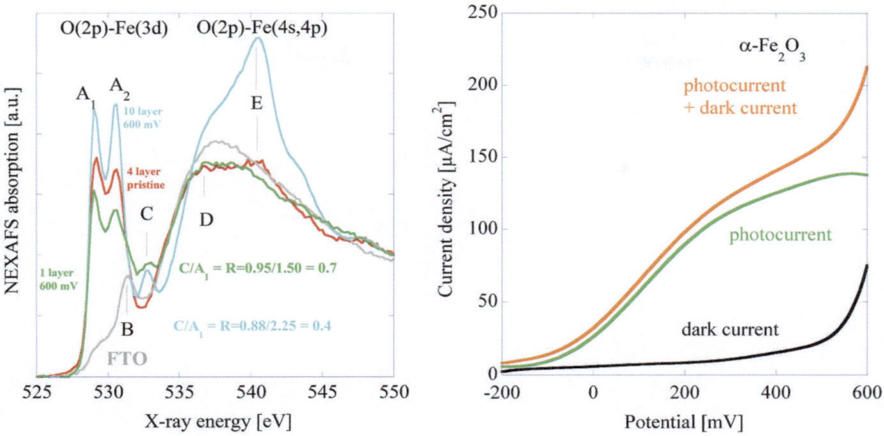

Fig. 8 Oxygen NEXAFS spectra of FTO, pristine hematite (four layer thick), and 600 mV anodized hematite with one-layer and ten-layer thickness. Photocurrent density of hematite films (*green middle curve*) obtained from subtraction of the dark current (*bottom spectrum*) from the light current (*top spectrum*). *Bottom* shows a schematic of the different layers that constitute a hematite photoanode. Reprinted with permission from J. Phys. Chem. C, 2011, 115 (13), 5619–5625. Copyright 2011 American Chemical Society

We therefore continue here for the time being with ex situ analysis of electrodes and look at the oxygen NEXAFS spectra of hematite electrodes that had been operated [14]. Figure 8 shows the spectra of two hematite PEC films that were anodized in KOH at 600 mV vs. Ag/AgCl reference and one pristine hematite film. In addition is an oxygen spectrum from fluorine-doped tin oxide (FTO) film shown, which in PEC is typically used as a transparent conducting oxide (TCO) underlayer. The pristine film is four-layer thick (which is around 0.5 μm) and shows the pronounced double-peak structure plus and a distinct intensity minimum at around 532 eV, followed be the readily known Fe(4sp)–O(2p) hybridized states. The hematite is deposited on a 20 nm thin FTO layer, the oxygen spectrum of which is shown as well (peak B).

The FTO spectrum will "shine through" particularly in any thin overlayer, such as the one-layer-thin (125 nm) hematite, which has an intensity shoulder at around 532 eV, exactly the position where FTO has peak B. This FTO signature is not visible anymore in the thicker, four-layer iron oxide film. The next observation we make is a newly emerging peak in the intensity minimum at position C, when the films were anodized at 600 mV. The ten-layer-thick film (around 1.3 μm) shows a very pronounced such peak, whereas the one-layer-thin film shows a more diffuse such peak. However, we have to relate the spectral weight of this peak intensity to the height of the A1 and A2 doublet. Then we learn that the new peak in C is specific to the surface. This surface species originates likely from an oxyhydroxyl-like species connected with an iron ion in a very high oxidation state, such as Fe^{4+} or higher (see Pourbaix diagram in Fig. 9 in ref. [15]). This observation is a very important one. The iron oxide photoelectrode needs a constant electrical bias during the

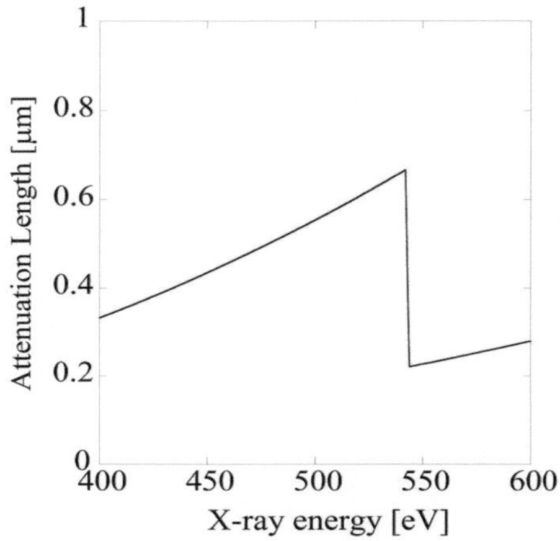

Fig. 9 Calculated [6] X-ray attenuation length for X-rays with energy from 400 to 600 eV in hematite with 5 g/cm³ mass density

photoelectrochemical operation [16, 17] which basically constitutes an anodization of the iron oxide. In a strongly alkaline electrolyte like the pH 13 KOH, the Pourbaix diagrams predict the existence of a highly oxidized Fe phase exceeding even Fe^{4+}. The new structures that we have observed in position C in the NEXAFS spectrum of anodized hematite do not show up when the bias potential is only 200 mV, for example. The observations and consideration made herein show us that the surfaces of the photoelectrodes are in any way influenced by the operation of the PEC device.

The fact that the oxygen NEXAFS experiment probes the electrode with an information depth of around 0.4–1 µm (Fig. 9) shows that while we clearly observe new structures that must be originating from changes at the surface, also the bulk electronic structure adds a significant background to the spectra. This is why X-ray absorption spectroscopy (XAS) is not a distinct surface-sensitive method.

For a clearer inspection of the surface of photoelectrodes, X-ray photoelectron spectroscopy (XPS) is the more suitable method [18] or photoemission spectroscopy (PES). It is particularly the very small mean free path of electrons in solids, depending on their kinetic energy, which makes XPS a surface analytical method.

We have subjected hematite electrodes to a systematic anodization protocol for various bias potentials and for various times. These samples were then analyzed with synchrotron-based XPS. Figure 10 shows the XPS valence band spectra of four samples that were *anodized* at 600, 200, 0 mV, and open circuit potential (OCV, typically around –0.2 V), i.e., kept under these four different conditions for 1 min in 1 M KOH, rinsed in distilled water, dried, and brought into the UHV chamber of the synchrotron end station (Beamline 7, Advanced Light Source, Berkeley).

The first observation is that the spectra shift toward smaller binding energies (toward the Fermi level) with increasing potential. This means that that the anodization causes hole doping in the electrode structure. The XPS valence band spectra are at large showing the hybridized states from Fe3d and O2p contributions. Beforehand, it is not possible to determine which ions or orbitals contribute to the suggested hole-doping effect. Therefore, we have tuned the X-ray photon energy to the Fe2p

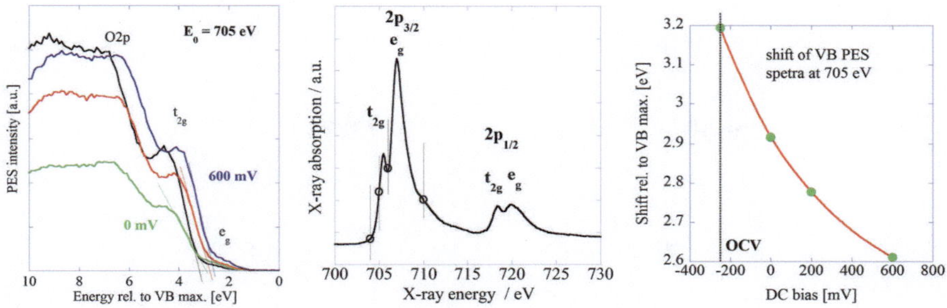

Fig. 10 Resonant Fe2p XPS valence band spectra of differently anodized iron oxide electrodes. Fe L-edge spectrum of iron oxide electrode with $2p_{3/2}$–$2p_{1/2}$ multiplet and t_{2g}, e_g crystal field splitting. The *green circles* with vertical lines indicate the X-ray energies at which the resonant VB spectra were recorded. On the *left*, we see the systematic energy shift of the VB XPS spectra as a function of DC bias [19]. Reprinted with permission from ChemPhysChem 2012, 13, 2937–2944. Copyright 2012 Wiley-VCH Verlag GmbH& Co. KGaA, Weinheim

absorption threshold of around 705 eV, which allows enhancement of the X-ray optical contrast between iron and oxygen. We clearly observe how the oxygen bonding peak (O2p) at around 6–7 eV changes its intensity with increasing photon energy. Moreover, the relative spectral weight of the e_g peak (at 2 eV, indicative to Fe^{2+}) and t_{2g} peak (at 4 eV, indicative of Fe^{3+}) changes significantly with photon energy.

The shift of the VB spectra is clearest when determined with Fe-resonant photon energy. This suggests that the holes which are formed during anodization may arise from Fe 3d levels. As we will see in the next section, such Fe 3d-type holes were recently observed together with O2p holes with in situ NEXAFS experiment [20]. This is not withstanding the possibility that it could be possible to identify a resonant VB PES effect with a shift when resonant measurements of the O 1 s transition energy are performed.

What we also learn from this XPS study is that the iron oxide films are not entirely Fe_2O_3. Rather, the small intensity of the e_g peak at around 2.5 eV in Fig. 10 indicates presence of Fe^{2+} at the sample surface. For further investigation of this finding, we chose another iron-resonant synchrotron experiment with XPS. Figure 11 shows the Fe3p-resonant valence band spectra of a hematite sample which had been anodized for 10 min at 500 mV. Spectra were recorded with X-ray photon energies ranging from 50 to 60 eV, thus scanning through the Fe3p absorption threshold. The 50 eV photon energy seems to be resonant for the O2p peak, because it has the highest intensity at this energy. At 60 eV, the peak from the orbitals with t_{2g} symmetry (around 3 eV) has the highest intensity and is very sharp. The signature of the e_g orbital (around 1.5 eV) has a maximum intensity at 56 eV. Because of our interest in this spectral feature which is indicative to Fe^{2+}, we chose to conduct the further studies with 56 eV photon energy. The graph on the right side in Fig. 11 shows the thus obtained VB spectra from hematite, which has been anodized for five different times at 700 mV, i.e., from 0 to 1440 min. Here, we notice systematic changes in the spectral weight of the e_g peak, which is decreasing with increasing

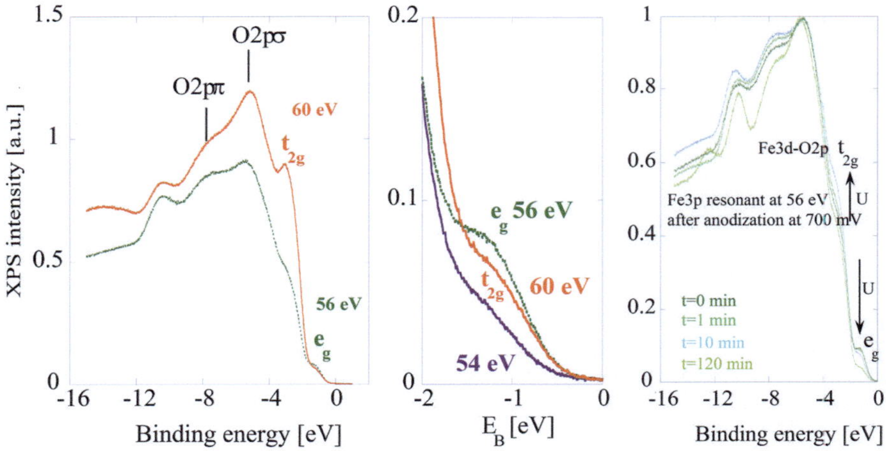

Fig. 11 Photoemission spectra of hematite recorded in Fe3p-resonant mode with 56 and 60 eV photon energy. The *middle panel* shows the magnified region near E_F to highlight the resonant effect for 56 eV photon energy. *Right panel* shows the spectra of hematite anodized for 0, 1, 10, and 1440 min at 700 mV. Reproduced from Phys. Chem. Chem. Phys. 2013, 15, 1443–1451 with permission from the PCCP Owner Societies

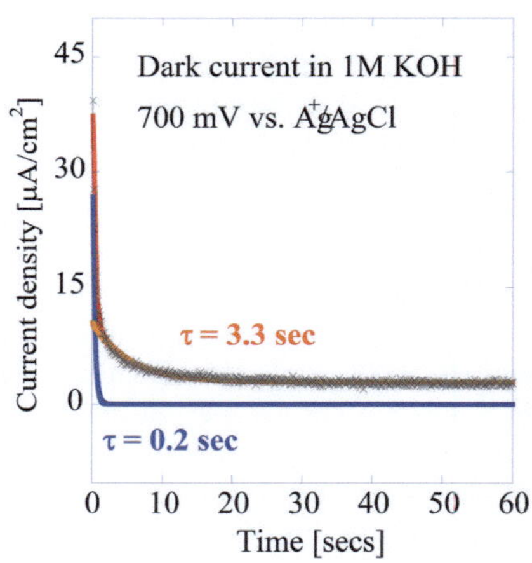

Fig. 12 Current density transient of hematite during the first minute of anodization at 700 mV in 1 M KOH electrolyte in dark condition [15]. The transient is modeled with two exponentials (*solid lines*) which yield time constants of 0.2 s and 3.3 s

anodization time. In return, the t_{2g} peak is increasing in intensity. We find therefore confirmation that the pristine hematite film is at the surface enriched with Fe^{2+}, and the concentration of this Fe^{2+} becomes depleted during anodization.

It is therefore obvious that the as-prepared hematite electrode is oxygen deficient at the surface. This is why we observe Fe^{2+} in the X-ray spectra. The current transient in Fig. 12 shows a rapid decay during the first few seconds, which we can attribute to a fast process likely from electrochemical double-layer charging and a somewhat slower galvanic process where the Fe^{2+} is converted to Fe^{3+}. Likely this

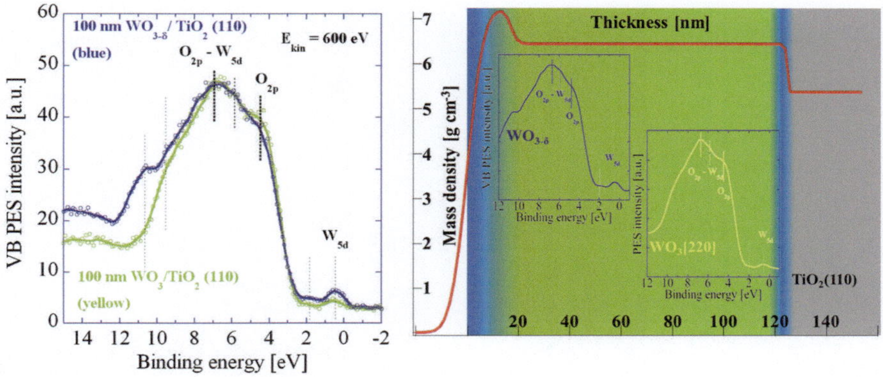

Fig. 13 Schematic of a 100 nm thick tungsten oxide film deposited with laser ablation on a TiO2 single-crystal substrate. The *bluish color* at the film surface and interface denote oxygen-deficient regions. The *solid red line* represents the variation of the mass of the film, as determined from the least square fit of X-ray reflectometry data and model. The VB PES spectra were recorded from the as-deposited and the oxidized film and show two defect structures in the VB region near the Fermi energy. Reprinted with permission from J. Phys. Chem. C, 2011, 115 (33), 16411–16417. Copyright 2011 American Chemical Society

process extends over time, as we notice from the decrease of the Fe^{2+} and increase of the Fe^{3+} structure in the four VB spectra in Fig. 11, right panel.

The oxygen deficiency on a metal oxide surface is not restricted to iron oxide. Rather, oxygen deficiency seems to be the rule at the surface. In Fig. 13, we see the schematic of a 100 nm thick tungsten oxide film, deposited with pulsed laser ablation on a TiO_2 single-crystal substrate. We have studied such films of 10 nm and 100 nm thickness on TiO_2 with various crystal plane orientations with synchrotron-based XPS spectroscopy. These films came with bluish color after synthesis, a clear indication for oxygen substoichiometry, i.e., the films stoichiometry is rather WO3-δ, $\delta < 1$. Note that the as-deposited, blue film is thus more conducting, has a much smaller bandgap, and is thus not suited as an absorber material for PEC applications. Oxidative treatment will fix this and turn the film into a 2.7 eV bandgap absorber material.

We have oxidized an aliquot of these films in oxygen atmosphere at 400 °C for 48 h, after which the color had changed to pale yellow without any traces of blue. How does the defect structure of the film look like? To answer this question, we record the valence band spectra. Oxygen-deficient tungsten oxide is in this respect a grateful system because the defect states from oxygen vacancies show up nicely in the VB spectra [21].

The right panel in Fig. 13 shows the VB spectra for as-deposited and oxidized condition. Between 0 and 3 eV, we see two broad peak structures, one with high intensity at around 0.5 eV and one with low flat intensity at around 2 eV. The overall intensity of these peaks is higher for the reduced blue sample and originates from oxygen vacancies in the tungsten oxide. We learn that even the oxidized sample

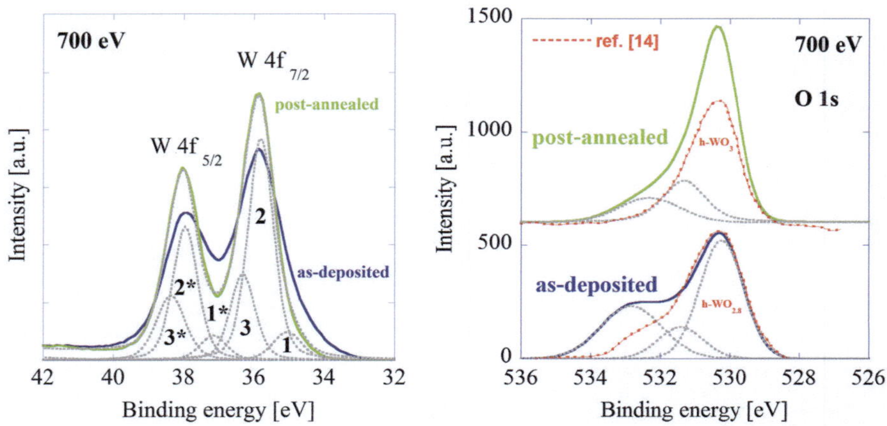

Fig. 14 W4f and oxygen XPS core-level spectra of the as-deposited film and the oxidized film, recorded in UHV at 300 K. Reprinted with permission from J. Phys. Chem. C, 2011, 115 (33), 16411–16417. Copyright 2011 American Chemical Society

with distinct yellow color has still such defect peak intensity. Excitation energy-dependent XPS studies [21] show that for the deeper regions of the film, the defect state signature is less pronounced in the post-treated film. For the tungsten oxide interfacing the TiO_2 substrate, we notice an increase in the defect peak intensity. We have investigated this film prior to and after oxidation with X-ray reflectometry and find that oxidation improves the smoothness of the film surface, as indicated by the onset of oscillations.

Finally, we want to inspect the W4f core-level region because this is specific to the element tungsten in the film. Figure 14 shows the XPS core-level spectra with the W4f doublet, recorded with 700 eV excitation energy [21]. The as-deposited film has a doublet which is slightly shifted to lower binding energies when compared to the one annealed in oxygen. This one is also broader than the oxidized one, suggesting that this film is more covalent, whereas the oxidized one is more ionic bonding. This suggestion is corroborated by the corresponding oxygen core-level spectra.

The doublet is deconvoluted into actually six peaks [21], which determine the width of the entire spectrum. The intensity of the central peaks labeled 2 and 2* is increasing upon oxidation. The corresponding oxygen core-level spectra show a more distinct change. Particularly, the defect peak at 533 eV is depleted upon oxidation.

4 In Situ Investigation of the Semiconductor Subsurface and Depletion Layer

As we have illustrated in Fig. 1, a multilayer film may be composed of several dipped layers, for example. In this particular case, the first iron oxide layer (layer #1) has been heated for 30 min before the next layer (layer #2) was deposited, which

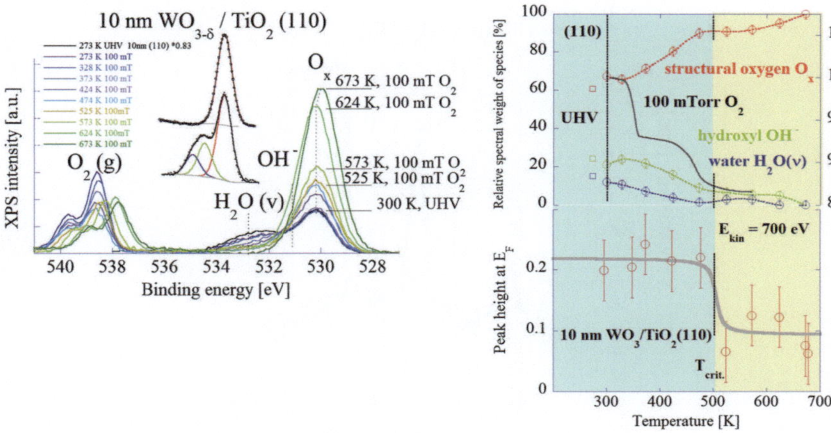

Fig. 15 Summary of in situ XPS study: Evolution of the oxygen core-level XPS spectra in a 10 nm tungsten oxide film on TiO_2 (110) during heating to 673 K and exposure to 100 mTorr oxygen in an ambient pressure XPS chamber. The *right panel* shows the evolution of spectral weight of peaks from structural oxygen, hydroxyl groups, water vapor, and the weight loss from thermogravimetry during temperature increase. The *bottom part* in the *right panel* shows the behavior of the VB XPS defect-state intensity. Reprinted with permission from Chem. Mater., 2012, 24 (17), 3473–3480. Copyright 2012 American Chemical Society

then is also heated for 30 min. Hence, layer #1 has been heated 2×30 min. An electrode with four layers has then layer #1 exposed 120 min to 500 °C, whereas the top layer #4 was heated only 30 min at that temperature. One possible scenario is then that Sn (tin) from the FTO layer or even Si (silicon) from the glass substrate may diffuse into the iron oxide absorber layers. It may depend on time, thickness, and microstructure how much of these dopant ions arrive in the absorber layer or even on their surface. Depending on the actual situation, this not-intended doping may be so low that it is difficult to analyze or even verify.

We have deposited with pulsed laser ablation 10 nm of tungsten oxide on TiO_2 substrates with different crystallographic orientations, as explained in the previous section. These oxygen-deficient films were then subjected to XPS spectroscopy while being heated in the XPS measurement chamber under 100 mTorr oxygen gas partial pressure (near-ambient pressure XPS, [22]). We observe that with increasing temperature up to 400 °C, the defect states which show up in the valence band spectra near the Fermi energy become depleted. Inspection of the corresponding oxygen core-level spectra shows systematic and significant changes that arise from the interaction of the oxygen-deficient tungsten oxide with the supplied oxygen, the readily prevalent adsorbed humidity in the film, activated by the temperature ramp. The filling of the oxygen vacancies is preceded by an intermediate formation of hydroxyl species. This parallelism of processes is illustrated in Fig. 15, where we summarize the evolution of spectral weight of defect peaks and oxygen species and the change of the weight during TGA experiments.

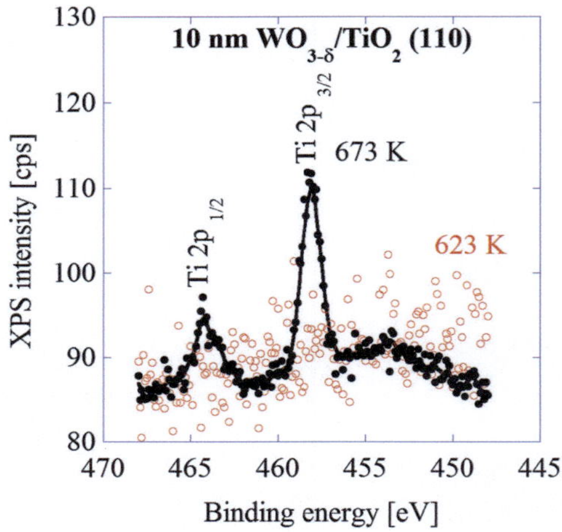

Fig. 16 Ti2p XPS spectra of a 10 nm tungsten oxide film deposited on TiO2 (110). Spectra were recorded with ambient pressure XPS at 100 mTorr oxygen partial pressure during heating from 300 to 673 K. Ti2p signal from interdiffused Ti from the substrate to the film surface cannot be identified at 623 K, but is clearly present at 673 K. Reprinted with permission from Chem. Mater., 2012, 24 (17), 3473–3480. Copyright 2012 American Chemical Society

Depending on the crystallographic orientation of the substrate, we find that another process is setting on, this is, the interdiffusion of Ti from the substrate into the tungsten oxide film. This is a possibility always to consider when multilayered or heterostructured electrodes are built. It is possible to identify such a progress by looking at the core-level XPS or X-ray absorption spectra of the suspected diffusing element. Figure 16 displays the Ti 2p core-level spectrum recorded at 623 K and at 673 K. There is no intensity that can be associated with Ti when recorded at the lower temperature at 623 K. After the temperature is increased to 673 K, a clear Ti 2p multiplet is visible, revealing that a substantial amount of Ti ions has diffused to the surface region of the WO_3 film, to an extent that it can be detected with the NEXAFS spectroscopy.

We have shown in the previous section that anodization of the iron oxide electrode (this is the electrochemical oxidation) can cause hole doping, as is shown by a shift of the valence band spectra toward the Fermi level. These are remnant holes because they remain in the sample even when the bias potential, which is necessary to carry out the anodization, is removed.

One question that could arise from this observation is, what is the influence of the light on the hole-doping process? If there is any.

We thus carried out a resonant photoemission experiment on a hematite single crystal in dark condition and when exposed to the light from a solar simulator. We have thus a series of PES spectra recorded along the Fe3p-resonant X-ray energies receded in light and in the dark. The spectral characteristics do change depending on the X-ray energy, as expected. We have plotted the intensity of the O2p bonding peak vs. the excitation energy and notice a sharp peak of the O2p intensity at 57.3

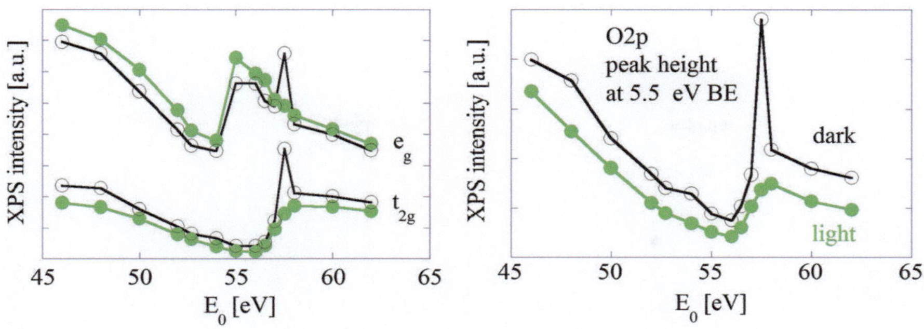

Fig. 17 Variation of the spectral weight of the e_g and t_{2g} shoulders and O2p bonding peak in the Fe3p-resonant VB spectra in dark (*open symbols*) and illuminated (*filled symbols*) hematite. Reprinted with permission from Journal of Electron Spectroscopy and Related Phenomena 190 (2013) 93–105. Copyright 2012 Elsevier B.V

eV resonant excitation as shown in Fig. 17. The intensity profile of the O2p peak vs. energy under illumination is virtually identical with the profile under the dark, except for this sharp peak at 57.3 eV. Under illumination, this peak transition is vanished. We thus conclude that the illumination induces photoconductivity which overcomes an electrostatic barrier originating from or at least being related with the oxygen bonding peak.

The same holds for the intensity shoulders from the e_g and t_{2g} symmetry orbitals in the VB spectra. Hence, the effect of suppressed overall intensity of O2p and $t_{2g}e_g$ upon illumination can be attributed to the entire superexchange unit Fe3d–O2p–Fe3d. A further observation is that the absolute intensity between 45 and 62 eV decreases upon illumination for the t_{2g} but increases during illumination for the e_g peak. This shows that Fe^{2+} signature from e_g becomes enhanced, on the cost of the Fe^{3+} signature from t_{2g}.

Now that we have learned about the influence of light and DC bias on the hematite photoelectrode, we want to advance to a combined photoelectrochemical in situ or *operando* study with NEXAFS spectroscopy. The condition for this experiment is as follows. We coat an around 30 nm iron oxide film on a 100 nm thick Si_3N_4 window, which is etched into a Si wafer with a Si_3N_4 layer. A metal bilayer from 1 nm Cr to 5 nm Au serves as buffer and current collector between hematite and Si_3N_4 support. This electrode assembly is glued on a PEEK micro spectro-photoelectrochemical cell which can be inserted into an UHV chamber. The Si_3N_4 is an X-ray window which supports the hematite film, which from the back is supplied with liquid electrolyte with a peristaltic pump. The cell contains a reference and counter electrode and thus provides an accurate 3-electrode system [23]. The X-ray beam enters the cell from the Si_3N_4 window, the metal buffer layers, and the hematite film and then reaches the electrolyte volume. The fluorescence signal is recorded as NEXAFS spectrum and contains the spectral signatures from all chemical elements which are in the X-ray optical path of the beam, that is, Si, N, Cr, Au, Fe, and O from the hematite and O, K, and H from the aqueous electrolyte. The cell

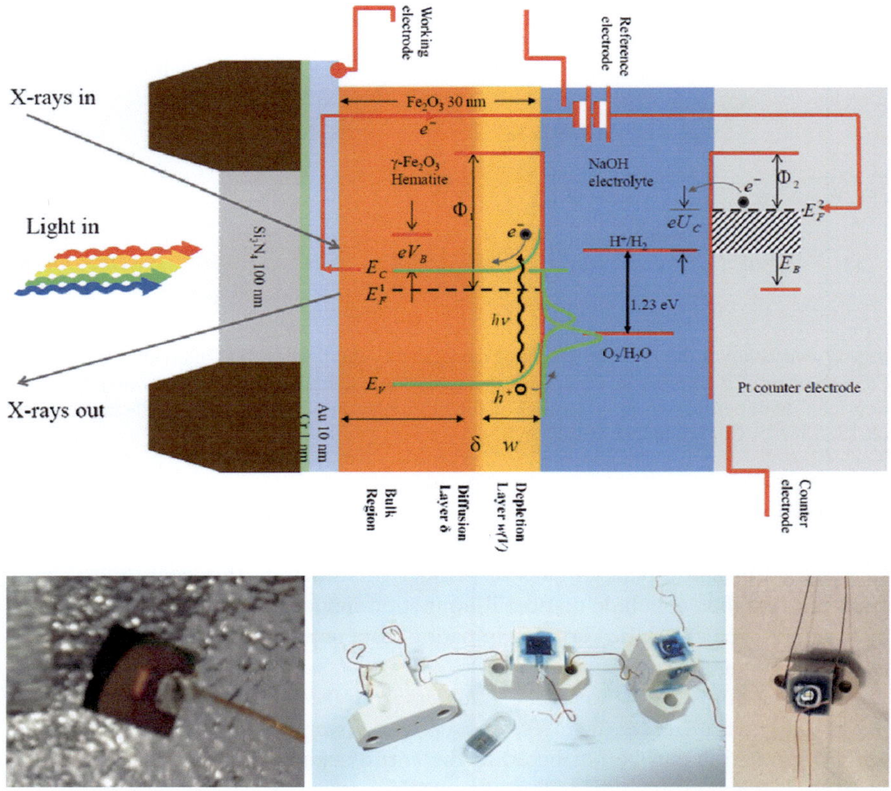

Fig. 18 Schematic representation of micro in situ spectro-photoelectrochemical cell at Advanced Light Source (ALS) in Berkeley with PEC electrode assembly during operando water-splitting NEXAFS experiment. The photograph shows a set of PEEK cells and one coated Si/Si_3N_4 frame with contact wire. Reprinted with permission from J. Phys. Chem. C, 2012, 116 (32), 16870–16875. Copyright 2012 American Chemical Society

is schematically sketched in Fig. 18 above. When the metal oxide film is thick, a lesser portion of the electrolyte will add to an oxygen spectrum than when the metal oxide film is thin.

We have recorded oxygen NEXAFS spectra in this manner, while the iron oxide film was under bias potentials from 0 to 900 mV in the dark and in light (Fig. 19). Close inspection of all spectra shows that upon illumination, two new spectral features evolve at around 526–527 eV in the bias range from around 300 mV to 900 mV. We can switch off these features by either removing the bias potential or switching off the light. These two features are formed by electron holes which are instrumental for the water oxidation according to the reaction

$$2Fe^{3+} \xrightarrow{h\nu} Fe^{2+}(Fe^{3+} - e^{-}) + Fe^{4+}(Fe^{3+} - h^{+})$$

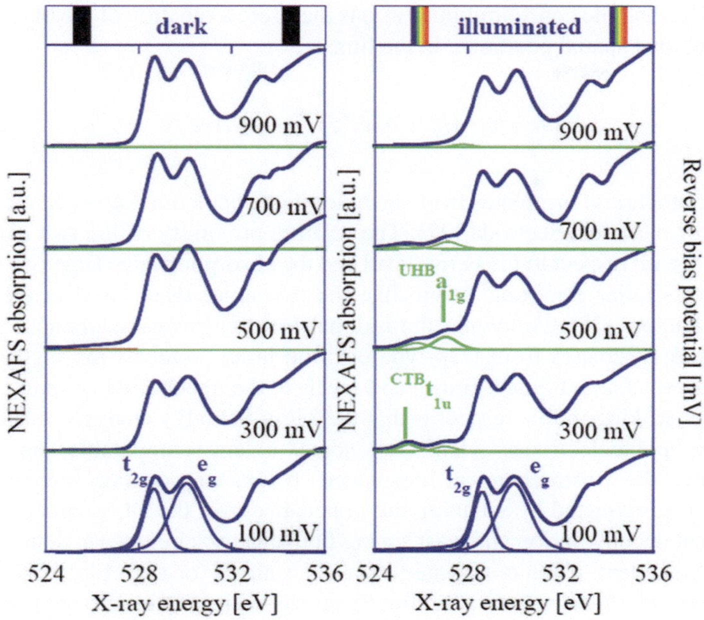

Fig. 19 Oxygen NEXAFS spectra from hematite film in KOH under DC bias from 0 to 900 mV in *dark* and in *light* recorded fully operando at the ALS in Berkeley. Reprinted with permission from J. Phys. Chem. C, 2012, 116 (32), 16870–16875. Copyright 2012 American Chemical Society

Note that this process can be interpreted as a photo-induced disproportionation of Fe^{3+}, where the corresponding Fe^{2+} is formed by an Fe^{3+} plus an electron e^- and the conjugated Fe^{4+} by an Fe^{3+} plus an electron hole h^+.

The existence of such two holes has been speculated in the past [24]. It has also been suggested that this photo-induced process could be the origin for the intrinsic conductivity in α-Fe_2O_3. This is an interesting suggestion because Toledano et al. [25] found a light-dependent effect on the UPS VB spectra of hematite with a reversible occurrence of an Fe^{2+} state. The extent of this reversible feature depended on the electron hole pair lifetime. It has not yet been shown that this suggested Fe^{2+} state is the one that can be seen in the Fe3p-resonant photoemission spectra.

Further analysis of the two hole states in the oxygen NEXAFS spectra suggests that the first state relates to a transition into the charge transfer band [20]. The spectral weight of this peak follows a sigmoidal characteristic when plotted vs. the DC bias potential. This is interesting because this sigmoidal profile provides a link to semiconductor photoelectrochemistry. Specifically, we can interpret the hematite film in electrolyte with the Schottky barrier theory and work by [26]. The thickness w of the charge carrier depletion layer in the film scales with the bias potential V in a parabolic growth law and is pinned by the flat band potential V_{fb} according to

$$w = w_0\sqrt{V - V_{fb}}$$

The electron hole concentration *h(w,x)* is therefore a function of the bias potential and the relative spatial position *x* in the film.

$$h = h_0 - (h_0 + A \cdot e^{-aw}) \cdot \frac{e^{w-x}}{L_h} + A \cdot e^{-ax}$$

The photocurrent originates from the holes in the depletion layer and the holes in the bulk of the photoelectrode [27]. The relative proximity of the two discovered hole peaks with respect to the Fermi level and the absorption edge suggests that both holes have similar electronic origin like the two holes discovered in the cuprate superconductors [28]. Following the rational for their peak assignment, the low-energy peak originates from O2p-type electron holes, whereas the higher energy peak is due to charge transfer from Fe3d levels to the upper Hubbard band.

What is striking in this respect is that the Mott–Schottky analysis of such semiconductor photoelectrodes yields to some extent comparable information. Specifically, the electrochemical impedance spectra may contain hidden features which can be extracted by scanning the impedance over the DC potential [29–31]. Subsequent deconvolution and least square fit to electric circuits can then produce a capacitive element which constitutes a surface state C_{ss} or C_{trap}. Figure 20 shows a set of impedance spectra from a differently processed iron oxide electrode at 1.0 and 1.25 V vs. RHE bias potential. The electric circuit is shown in section b of Fig. 20. Depending on the bias potential, the value for C_{trap} may change. This is exercised in [29–31] for various electrodes including TiO_2 and α-Fe_2O_3. Typically, we notice then a Gaussian-shaped profile of this defect peak vs. the DC bias. We compare the aforementioned NEXAFS study in Fig. 19 with the impedance study in Fig. 20 [32] and notice that the hole state which is related with the UHB is at the same water-splitting onset potential like the surface trap state as determined from impedance spectroscopy (Fig. 21).

We can now summarize this Chapter about X-ray spectroscopy on photoelectrodes. The core-level spectroscopy methods allow for a virtually unambiguous assignment of the electronic structure to particular chemical elements in the photoelectrodes. Since the absorber in a photoelectrode assembly should be typically thick in the range of ½ to 3 μm, around one optical absorption length, surface-sensitive photoelectron methods are not necessarily the first choice for probing this deep part of the electrode. Hard X-ray absorption spectroscopy would be here the method of choice. Also, hard X-rays are easier to use for in situ experiments. The core-level spectra of the ligands from oxygen, but also nitrogen, sulfur, phosphor, and carbon, allow accessing the valence band properties of the electrodes. These low-*Z* elements have their core levels in the soft X-ray regime with information and probing depth of few nanometers only and thus cannot probe the bulk of the electrode. However, hard X-ray inelastic scattering is an established and increasingly emerging method for getting bulk representative information in the soft X-ray range. The electrochemical and electrocatalytic properties of the electrodes are relevant at their surfaces only, where they will interface with the electrolyte. The surface region can be probed with X-ray photoelectron spectroscopy, which will

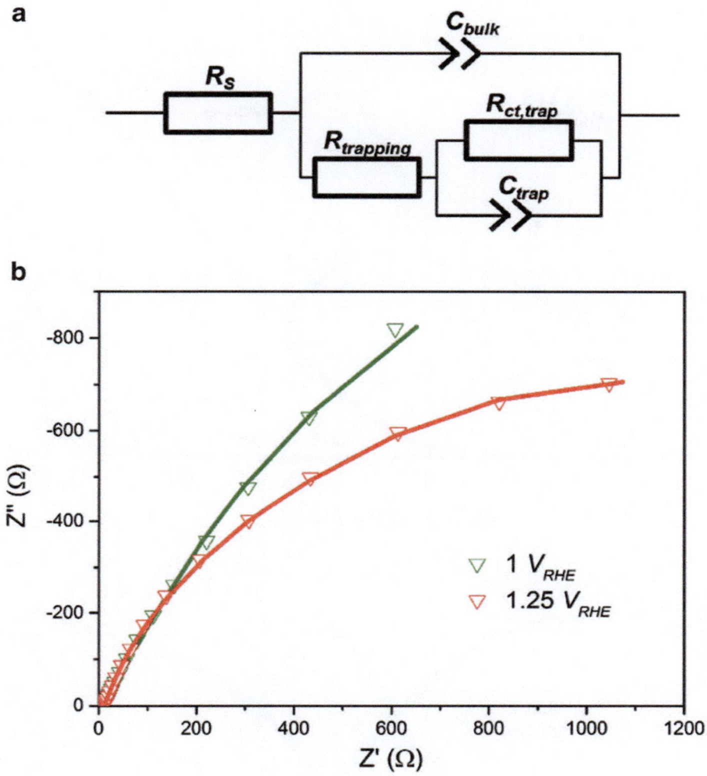

Fig. 20 Electrochemical impedance spectra of hematite in 1 M KOH recorded in *dark* at 1 V and 1.25 V vs. RHE from 100 kHz to 1 Hz. The samples were pristine ("0") or subject to an oxygen plasma for 20 min ("20")

provide core-level spectra and valence band spectra. The influence of synthesis and processing conditions on the electronic structure of photoelectrodes can be well investigated with these methods. It may be important to assign changes to particular elements, for instance, in doping and substitution studies or when oxygen vacancies are formed or filled. To tell apart which of these have influence on the electronic structure, valence band spectra can be recorded in the resonant mode for chemical contrast variation by tuning the X-ray photon energy to an X-ray absorption threshold of a chemical element of interest. This can help to elucidate the structural origin of defect states, which can be important for the transport properties and optical properties of the photoelectrode. The tunability of the X-ray photon energy allows also for depth-dependent studies, which is important when the subsurface of electrodes or heterostructures is investigated. Insofar, we have investigated the solid and inorganic electrode materials. This can be combined, as we have shown, with complementary electroanalytical methods.

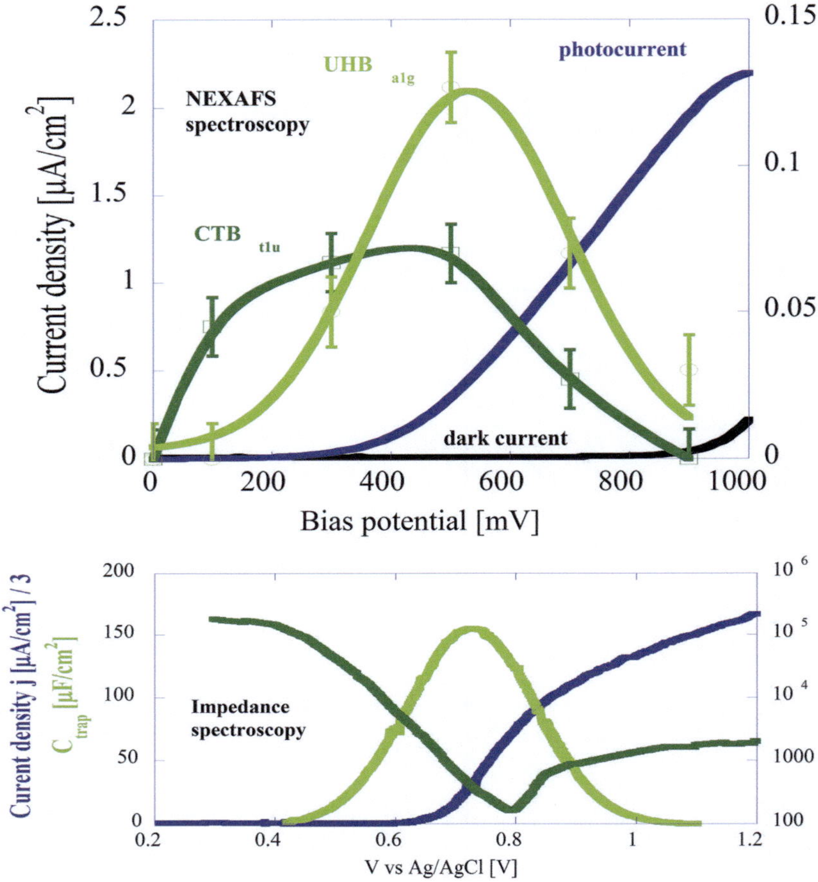

Fig. 21 The *top panel* shows the evolution of the spectral weight of the NEXAFS hole peaks as obtained from the in situ experiment shown in Fig. 19 and the dark current and photocurrent. The *panel below* shows the Gaussian profile of the surface trap state C_{ss} as obtained from impedance and the corresponding resistive element $R_{C,trap}$, together with the photocurrent. Reproduced from [31]

The opportunities that arise from the use of X-ray spectroscopy are certainly much wider. A whole variety of X-ray methods has been explored and developed in the field of photosynthetic protein spectroscopy since several decades. XANES and EXAFS have been extensively used for the determination of oxidation states and bond lengths of the metal centers [33] in hydrogenases, oxygenases, and nitrogenases [34, 35]. K-beta spectroscopy has been used for the determination of spin states in the Mn cluster in the oxygen-evolving complex of photosystem II [36, 37] and Fe porphyrins [38]. Since about 25 years, soft X-rays are being used for protein spectroscopy. Specifically, the L-edges of the transition metal centers in proteins are investigated [39]. This kind of protein studies has kept pace with the development of synchrotron spectroscopy in general, for example, the application of X-ray circular dichroism [40]. The L-edge multiplets are quite sensitive to the spin states and originate from dipole-allowed transitions and thus may sometimes outperform hard X-ray methods [41].In some cases, X-ray studies have been instrumental for the elucidation of fundamental steps in natural photosynthesis [42]. Since about 10 years, nuclear vibration-resonant spectroscopy (NVRS) is being used on Fe-containing compounds in photosynthesis research [43].

It would be desirable that X-ray spectroscopy is applied to bioorganic photosynthetic electrode assemblies *operando* or in situ. This requires technical progress in the sample environment infrastructure at synchrotron sources and certainly also at free-electron lasers. Such progress has indeed been made in the last 5 years, for example, with ambient pressure XPS technology and with ultrafast spectroscopy.

Acknowledgment A.B is indebted to Dr. Andri Vital (Empa) and Prof. Dr. Andreas Luzzi (Schweizer Solarzentrum Rapperswil) for giving him the opportunity to start working again on artificial photosynthesis. Since, funding for this research was granted by the Swiss Federal Office of Energy, the Empa Board of Directors, the Empa Laboratory for High Performance Ceramics, the Swiss National Science Foundation, the European Commission, the VELUX Foundation, the CRUS Sciex Program, and the Swiss COST Office and the EU COST Program.

References

1. L. Weinhardt, M. Blum, M. Bär, C. Heske, B. Cole, B. Marsen, E. L. Miller, Electronic Surface Level Positions of WO3 Thin Films for Photoelectrochemical Hydrogen Production, J. Phys. Chem. C 2008, 112, 3078-3082.
2. N. Gaillard, B. Cole, B. Marsen, J. Kaneshiro, E.L. Miller, L. Weinhardt, M. Bär, C. Heske, K. -S. Ahn, Y. Yan, and M. M. Al-Jassim, Improved current collection in WO3:Mo/WO3 bilayer photoelectrodes. J. Mater. Res. 25, 45 (2010).
3. Pieter Glatzel, Junko Yano, Uwe Bergmann, Hendrik Visser, John H. Robblee, Weiwei Gu, Frank M.F. de Groot, Stephen P. Cramer, Vittal K. Yachandra, Resonant inelastic X-ray scattering (RIXS) spectroscopy at the Mn K absorption pre-edge - a direct probe of the 3d orbitals, Journal of Physics and Chemistry of Solids 66 (2005) 2163–2167.
4. Uwe Bergmann, Pieter Glatzel, Stephen P. Cramer, Bulk-sensitive XAS characterization of light elements: from X-ray Raman scattering to X-ray Raman spectroscopy; Microchemical Journal 2002, 71, (2-3), 221-230.
5. D.K. Bora, A. Braun, S. Erat, O. Safonova, T. Graule, E. C. Constable, Evolution of structural properties of iron oxide nanoparticles during temperature treatment from 250 °C – 900 °C: X-ray diffraction and Fe K-shell pre-edge x-ray absorption study, Current Applied Physics 2012, 12 (3), 817–825.
6. B.L. Henke, E.M. Gullikson, J.C. Davis. X-ray interactions: photoabsorption, scattering, transmission, and reflection at E=50-30000 eV, Z=1-92, Atomic Data and Nuclear Data Tables Vol. 54 (no.2), 181-342 (July 1993).
7. X-Ray Absorption: Principles, Applications, Techniques of EXAFS, SEXAFS and XANES, by D. C. Koningsberger (Editor), R. Prins (Editor), Wiley-Interscience; 1 edition (January 18, 1988); ISBN-13: 978-0471875475 ISBN-10: 0471875473.
8. John B. Goodenough (2008), Scholarpedia, 3(10):7382.
9. A. Braun, S. Erat, D. Bayraktar, A. Harvey, T. Graule, Electronic origin of conductivity changes and isothermal expansion of Ta- and Ti-substituted $La_{1/2}Sr_{1/2}$Fe-oxide in oxidative and reducing atmosphere, Chem. Mater., 2012, 24 (8), 1529–1535.

10. A Braun, D. Bayraktar, A.S. Harvey, D. Beckel, J.A. Purton, P. Holtappels, L.J. Gauckler, T. Graule, Pre-edges in oxygen (1 s) x-ray absorption spectra: A spectral indicator for electron hole depletion and transport blocking in iron perovskites, Applied Physics Letters 2009, 94, 202102.
11. A. Braun, K.K. Akurati, G. Fortunato, F.A. Reifler, A. Ritter, A.S. Harvey, A. Vital, T. Graule, Nitrogen doping of TiO2 photocatalyst forms a second eg state in the Oxygen (1 s) NEXAFS pre-edge, J. Phys. Chem. C, 2010, 114 (1), 516–519.
12. Y. Hu, D. K. Bora, F. Boudoire, F. Häussler, M. Graetzel, E. C. Constable, A. Braun, A dip coating process for large area silicon-doped high performance hematite photoanodes, J. Renewable Sustainable Energy 5, 043109 (2013).
13. Evgenij Barsoukov (Editor), J. Ross Macdonald (Editor), Impedance Spectroscopy: Theory, Experiment, and Applications, 2nd Edition, Wiley, ISBN: 978-0-471-64749-2, April 2005.
14. D.K. Bora, A. Braun, S. Erat, R. Löhnert, A.K. Ariffin, R. Manzke, K. Sivula, T. Graule, M. Grätzel, E. Constable, Evolution of an oxygen NEXAFS transition in the upper Hubbard band in α-Fe2O3 upon electrochemical oxidation, J. Phys. Chem. C, 2011, 115 (13), 5619–5625.
15. K. Gajda-Schrantz, S. Tymen, F. Boudoire, R. Toth, D.K. Bora, W. Calvet, M. Grätzel, E. C. Constable, A. Braun, *Formation of an electron hole doped film in the α-Fe2O3 photoanode upon electrochemical oxidation*, Phys. Chem. Chem. Phys., 2013, 15, 1443-1451.
16. Dare-Edwards,M.P.;Goodenough,J.B.;Hamnett,A.;Trevellick,P.R.,ELECTROCHEMISTRY AND PHOTOELECTROCHEMISTRY OF IRON(III) OXIDE. Journal of the Chemical Society-Faraday Transactions I 1983, 79, 2027-2041.
17. Laurence M. Peter, K. G. Upul Wijayantha, Asif A. Tahir, Kinetics of light-driven oxygen evolution at α-Fe2O3 electrodes, Faraday Discuss., 2012,155, 309-322.
18. Stephan Hüfner, Photoelectron Spectroscopy, Series: Advanced Texts in Physics, Publisher: Springer; 3rd revised and enlarged ed. 2003 edition (April 8, 2003), ISBN-10: 3540418024, ISBN-13: 978-3540418023.
19. A. Braun, Q. Chen, D. Flak, G. Fortunato, K. Gajda-Schrantz, M. Grätzel, T. Graule, J. Guo, T.-W. Huang, Z. Liu, A. Popelo, K. Sivula, H. Wadati, P. P. Wyss, L. Zhang, J. Zhu; Iron resonant photoemission spectroscopy on anodized hematite points to electron hole doping during anodization, invited for Special Issue: Electrochemistry and Energy in ChemPhysChem 2012, 13(12), 2937-2944.
20. A. Braun, K. Sivula, D. K. Bora, J. Zhu, L. Zhang, M. Grätzel, J. Guo, E. C. Constable, Direct observation of two electron holes in hematite during photo-electrochemical water splitting, J. Phys. Chem. C 2012, 116 (23) 16870-16875.
21. A. Braun, S. Erat, X. Zhang, Q. Chen, F. Aksoy, R. Löhnert, Z. Liu, S.S. Mao, T. Graule, Surface and bulk oxygen vacancy defect states near the Fermi level in 100 nm $WO_{3-}\delta/TiO_2$ (110) films: A resonant valence band photoemission spectroscopy study, J. Phys. Chem. C, 2011, 115 (33), 16411–16417.
22. Michael E. Grass, Patrik G. Karlsson, Funda Aksoy, Måns Lundqvist, Björn Wannberg, Bongjin S. Mun, Zahid Hussain, Zhi Liu Rev. Sci. Instrum. (2010) May.
23. Jiang, P.; Chen, J.-L.; Borondics, F.; Glans, P.-A.; West, M. W.; 446 Chang, C. L. Electrochem. Commun. 2010, 12, 820 – 822.
24. Kennedy J.H. and Frese K.W. Photooxidation of water at α-Fe2O3 electrodes. J. Electrochem. Soc. 125(5), 709-714 (1978).
25. D.S. Toledano, E.R. Dufresne, V.E. Henrich, J. Vac. Sci. Technol. A. 16 (1998), 1050-1054.
26. Gärtner, PRB 1959, 116(1) 84-87.
27. Butler, J. Appl. Phys. 1977, 48(5), 1914-1920.
28. Chen, C.T., et al. Electronic states in La2-xSrxCuO4 + δ probed by soft-x-ray absorption. Phys. Rev. Lett. 66, 104–107 (1991).
29. Tomkiewicz, J. Electrochem. Soc. 1980, 127, 1518.
30. Shen, W. M.; Tomkiewicz, M.; Cahen, D., Impedance study of surface optimization on n-$CuInSe_2$ in photoelectrochemical solar-cells. *J. Electrochem. Soc.***1986**, 133, (1), 112-116.

31. Klahr, B.; Gimenez, S.; Fabregat-Santiago, F.; Hamann, T.; Bisquert, J., Water Oxidation at Hematite Photoelectrodes: The Role of Surface States. *J. Am. Chem. Soc.***2012**, 134, (9), 4294-4302.
32. D. K. Bora, Y. Hu, S. Thiess, S. Erat, X. Feng, S. Mukherjee, G. Fortunato, N. Gaillard, R. Toth, K. Gajda-Schrantz, W. Drube, M. Grätzel, J. Guo, J. Zhu, E.C. Constable, D.D. Sarma, H. Wang, A. Braun, Between Photocatalysis and Photosynthesis: Synchrotron spectroscopy methods on molecules and materials for solar hydrogen generation, J. Electron Spectr. Rel. Phenom. 2013, 190 A, 93-105.
33. Cramer, S. P., in Synchrotron Radiation Applied to Biophysical and Biochemical Research: Structure Determination by X-ray Absorption Spectroscopy; Castellani, A. and Quercia, I. F., Ed.; Plenum: New York, 1979, pp. 291-322.
34. Cramer, S. P.; Eccles, T. K.; Kutzler, F.; Hodgson, K. O.; Mortenson, L. E. Molybdenum X-ray Absorption Edge Spectra - The Chemical State of Molybdenum in Nitrogenase, J. Am. Chem. Soc., 1976, 98, 1287-1288.
35. Cramer, S. P.; Eidsness, M. K.; Pan, W.-H.; Morton, T. A.; Ragsdale, S. W.; DerVartanian, D. V.; Ljungdahl, L. G.; Scott, R. A. X-Ray Absorption Spectroscopic Evidence for a Unique Nickel Site in Clostridium thermoaceticum Carbon Monoxide Dehydrogenase, Inorg. Chem., 1987, 26, 2477-2479.
36. Prince, R. C.; Cramer, S. P.; George, G. N., in Current Research in Photosynthesis: The Manganese Cluster of the Water-Splitting Enzyme, Baltscheffsky, M., Ed.; Kluwer, 1990; Vol. 1, pp. 685-692.
37. Bergmann, U.; Grush, M. M.; Horne, C. R.; DeMarois, P.; Penner-Hahn, J. E.; Yocum, C. F.; Wright, D. W.; Dube, C. E.; Armstrong, W. H.; Christou, G.; Eppley, H. J.; Cramer, S. P. Characterization of the Mn Oxidation States in Photosystem II by K-β X-Ray Fluorescence Spectroscopy, J. Phys. Chem. B, 1998, 102, 8350-8352.
38. Wang, X.; Randall, C. R.; Peng, G.; Cramer, S. P. Spin-Polarized and Site-Selective X-Ray Absorption - Demonstration with Fe Porphyrins and K-Beta Detection, Chem. Phys. Lett., 1995, 243, 469-473.
39. Cramer, S. P.; Degroot, F. M. F.; Ma, Y.; Chen, C. T.; Sette, F.; Kipke, C. A.; Eichhorn, D. M.; Chan, M. K.; Armstrong, W. H.; Libby, E.; Christou, G.; Brooker, S.; Mckee, V.; Mullins, O. C.; Fuggle, J. C. Ligand Field Strengths and Oxidation States from Manganese L-Edge Spectroscopy, J. Am. Chem. Soc., 1991, 113, 7937-7940.
40. van Elp, J.; George, S. J.; Chen, J.; Peng, G.; Chen, C. T.; Tjeng, L. H.; Meigs, G.; Lin, H. J.; Zhou, Z. H.; Adams, M. W. W.; Searle, B. G.; Cramer, Soft X-Ray Magnetic Circular Dichroism - A Probe for Studying Paramagnetic Bioinorganic Systems, S. P. Proc. Nat. Acad. Sci. U. S. A., 1993, 90, 9664-9667.
41. Wang, H.; Patil, D. S.; Ralston, C. Y.; Bryant, C.; Cramer, L-edge X-ray Magnetic Circular Dichroism of Ni Enzymes: Direct Probe of Ni Spin States, S. P. J. Electron Spec., 2001, 114, 865-871.
42. Messinger, J.; Robblee, J. H.; Bergmann, U.; Fernandez, C.; Glatzel, P.; Visser, H.; Cinco, R. M.; Holman, K., M.; Bellacchio, E.; Pizarro, S. A.; Cramer, S. P.; Sauer, K.; Klein, M. P.; Yachandra, V. K. Absence of Mn-Centered Oxidation in the S2 → S3 Transition: Implications for the Mechanism of Photosynthetic Water Oxidation, J. Am. Chem. Soc., 2001, 123, 7804-7820.
43. Xiao, Y.; Fisher, K.; Smith, M. C.; Newton, W.; Case, D. A.; George, S. J.; Wang, H.; Sturhahn, W.; Alp, E. E.; Zhao, J.; Yoda, Y.; Cramer, S. P. How Nitrogenase Shakes - Initial Information about P-Cluster and FeMo-Cofactor Normal Modes from Nuclear Resonance Vibrational Spectroscopy (NRVS), J. Am. Chem. Soc., 2006, 128, 7608-7612.

MIX
Papier aus verantwortungsvollen Quellen
Paper from responsible sources
FSC® C105338

If you have any concerns about our products,
you can contact us on
ProductSafety@springernature.com

In case Publisher is established outside the EU,
the EU authorized representative is:
Springer Nature Customer Service Center GmbH
Europaplatz 3, 69115 Heidelberg, Germany

Printed by Libri Plureos GmbH
in Hamburg, Germany